MICROORGANISMS IN SUSTAINABLE AGRICULTURE, FOOD, AND THE ENVIRONMENT

MICROORGANISMS IN SUSTAINABLE AGRICULTURE, FOOD, AND THE ENVIRONMENT

Edited by
Deepak Kumar Verma
Prem Prakash Srivastav, PhD

APPLE
ACADEMIC
PRESS

Apple Academic Press Inc.
3333 Mistwell Crescent
Oakville, ON L6L 0A2 Canada

Apple Academic Press Inc.
9 Spinnaker Way
Waretown, NJ 08758 USA

Library and Archives Canada Cataloguing in Publication

Microorganisms in sustainable agriculture, food, and the environment/edited by Deepak Kumar Verma and Prem Prakash Srivastav, PhD.

Includes bibliographical references and index.
Issued in print and electronic formats.

ISBN 978-1-77188-479-2 (hardcover).—ISBN 978-1-315-36582-4 (PDF) 1. Agricultural microbiology. 2. Sustainable agriculture. 3. Food--Microbiology. 4. Microbial ecology. I. Srivastav, Prem Prakash, editor II. Verma, Deepak Kumar, editor

| QR51.M57 2017 | 630.2795 | C2017-902000-5 | C2017-902001-3 |

Library of Congress Cataloging-in-Publication Data

Names: Verma, Deepak Kumar, 1986- editor. | Srivastav, Prem Prakash, editor.
Title: Microorganisms in sustainable agriculture, food, and the environment / editors: Deepak Kumar Verma, Prem Prakash Srivastav, PhD.
Description: Waretown, NJ : Apple Academic Press, 2017. | Includes   bibliographical references and index.
Identifiers: LCCN 2017012334 (print) | LCCN 2017015162 (ebook) | ISBN 9781315365824 (ebook) | ISBN 9781771884792 (hardcover : alk. paper)
Subjects: LCSH: Agricultural microbiology.
Classification: LCC QR51 (ebook) | LCC QR51 .M458 2017 (print) | DDC 579--dc23
LC record available at https://lccn.loc.gov/2017012334

Apple Academic Press also publishes its books in a variety of electronic formats. Some content that appears in print may not be available in electronic format. For information about Apple Academic Press products, visit our website at **www.appleacademicpress.com** and the CRC Press website at **www.crcpress.com**

This book is dedicated
to
the memory of remarkable people and
The Father of Microbiology

Antonie Philips van Leeuwenhoek
(24th October, 1632 to 26th August, 1723)

ABOUT THE EDITORS

Deepak Kumar Verma

Deepak Kumar Verma is an agricultural science professional and is currently a PhD Research Scholar in the specialization of food processing engineering in the Agricultural and Food Engineering Department, Indian Institute of Technology, Kharagpur (WB), India. In 2012, he received a DST-INSPIRE Fellowship for PhD study by the Department of Science & Technology (DST), Ministry of Science and Technology, Government of India. Mr. Verma is currently working on the research project "Isolation and Characterization of Aroma Volatile and Flavoring Compounds from Aromatic and Non-Aromatic Rice Cultivars of India." His previous research work included "Physico-Chemical and Cooking Characteristics of Azad Basmati (CSAR 839-3): A Newly Evolved Variety of Basmati Rice (*Oryza sativa* L.)". He earned his BSc degree in agricultural science from the Faculty of Agriculture at Gorakhpur University, Gorakhpur, and his MSc (Agriculture) in Agricultural Biochemistry in 2011. He also received an award from the Department of Agricultural Biochemistry, Chandra Shekhar Azad University of Agricultural and Technology, Kanpur, India. Apart of his area of specialization as plant biochemistry, he has also built a sound background in plant physiology, microbiology, plant pathology, genetics and plant breeding, plant biotechnology and genetic engineering, seed science and technology, food science and technology etc. In addition, he is member of different professional bodies, and his activities and accomplishments include conferences, seminar, workshop, training, and also the publication of research articles, books, and book chapters.

Prem Prakash Srivastav

Prem Prakash Srivastav, PhD, is Associate Professor of Food Science and Technology in the Agricultural and Food Engineering Department at the Indian Institute of Technology, Kharagpur (WB), India, where he teaches various undergraduate, post-graduate, and PhD level courses and guides research projects at PhD, masters and undergraduate levels. His research interests include the development of specially designed convenience, functional, and therapeutic foods; the extraction of nutraceuticals; and the development of various low-cost food processing machineries. He has organized many sponsored short-term courses and completed sponsored research projects and consultancies. He has published various research papers in peer-reviewed international and national journals and proceedings and many technical bulletins and monographs as well. Other publications include books and book chapters along with many patents. He has attended, chaired, and presented various papers at international and national conferences and delivered many invited lectures in various summer/winter schools. Dr. Srivastav has received several best poster paper awards for his presentations. He is a member of various professional bodies, including the International Society for Technology in Education (ISTE), the Association of Food Scientists and Technologists (India), the Indian Dairy Association (IDA), the Association of Microbiologists of India (AMI), and the American Society of Agricultural and Biological Engineers and the Institute of Food Technologists (USA).

CONTENTS

LIST OF CONTRIBUTORS

Ramesh Chander Anand
Senior Scientist, Department of Microbiology, Chaudhary Charan Singh Haryana Agricultural University, Hisar 125004, Haryana, India.

Sudhanshi Billoria
Department of Agricultural and Food Engineering, Indian Institute of Technology, Kharagpur 721302, West Bengal, India.

Bindu Devi
Department of Microbiology, College of Basic Sciences, CSK, Himachal Pradesh Agricultural University, Palampur 176062, Himachal Pradesh, India.

Sunita Devi
Assistant Professor (Microbiology), Department of Basic Sciences, College of Forestry, University of Horticulture and Forestry—Nauni, Solan 173230, Himachal Pradesh, India.

Y. K. Jhala
Research Associate, Department of Agricultural Microbiology & Bio-fertilizers Projects, B. A. College of Agriculture, Anand Agricultural University, Anand 388110, Gujarat, India.

Deepika Kadian
Department of Microbiology, Chaudhary Charan Singh Haryana Agricultural University, Hisar 125004, Haryana, India.

Mandira Kapri
Department of Food Technology, Maharshi Dayanand University, Rohtak 124001, Haryana, India.

Rakesh Kumar
District Extension Specialist, CCS Haryana Agricultural University, Krishi Vigyan Kendra, Fatehabad 125120, Haryana, India.

Dipendra Kumar Mahato
Senior Research Fellow (SRF), Division of Food Science and Post-harvest Technology, Indian Agricultural Research Institute (IARI), New Delhi, India.

Kamla Malik
Assistant Scientist, Department of Microbiology, Chaudhary Charan Singh Haryana Agricultural University, Hisar 125004, Haryana, India.

S. K. Mehta
District Extension Specialist, CCS Haryana Agricultural University, Krishi Vigyan Kendra, Bhiwani 127021, Haryana, India.

Amrita Narula
Department of Microbiology, Chaudhary Charan Singh Haryana Agricultural University, Hisar 125004, Haryana, India.

Alaa Kareem Niamah
Department of Food Science, College of Agriculture, University of Basrah, Basra City, Iraq.

Abhay K. Pandey
Research Associate, Plant Pathology, Plant Health Management Division, National Institute of Plant Health Management, Department of Agriculture and Cooperation, Ministry of Agriculture, Government of India, Rajendranagar, Hyderabad 500030, Telangana, India.

Ami Patel
Division of Dairy and Food Microbiology, Mansinhbhai Institute of Dairy & Food Technology, MIDFT, Dudhsagar Dairy Campus, Mehsana 384002, Gujarat, India.

D. V. Pathak
CCS Haryana Agricultural University, Regional Research Station, Bawal, Rewari 123501, Haryana India.

Amrita Poonia
Centre of Food Science & Technology, Institute of Agricultural Science, Banaras Hindu University, Varanasi 221005, Uttar Pradesh, India.

P. K. Prabhakar
Department of Agricultural and Food Engineering, Indian Institute of Technology, Kharagpur 721302, West Bengal, India.

Satish K. Sain
Assistant Director, Plant Health Management, Plant Health Management Division, National Institute of Plant Health Management, Department of Agriculture and Cooperation, Ministry of Agriculture, Government of India, Rajendranagar, Hyderabad 500030, Telangana, India.

Nihir Shah
Division of Dairy and Food Microbiology, Mansinhbhai Institute of Dairy & Food Technology, MIDFT, Dudhsagar Dairy Campus, Mehsana 384002, Gujarat, India.

Harsha N. Shelat
Associate Research Scientist, Department of Agricultural Microbiology & Bio-fertilizers Projects, B. A. College of Agriculture, Anand Agricultural University, Anand 388110, Gujarat, India.

Prem Prakash Srivastav
Department of Agricultural and Food Engineering, Indian Institute of Technology, Kharagpur 721302, West Bengal, India.

Ajesh Kumar V.
Department of Agricultural and Food Engineering, Indian Institute of Technology, Kharagpur 721302, West Bengal, India.

Deepak Kumar Verma
Department of Agricultural and Food Engineering, Indian Institute of Technology, Kharagpur 721302, West Bengal, India.

Seema Verma
Technical Assistant, Department of Basic Sciences, College of Forestry, University of Horticulture and Forestry—Nauni, Solan 173230, Himachal Pradesh, India.

R. V. Vyas
Research Scientist, Department of Agricultural Microbiology & Bio-fertilizers Projects, B. A. College of Agriculture, Anand Agricultural University, Anand 388110, Gujarat, India.

LIST OF ABBREVIATIONS

2-AP	2-acetyl-1-pyrroline
A.D.	Anno Domini
AAB	acetic acid bacteria
ABA	abscisic acid
ACC	aminocyclopropane-1-carboxylate
AFB1	aflatoxin B1
AFDA	American Food and Drug Administration
AFM1	aflatoxin M1
ALF	alcoholic fermentation
AM	arbuscular mycorrhizal
AMD	acid mine drainage
ANF	santinutritional factors
Aq	*Ampelomyces quisqualis*
ATP	adenosine triphosphate
a_w	water activity
BAs	biogenic amines
BGA	blue green algae
BNF	biological nitrogen fixation
Bt	*Bacillus thuringiensis*
cDNA	complementary DNA
CE	capillary electrophoresis
CLA	conjugated linoleic acid
CMV	cucumber mosaic virus
CO_2	carbon dioxide
CSS	cold-smoked salmon
CTV	citrus tristeza virus
Cu_2S	copper sulfide
$CuSO_4$	copper sulfate
CVD	cardiovascular disease
DAP	diammonium phosphate
DEEMM	diethyl ethoxymethylenemalonate
DHA	docosahexaenoic acid
DNA	deoxyribonucleic acid
dsRNA	double standard RNA

EDTA	ethylenediaminetetraacetic acid
EFSA	European Food Safety Authority
Eh	redox potential
EHEC	enterohemorrhagic *E. coli*
EPA	wiocosapentaenoic acid
EPN	entomopathogenic nematodes
EPS	exopolymeric substance
EPS	exopolysaccharide
EPS	extracellular polysaccharides
ET-A	exfoliative toxin A
ET-B	exfoliateve toxin B
FAO	Food and Agricultural Organization
FCO	fertilizer control order
FDA	Food and Drug Administration
FDA	Food and Drug Agency
FFMF	fermented foxtail millet flour
FMF	foxtail millet flour
FMG	fermented milk product
FOD	foodborne outbreak database
FOSHU	foods for specified health uses
FSIS	food safety inspection service
FSSAI	Food Safety and Standards Authority of India
GABA	gamma γ-aminobutyric acid
Gal	galactose
GC	gas chromatography
GC–MS	gas chromatography–mass spectrometry
GDP	gross domestic product
GI	gastrointestinal
Glc	glucose
GMO	genetically modified organisms
GMP	good manufacturing practices
GRAS	generally regarded as safe
GV	granuloviruses
H_2O_2	hydrogen peroxide
HACCP	hazard analysis and critical control point
HCN	hydrogen cyanide
HDL	high density lipoprotein
HHP	high hydrostatic pressure
HIPEF	high intensity pulsed-electric field

HMT	heat moisture treatment
HPH	high pressure homogenization
HPLC	high-performance liquid chromatography
HSP	heat shock response
HUSH	emolytic uremic syndrome
IAA	indole-3-acetic acid
IEC	ion-exchange chromatography
IPM	Integrated Pest Management Programme
ISL	*in-situ* leaching
ISR	induced systemic resistance
JA	jasmonic acid
KDO	keto-deoxyoctulosonate
KMB	potash mobilizing bacteria
LAB	lactic acid bacteria
LAF	lactic acid fermentation
LBD	late blowing defect
LDL	low density lipoprotein
LDPE	low-density polyethylene
LHCs	light harvesting complexes
LPS	lactoperoxidase system
LPSl	lipopolysaccharide
MAG	monoammonium glyctyrrhizinate
MAP	modified atmosphere packaging
MFP	microbial fermentation process
MLF	malolactic fermentation
MMT	million metric tonnes
MoFe	molybdenum–iron
MPC	milk protein concentrates
NGa	*N*-acetyl-galactosamine
NGc	*N*-acetyl-glucosamine
NPV	nucleopolyhedroviruses
NSC	natural starter cultures
NSLAB	nonstarter lactic acid bacteria
NSPs	nonstarch polysaccharides
PAL	phenylalanine ammonialyase
PBRs	photobioreactors
PCR	polymerase chain reaction
PCR-DGGE	polymerase chain reaction denaturing gradient gel electrophoresis

PEF	pulsed electric field
PGPR	plant-growth promoting rhizobacteria
PHB	poly-3-hydroxy butyrate
PR	pathogen related
PSB	phosphate solubilizing bacteria
PUFAs	polyunsaturated fatty acids
RAPD	random amplified polymorphic DNA
rDNA	ribosomal DNA
rRNA	ribosomal ribonucleic A
RTE	ready to eat
RTF	ready-to-feed
RT-PCR	real time polymerase chain reaction
SA	salicylic acid
SEls	Staphylococcal enterotoxin-like
SEs	Staphylococcal enterotoxins
SFP	*S. aureus* food poisoning
SFP	Staphylococcal food poisoning
SiO$_2$	silicon dioxide
SMFGM	skim milk rich with fat globule membrane
SMP	skimmed milk protein
SSB	silicate solubilizing bacteria
SSC	selected starter cultures
SSF	solid state fermentation
SSP	single super phosphate
STEC	shigatoxigenic *E. coli*
Stx	shiga-like toxins
TAGs	triacylglycerols
TFC	total flavonoid content
TLC	thin layer chromatography
TPC	total phenolic content
TST	toxic shock syndrome toxin-1
U.S.	United States
UHPLC	ultra-high-pressure liquid chromatography
UHT	ultrahigh temperature
USA	United State of America
USSR	Union of Soviet Socialist Republics
VAM	vesicular arbuscular mycorrhizae
VP	voges proskauer
VT	verotoxins

VTEC	verocytotoxic-producing *E. coli*
WBCs	white blood cells
WHO	World Health Organization
WPC	whey protein concentrates
ZSB	zinc solubilizing bacteria

PREFACE

Microbiology is the study of microscopic living organisms, which are recognized for their ubiquitous presence, diverse metabolic activity, and unique survival strategies under extreme conditions. These microscopic organisms include fungi, bacteria, viruses (infectious agents at the borderline of life) algae, and protozoa. These microorganisms are poorly explored in our surrounding environments with their diversity and abundance in universe. In microbiology, these organisms are concerned with their form, structure, classification, reproduction, physiology, and metabolism. It includes the study of (1) their distribution in nature, (2) their relationship with each other and other living organisms, (3) their effects on human beings and on other animals and plants, (4) their abilities to make physical and chemical changes in our environment, and (5) their reactions to physical and chemical agents. Thus, microbiology is a part of science that has given birth to several branches, viz. mycology, bacteriology, virology, protozoology, parasitology, phycology or algology, microbial morphology, microbial physiology, microbial taxonomy, microbial genetics, molecular biology, microbial ecology, food microbiology, dairy microbiology, aquatic microbiology, industrial microbiology, etc., each being pursued as a specialty in itself.

Agricultural microbiology is the branch of science concerned with the relationships between microbes and agricultural crops, with an emphasis on improving yields and combating plant diseases. This branch presents as a synthetic research field that is responsible for transfer of knowledge from general microbiology and microbial ecology to the agricultural biotechnology. Agricultural microbiology has a major goal comprehensive analysis of symbiotic microorganisms, viz. bacteria and fungi, interacting with agriculture. In agriculture, the entire food and agricultural processes are permeated by these organisms.

While the most visible and important role of agriculture is probably producing and delivering food, microbiology is critical to other agricultural sectors as well, for example, for bioremediation of agricultural wastes and for production of energy. Some microorganisms are an integral part of successful food production, whereas others are a constant source of trouble for agricultural endeavors. Microbial influences on food and agriculture have produced both advancements and disasters that have punctuated human history. Some examples of microbe-driven outcomes set the stage for describing how important it is to seize research opportunities in food and agriculture microbiology.

In agricultural education and research, the study of microbiology has undergone tremendous changes in the past few decades in all the above-mentioned areas. This covers all human endeavors in the broad sense as the transmission, absorption, and acquisition of knowledge for the better means of understanding the processes that lead to the scientific farming. We all know that agriculture is a backbone of economy all over the globe. This book is a general consensus on the need for a comprehensive study of recent advances and innovations in microbiology.

This book, *Microorganisms in Sustainable Agriculture, Food, and the Environment*, contains 12 chapters that are further divided into four main parts.

Part 1: Food Microbiology is devoted to research opportunities in microbiology as the gateway to sustaining and improving agriculture and food production, quality, and safety. In food, microorganisms are teeming with threats and benefits to abundant, healthy food and associated environments. Threats come from microbial pathogens that perpetrate a wide range of plant and animal diseases, destroying agricultural productivity. The constant spread and evolution of agricultural pathogens provide a continually renewed source of challenges to productivity and food safety. Pathogens continue to cause harm once food has left the farm, causing spoilage, and in some cases poisoning and diseases of humans and animals. New vulnerabilities are generated for agriculture by the global movement of agricultural products, trading policies, industrial agricultural practices, and the potential for malicious releases of pathogens by "bioterrorists." In addition to the threats, benefits also come from the many microorganisms associated with, or introduced into, our food supply where they serve important roles in bioprocessing, fermentation, or as probiotics.

Part 2: Soil Microbiology deals extensively with studies on the isolation, culture, and use of *Rhizobium* spp. and mycorrhizae to improve soil fertility, plant growth, and yield. This part of the book includes research progress on biogeochemical cycles, plant growth promoting rhizobacteria (PGPR), microbial interactions in soil and other soil activities, microbial diversity in soil, biological control and bioremediation, and improvement of beneficial microorganisms (N_2 fixers, phosphate solubilizers, etc.).

Part 3: Environmental Microbiology deals with the study of the composition and physiology of microbial communities in the environment *(viz.*

soil, water, and air). This third part of this book is devoted to research opportunities and describes the environment, water, soil, air pollution, and bioremediation, microbiological control of agricultural enemies and pathogens of agricultural important crop plants.

Part 4: Industrial Microbiology and Microbial Biotechnology both are associated with the commercial exploitation of microorganisms, which involve achieving specific goals for the processes and production of useful products, whether creating new products with monetary value that are of major economic, environmental, or social importance. There are two key aspects of industrial microbiology, the first relating to production of valuable microbial products via fermentation processes, whereas the second aspect is the role of microorganisms in providing services, particularly for waste treatment and pollution control, which utilizes their abilities to degrade virtually all natural and man-made products. This part of the book explores the wide range of industrial microbial processes and products, including traditional fermented foods and beverages (using yeast technology, citric acid, and lactic acid fermentations), such as bread, beer, cheese, and wine. In addition, this part of the book is further employed in the production of numerous chemical feedstocks (primary and secondary metabolites like enzymes, amino acids) and products for application in human and animal health), the provision of animal feed production, alternative energy sources, biofertilizers production (N_2 fixers, phosphate solubilizers, PGPR, BGA, composting, etc. for maintaining the soil and plant health for sustaining crop productivity and their importance in organic farming).

With contributions from a broad range of leading researchers, this book focuses on such areas of microbiology as we discussed above that will provide a guide to students, instructors, and researchers. It covers the topics expected of students, instructors, and researchers taking agricultural microbiology courses as major. The text in the chapters is suitable for undergraduate and graduate students. In addition, microbiology professionals seeking recent advanced and innovative knowledge in agriculture will find this book helpful. It is envisaged that this book will also serve as a reference source for individuals engaged in microbiology research, processing, and product development.

With great pleasure, we would like to extend our sincere thanks to all the learned contributors for the magnificent work they did in making sure that their timely response, excellent devoted contributions to detail and accuracy of information presented in this text, along with their consistent support and cooperation, has made our task as editors a pleasure.

We feel that we covered most of the topics expected for agricultural microbiology in this text. It is hoped that this volume will stimulate discussions and generate helpful comments to improve upon future volume editions. Efforts are made to cross reference the chapters as such.

Finally, we acknowledge Almighty God, who provided all the inspirations, insights, positive thoughts, and channels to complete this book project.

—Deepak Kumar Verma
Prem Prakash Srivastav

PART I
Food Microbiology

CHAPTER 1

FERMENTED FOODS: AN OVERVIEW

AMI PATEL* and NIHIR SHAH

Division of Dairy and Food Microbiology, Mansinhbhai Institute of Dairy & Food Technology (MIDFT), Dudhsagar Dairy Campus, Mehsana 384 002, Gujarat, India

Corresponding author. E-mail: amiamipatel@yahoo.co.in

CONTENTS

ABSTRACT

Fermented foods have become an essential part of our daily diet. Various milk, vegetables, cereals-legumes, meat, and fish-based fermented foods might have evolved in the dietary, religious, and cultural ethos of individuals living in the regions of the world and owe their origin. Employing selected starter strains such as that have an ability to produce flavor components, vitamins, enzymes, antimicrobial compounds or exopolysaccharides, or health beneficial microbes, that is, probiotics within the matrix of fermented foods will help to develop novel functional fermented foods by utilizing the traditional indigenous technologies. It may help to spread fermented foods of different regions and provide a range of choice of food to consumers worldwide. Moreover, industrial manufacturing of fermented foods would serve to improve the livelihood of rural as well as urban population with generation of new source of income and employment particularly for poor and needed.

1.1 INTRODUCTION

Current food processing depends on a diverse preservation method that can ensure an acceptable level of food quality and safety from the time of production to the time of intake. Fermentation is one of the well-known oldest low-cost processes, and different civilizations of the world have employed it to enhance the shelf life and enrich the nutritional value of many perishable foodstuffs including fruits, vegetables, milk, fish, meat, legumes, and cereals (Chandan, 2013). Irrespective of age groups, traditional foods play a very vital role in the diet of a large proportion of global population. Fermented foods may be described as the foodstuff whose production obliges direct action of a single or mixed type of microflora or their enzymes to bring enviable biochemical changes with or without significant change in the physical properties of raw food. Fermented foods are enormously popular in numerous developing countries because of their unique sensory attributes, enhanced nutritional values, and digestibility, as well as extended shelf life under ambient conditions (Holzapfel, 2002; Hutkins, 2006). It is observed that the fermented products of specific region of world fit in to the dietary habits as well as their cultural and religious traditions.

Developed countries can afford to enrich its food with dietary fibers, amino acids, or vitamins, whereas the developing countries of the world must rely upon biological enrichment of food to fulfill their routine requirements.

Fermentation leads to improve the keeping quality, body and texture, aroma, and organoleptic properties of the final food product. It seems that this conventional biotechnology have evolved through inherent processes where existing nutrient together with ecological conditions selected particular microbe in specific food product (Tamang et al., 2009). At industrial level, now it is possible to select particular fermenting starter through strain improvement process, and modern gene technology methods. In general, lactic acid bacteria (LAB), acetic acid bacteria, yeasts, and molds play imperative role in the fermentation of large variety of substrates including milks, cereals, vegetables, fruits, and meats. In general, fermentation leads biosynthesis of organic acids like lactic acid, acetic acid; alcoholic; gaseous, or alkaline compounds during fermentation to preserve substantial amount of food which in turn develop variety of flavors, aroma, and textures within the food matrix (Soni and Sandhu, 1990).

The major benefits associated with fermentation include: biological fortification of original food substrates through easily digestible carbohydrates, proteins, vitamins, essential amino acids, minerals, and dietary fibers; destruction or reduction of toxic compounds naturally present in substrates specially in case of plant-based materials or cereals; and most importantly decrease in cooking time and fuel requirement. The presence of unwanted antinutritional factors like polyphenols, phytate, oligosaccharides, enzyme inhibitors, and lectins often limits the digestibility of several nutrients in food. Recent studies have confirmed the action of fermenting microflora during the preparation of fermented foods found to destruct or reduce antinutritional compounds and enhance the digestibility as well as nutritional quality of the foods (Rose, 2015; Steinkraus, 1997). In several regions, traditionally, the fermented food products are associated with curing of disease or an illness, for instance, a fermented product *gariss*, obtained from camel milk was used for the treatment of leishmaniasis (*kala-azar*) in Sudan; kumiss was used to cure pulmonary tuberculosis in Russia; and dahi like fermented milks were used to treat gastrointestinal (GI) disturbances in India (Yadav et al., 1993).

Fermented foods play a major role as dietary staples in various countries across Asia, Middle East, Africa, Latin America, etc. where small-scale fermentation industries contribute to quality and security of food, principally in areas vulnerable to food shortages (Prajapati and Nair, 2008). Till date, the most of the fermented foods are manufactured in traditional manner at cottage level and small industrial scale. It is evident that fermented foods are consumed as a staple meal or an accompaniment to a meal either in form of breakfast snack, beverage, dessert, condiment, or as a component of cooked dishes.

The implementation of standardization in the methods used, selection of starters, such as probiotics and/or those have an ability to synthesize vitamins, antimicrobials (bacteriocins or antimicrobial peptides), or enzymes in fermented foods, is essential for large-scale production of various fermented food products to enhance flavor, texture, or health-promoting properties. Recent technologies and researches carried away in different fermented foods opened the opportunities to improve the popularity, sales, and turnover of various fermented foods, chiefly of fermented milks globally.

1.2 HISTORY OF FERMENTED FOODS

Fermentation is one the ancient technology used to preserve various food products in desired modified form, and many of fermented foods have evolved more by accidently than by direct efforts. The production and consumption of fermented foods began date back thousands of years, from initial evidences of the fermentation of left over milk to fermented milk, barley to beer, and grapes to wine. Until the 19th century, fermentation processes were carried out without any prior knowledge or understanding of microorganisms or their activities.

From the archaeological findings allied with Indo-Aryans of the Indian subcontinent, the Sumerians and Babylonians of Mesopotamia and the Pharoes of northeast Africa provides early evidence of utilizing technology of fermented milks (Chandan and Nauth, 2012; Tamime and Robinson, 2007). Dating back about 5000 years, the Vedas—prehistoric Indian scriptures refer the use of Dadhi (modern dahi) and buttermilk (Aneja et al., 2002). Thus, food fermentation should be considered as one of the oldest known applications of biotechnology.

Bread like products finds their origin dates back to the Neolithic era. In ancient times, Asians developed the processes to hydrolyze complex carbohydrates like starch as well as proteins to enhance the digestibility and sensorial attributes of vegetables and cereals based dietary food. It is also stated that Asians were the first to develop techniques related to production of meat like flavors from fermented plant proteins; the process of souring and leavening cereal batters were developed by Indians, whereas Indonesians came out with methods to generate a meat-like body and texture to vegetables; and yeast leavened wheat breads were first developed by Egyptians (Soni and Sandhu, 1999).

The earliest fermentation processes included the manufacturing of beer from Babylonia, soy sauce from Japan and China, and fermented milk beverages from Balkans and Central Asia. Fermented beverages emerged in around 5000 B.C. in Babylon followed by 3150 B.C. in Egypt, 2000 B.C. in Mexico, and 1500 B.C. in Sudan, respectively, as mentioned by Chojnacka (2009).

It is believed that fermentation of vegetables was begun in China and Korea while the pickling of cucumbers and carrots probably originated in Southeast Asia. While recent history suggest that pickled olives were oriented from Mediterranean countries (Prajapati and Nair, 2008).

Prior to First World War, ethanol was the only product synthesized by fermentation at large-scale, but since acetone was used in production of explosives within the period of World War I, acetone–butanol fermentation was commercially established (Chojnacka, 2009). Afterward, the market of fermentation products increased sharply with organic acids manufacturing. The market for conventional fermentation products like antibiotics was established than after in between 1941 and 1946, and later on, it led to increased interest in industrial utilization of different microorganisms significantly. In last few years, the world market of fermented dairy and food products flourished greatly.

1.3 MICROORGANISMS ASSOCIATED WITH FERMENTED FOODS

In the majority of traditional food fermentation techniques, frequently a mixed population involving diverse species of bacteria, yeasts, and molds which finds origin from the indigenous microbiota of the raw substrates is found to be active (Table 1.1). Among these, LAB are most commonly associated with natural fermentation of different fermented foods through their harmless metabolic activities. During growth, they usually utilize available carbohydrates (generally a sugar) as an energy source and mainly produce organic acids with other metabolites in fewer amounts. Due to their natural occurrence in various types of raw food materials together with prolonged safe use give them GRAS (Generally Regarded As Safe) status. Apart from LAB, many other bacteria as well as yeasts and molds are also associated with different types of fermented foods such as *Propionibacteria, Brevibacterium, Acetic acid bacteria, Bifidobacteria, Bacillus, Micrococcus; Candida, Saccharomyces, Torulopsis, Geotrichum, Aspergillus, Penicillium, Rhizopus, Mucor,* etc.

During natural fermentation, a diverse group of microorganisms participate. However, at industrial scale, a defined starter culture developed under controlled conditions is of foremost choice. It will serve to maintain the qualities of the finished product consistently after each consignment. With recent methods of genetic engineering and gene technology, now it is possible to design and build starter cultures with specific desired attributes (Chandan, 2013).

TABLE 1.1   Major Species of Microorganisms Associated With Fermented Foods.

Microorganisms group	Major species	Fermented product
Bacteria	*Lactobacillus, Streptococcus, Leuconostoc, Pediococcus, Acetobacter, Propionibacterium, Brevibacterium, Bacillus, Micrococcus, Staphylococcus*, etc.	Yoghurt, dahi, cultured butter milk
Yeast	*Saccharomyces, Candida, Torulopsis, Zygosaccharomyces, Hansenula*	Kefir, kumiss
Mold	*Geotrichum, Aspergillus, Penicillium, Rhizopus, Mucor, Actinomucor, Monascus*	Cheese, soya based products
Algae	Cholera (*C. vulgaricus*), Spirulina (*S. platensis*)	Fermented milks

1.3.1   LACTIC ACID BACTERIA

On the basis of 16s rRNA sequencing, ribotyping, nucleic acid hybridization and other molecular techniques, the LAB are grouped into the phylum *Firmicutes*; Order, *Lactobacillales* and family *Lactobacteriaceae* in current taxonomy. This group consists of Gram-positive rods and cocci-shaped bacteria which are catalase negative, facultatively anaerobic, nonmotile, nonspore forming which are characterized by a low genomic G+C content. The genome contains a chromosome, the size of which ranged from 1.8 to 3.4 Mbp. The LAB group consists of 12 genera viz. *Lactobacillus, Lactococcus, Streptococcus, Leuconostocs, Pediococcus, Oenococcus, Aerococcus, Tetragenococcus, Carnobacterium, Enterococcus, Vagococcus,* and *Weissella*. Out of these 12, only first 7 genera are used directly during fermentation because of their ability to synthesize lactic acid from fermentation of carbohydrates as the key metabolic end-product (Hutkins, 2006). At present, the genus *Lactobacillus* is the largest LAB group that contains about 140 species and 30 subspecies and is based on the modern molecular biology techniques such as whole genome sequencing or ribotyping, the number is getting revaluated continuously (Bernardeau et al., 2008). Basically, three types of *Lactobacillus* spp. are found viz., obligatory homofermentative, facultatively heterofermentative, and obligatory heterofermentative based on their fermentation pattern (Axelsson, 2004).

The genus, *Lactococcus*, comprises homofermentative, mesophilic, Gram positive cocci which is usually found singly, in pairs, or in chains. Some species like *Lc. lactis* sub spp. *diacetylactis* play significant role in development of diacetyl flavor in fermented milks. *Stre. thermophilus* is the key bacterium from the genus *Streptococcus* which appears as Gram positive cocci, mainly in chains under microscope. The genus, *Pediococcus*, is purely homofermentative Gram positive cocci-shaped bacteria that are typically observed in pairs or tetrads under microscope. They have an ability to grow in a broad range of pH, temperature, osmotic pressure, and other related unfavorable conditions making them suitable candidates that can colonize the digestive or GI tract. They are commonly isolated from fermented milks, vegetables, cereals, and meat products. Another genus, *Leuconostocs*, generally appear as Gram positive cocci to irregular-sized cells that may found in pairs and in short chains on microscopic examination. They are heterofermentative bacteria producing carbon dioxide along with organic acids from glucose and other sugars.

Leuconostocs species are important flavor producers (diacetyl) in some fermented dairy products like butter, cultured butter milk, dahi, and cheese. Rogosa agar has found to be the most suitable media for the growth of various species of *Leuconostocs*. Species currently classified as *Weissella* species share similar phenotypic characteristics to that of *Lactobacillus* species and are frequently reported from diverse fermented foods like sour cream, soya, fermented cassava, sauerkraut, sourdough, and blood sausage, "Morcilla de Burgos" (Plengvidhya et al., 2007; Galle et al., 2010).

1.3.2 NONSTARTER LACTIC ACID BACTERIA

Several nonlactic microorganisms like *Propionibacterium shermani*, *Brevibacterium linen*, *Bacillus* spp.; yeasts like *Kluyveromyces*, *Saccharomyces*, *Candida*, *Torulopsis*, and *Zygosaccharomyces*; molds such as *Geotrichum*, *Aspergillus*, *Penicillium*, *Rhizopus*, and *Mucor* are employed as starters during the manufacturing of milk, cereals–legumes, soya beans, and vegetables based fermented foods.

In addition to these normal microorganisms, some cyanobacteria have got enormous attention in recent times because of being a remarkable source of high-valued proteins, dietary fibers, and vitamins as well as presence of several therapeutically significant compounds. Microalgae, such as chlorella and spirulina, is a valuable food supplement that has a wide range of beneficial nutritional effects. Spirulina typically contains 53–63% protein,

17–25% carbohydrate, 4–6% lipids, 3–7% moisture, 8–10% fiber, 8–13% ash, 1–1.5% chlorophyll α, and a wide range of vitamins and other bioactive components (Molnár et al., 2009). Many studies have been conducted to incorporate chlorella and spirulina in development of fermented milk in combination with LAB; however, addition of microalgae found to alter the organoleptic properties of fermented milks. Nevertheless, much work is required in this direction as there is not enough information available in literatures (Molnár et al., 2009; Beheshtipour et al., 2013).

Many studies have reported the isolation and identification of microorganisms from a variety of fermented foods. Popular fermented milks including yoghurt, cultured butter milk, and sour cream utilize different species of LAB as starter. Many of the lactose fermenting yeasts have been frequently isolated from naturally fermented traditional milk products like kumiss, kefir, zabady, dahi, and cheese in conjunction to LAB. Although fermented vegetables like sauerkraut, kimchi, and gundruk finds use of diverse species of the genus, *Leuconostocs*, *Lactococcus*, *Pediococcus*, and *Lactobacillus*, in soya-based fermented foods; for the most parts, molds of the different genera including *Rhizopus*, *Aspergillus*, *Geotrichum*, *Mucor*, *Actinomucor*, and *Monascus* are found to be associated (Surh et al., 2008; Hutkins, 2006). In Europe and the United States, mold-ripened cheese varieties like Roquefort cheese, Stilton cheese, camembert cheese, etc. are quite popular, and many nonstarter LAB (NSLAB) also found to be part of other cheese varieties such as emmental cheese, brick cheese, etc. Many of the African origin fermented foods such as gari, kenkey, and dawadawa reported presence of *Bacillus* spp., *Staphylococcus* spp., *Micrococcus* as well as *Aspergillus*, and *Penicillium* spp. as the major microflora (Blandino et al., 2003).

1.4   CLASSIFICATION OF FERMENTED FOODS

Classification of fermented foods can be done in numerous ways based on (1) the kind of microorganisms involved with the fermentation process, (2) substrate or raw material, (3) type of fermentation, (4) texture of end product, or (5) function of the food, the same is depicted in Figure 1.1.

The fermentation process can be of four kinds such as organic acids, alcoholic, alkali, and mixed fermentation (acid, gas, and alcohol) based on the end product biosynthesis (Soni and Sandhu, 1990). LAB and related genera are the principle group associated with lactic fermentation processes and play a significant role in manufacturing of fermented milks and cereal-based products. Acetic acid fermentation leads to synthesis of acetic acid,

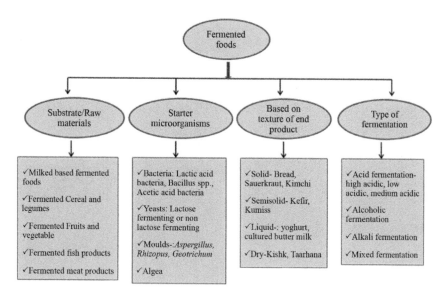

FIGURE 1.1   Classification of fermented foods.

and acetobacter species are the predominant organisms in the production of vinegar, i.e., palm wine vinegar, apple cider vinegar, and coconut water vinegar. In the presence of excess oxygen, these bacteria convert alcohol to acetic acid. Yeasts are the predominant organisms in alcohol fermentation that lead to production of ethanol and CO_2 in the products like, for example, bread, sourdough, wines, beers, and other alcoholic beverages. Generally, during the fermentation of fish and seeds that is popularly used as condiment or spread alkali fermentation proceeds (McKay and Baldwin, 1990). In the current text, the different fermented food products have been described based on their classification of raw materials or substrates.

1.4.1   FERMENTED MILKS

Fermented milk constitutes a crucial part of our diet since immemorial times in different parts of the world. Products obtained from the fermentation of milks resulted in preservation of precious nutrients; if not then milk obtained from various cattle would have deteriorated rapidly, particularly in the several South Asian and Middle East countries where throughout the year high ambient temperatures prevail. Thus, fermentation processes allowed us to preserve and consume milk constituents over a significantly longer period

TABLE 1.2 Major Fermented Milk-Based Products Consumed in Different Regions of the World.

Product name	Microorganisms involved	Major country/region
Yoghurt	*Lactobacillus delbrueckii* subsp. *bulgaricus*, *Streptococcus thermophilus*	USA
Cultured buttermilk	*Lactococcus lactis* subsp. *lactis*, *Lc. lactis* subsp. *diacetylactis*, and *Leuc. mesenteroides*	
Acidophilus milk	*L. acidophilus*	USA, Russia
Kefir	Mixed species of mesophilic and thermophilic lactobacilli and Lactococci, and yeasts	Russia, Central Asia
Koumiss/Kumys	*Lactobacillus delbrueckii* subsp. *bulgaricus*, *L. acidophilus*, and yeasts-*Torula*	
Donskaya/Varenetes/Kurugna/ Ryzhenka/Guslyanka	—	Russia
Ayran	*Lactobacillus delbrueckii* subsp. *bulgaricus*, *Streptococcus thermophilus*	Turkey, Azerbaijan, Bulgaria, Macedonia, Kazakhstan, Kyrgyzstan
Jugurt/Eyran/Ayran	*Lactobacillus delbrueckii* subsp. *bulgaricus*, *Streptococcus thermophilus*	Turkey
Ayrani	—	Cyprus
Busa	—	Turkestan
Chal	—	Turkmenistan
Skyr	*Lactobacillus delbrueckii* subsp. *bulgaricus*, *Streptococcus thermophilus*, lactose-fermenting yeast	Iceland
Filmjolk/Fillbunke/Fillbunk/ Surmelk/Taettemjolk/Tettemelk	*Lc. lactis* subsp. *lactis*, *Lactobacillus delbrueckii* subsp. *bulgaricus*, *Streptococcus thermophilus*, lactose-fermenting yeast	Sweden, Norway, Scandinavia

TABLE 1.2 *(Continued)*

Product name	Microorganisms involved	Major country/region
Pitkapiima, Viili	*Lc. lactis* subsp. *lactis*, *Leuc. mesenteroides* subsp. *cremoris*, and a fungus, *Geotrichumcandidum*	Finland
Ymer	*Lc. lactis* subsp. *lactisbiovar. diacetylactis* and *Leuc. mesenteroides* subsp. *cremoris*	Denmark
Cieddu	—	Italy
Yiaourti	—	Greece
Kurunga	—	Western Asia
Lassi, Mattha, Ghol, Chhas, Raita	*Lactococcuslactis* subsp. *lactis*, *Lc. lactis* subsp. *diacetylactis*, *Streptococcus thermophilus*, *Lactobacillus* spp.	India, Pakistan, Bangladesh, Nepal
Dahi/Dudhee/Dahee, Mishti-Dahi, Shrikhand	*Lactococcuslactis* subsp. *lactis*, *Lc. lactis* subsp. *diacetylactis*, or *Leuconostoc* spp. and *Lc. lactis* subsp. *cremoris*, *Streptococcus thermophilus*, *Lactobacillus* spp.	India
Shosim/Sho/Thara	—	Nepal
Dogh/Abdoogh/Mast	Yoghurt cultures	Afghanistan, Iran
Ergo	—	Ethiopia
Gioddu	—	Sardinia
Gruzovina	—	Yugoslavia
Iogurte	—	Brazil, Portugal
Mazun/Matzoon/Matsun/ Matsoni/MadzoonTan/Tahn	*L. mazun*, *L. bulgaricus*, *Stre. thermophilus*, spore-forming bacillus and lactose-fermenting yeast	Armenia
Katyk	—	Transcaucasia

TABLE 1.2 *(Continued)*

Product name	Microorganisms involved	Major country/region
KisselMaleka/Naja/Yaourt/Urgotnic	*Lactobacillus delbrueckii* subsp. *bulgaricus*, *Streptococcus thermophilus*, lactose-fermenting yeast	Balkans
Leben/Laban/Laban Rayeb	Thermophilic lactobacilli and mesophilic lactococci—*Lc. lactis* subsp. *lactis*, *Stre. thermophilus*, *L. delbrueckii* subsp. *bulgaricus*	Lebanon, Syria, Jordan
Mezzoradu	—	Sicily
Zabady/Zabade	*Lactobacillus bulgaricus* and *Streptococcus thermophilus*, *Candida krusei*	Egypt, Sudan
Rob	*Lactobacillus fermentum*, *Lc. lactis* subsp. *Lactis*	Sudan
Gariss/Hameedh	*Lactobacillus helveticus*, *Candida*, Lactococci or streptococci	
Roba/Rob	—	Iraq
Tarag	—	Mongolia
Tarho/Taho	—	Hungary

which was otherwise not possible for milk itself. Approximately, more than 400 various fermented milk products are obtained via milk fermentation, derived from different animals, are consumed around the world in present epoch (Chandan, 2013). In addition to conserve essential milk nutrients, fermentation process improves some milk components and leads to increase the nutritional values and health improving properties, provides active cultures in significant numbers. Such fermented milk products are sometimes referred as "functional foods" which constitute an important part of human diet (Chandan, 2013). The international market of yoghurt and related fermented dairy products stands for euro 63.2 billion and 31.6 million tons/ year production in that Asia, Europe, and North America accounts for 77%. During last 5 years, the sales value has been increased by 39% globally. The selling and turnover of fermented dairy products continue to increase worldwide and more profoundly in China, Russia, Latin America, Brazil, Middle East, and Africa (http://www.ylfa.org/about.php?classement=02).

The most popular types of traditional as well as modern fermented milks are shown in Table 1.2. In general, the variety of fermented dairy products may be ascribed to (1) milk procured from different domesticated animals such as cow, buffalo, sheep, goat, or camel, (2) appliance of assorted microorganisms as starter culture, (3) blending of sugar, spices, fresh fruits, and dry nuts to provide arrange of flavors and consistencies, and (4) making use of different preservative technologies such as condensing, drying, freezing, heating, thermization, etc.

1.4.1.1 POPULAR FERMENTED DAIRY PRODUCTS

1.4.1.1.1 *Yoghurt*

Yoghurt (also spelled yoghurt or jugurt) is a semisolid fermented milk product, which originated centuries ago in Bulgaria. In recent years' popularity, yoghurt has grown tremendously and is now consumed in almost all parts of the world either as set or stirred product with or without additional flavors and fruits. Although the consistency, flavor, and aroma of yoghurt may vary from one region to another, the basic ingredients, microflora, and manufacturing processes are consistent (Chandan and Nauth, 2012). According to FAO/WHO, "yoghurt is strictly defined as a coagulated milk product obtained by lactic acid fermentation through the action of two bacteria—*Streptococcus thermophilus* and *L. delbrueckii* spp. *bulgaricus*." Further, yoghurt may contain bifidobacteria and supplementary flora like

Lactobacillus acidophilus to improve its therapeutic value. The microflora of yoghurt is found to inhibit the growth of many putrefactive bacteria including anaerobic sporeformers as well as several food borne pathogens like *Staphylococcus aureus* and *E. coli* (Yadav et al., 1993). Diverse varieties of yoghurt such as set type, stirred type, fruit flavored, frozen, concentrated and dry powder forms are available in world market.

1.4.1.1.2 *Cultured Buttermilk*

Cultured buttermilk is fermented milk beverage produced from pasteurization of skim milk or double tone milk followed by culturing with mesophilic species of Lactococci such as *Lc. lactis* subsp. *lactis* and aroma-producing *Lc. lactis* subsp. *diacetylactis* or leuconostocs and is ripened for 14–16 h at 25 °C. As soon as the acidity of the curd reaches to 0.80–0.85%, the coagulum is broken, blended with adequate amount of salt or spices and butter flakes, and packaged in plastic or paper containers (Chandan, 2013). Buttermilk is mostly consumed as a fermented milk beverage with meal.

1.4.1.1.3 *Sour/Cultured Cream*

Sour cream is an extremely viscous fermented milk product of America and several European countries with the flavor and aroma like butter milk (Chandan, 2013). It is manufactured by inoculating pasteurized cream making use of mesophilic lactococci such as *Lc. lactis* subsp. *lactis* and aroma-producing *Lc. lactis* subsp. *diacetylactis* or leuconostocs. The product has a butter-like aroma and flavor. The product is made in a similar way to that of cultured butter milk. It is packaged in cartons or serving cups/pack ages. Sour cream is consumed as a dressing or topping on salads, fresh vegetables and fruits, fish, and meat products. It is also used in soups, cooked dishes, as a dip, and filler in pastries and cakes.

Crème fraiche bear a resemblance to sour cream and is well liked in European countries like France. The product contains up to 50% fat with a higher pH of about 6.2–6.3 and mild acidic taste and flavor (Goddik, 2012).

1.4.1.1.4 *Cheese*

Cheese is obtained by coagulation of the milk protein casein through action of rennet and/or microorganisms. Around 500 different varieties of cheese are available in world as recognized by the International Dairy Federation (Fox and Fox, 2000). As stated by FAO (Food and Agricultural Organization) of the United Nations, in 2011, more than 20 million metric tons of cheese was manufactured globally; in that the United States being the major cheese producer accounted for 26% of the total production, followed by Germany and France. Mozzarella, Gouda, Cheddar, and Emmental are by far the top cheese types.

The common steps involved with the production of cheese are selection of milk that is usually acidified either by using LAB or food grade acid (acetic acid) or combination of thereof and addition of enzyme rennet causes coagulation. The starter cultures develop lactic acid and so reduction of milk pH. Further, cooking and stirring of curd promote syneresis and whey expulsion. Concomitantly, the solid mass is separated from the whey and given a shape by pressing. This aging or ripening period may last from several days to months. During this period, casein and milk fat transform into simpler form like amino acids, peptides, and fatty acids by the action of microorganisms and enzymes that ultimately develops typical texture of cheese (Hutkins, 2006). Some cheeses have molds on the rind or throughout. The primary starters together with the secondary microflora develop both flavor and texture in cheese through promoting complex biochemical reactions (Heller et al., 2008; Yadav et al., 1993).

Microorganisms play very essential role throughout cheese manufacturing and ripening, and are categorized into two major groups; primary microflora and secondary flora. The primary starters, *Lc. lactis*, *Lc. lactis* subsp. *diacetylactis*, *Stre. Thermophilus*, and *L. delbrueckii* subsp. *bulgaricus*, are used either individually or in combinations depending on the type of cheese variety, whereas the secondary flora comprises NSLAB such as *Propionibacterium* spp., acetic acid bacteria, *Brevibacterium* spp., *Saccharomyces* spp., *Candida* spp., etc. which may grow either internally or externally and impart typical flavor, aroma, and body and texture to specific cheese varieties.

1.4.1.2 ASIAN FERMENTED MILKS

In India, Nepal, Bangladesh, Pakistan, Sri Lanka, and other related countries, following fermented milk products are of cultural and commercial importance.

1.4.1.2.1   Dahi

Dahi is very popular fermented milk of Asian countries like India, Nepal, Bangladesh, Pakistan, Sri Lanka, etc. It is also called curd, dahee, dadhi, doi, etc. in different regions of these countries. Dahi is widely consumed and accounts for about 90% of the total production of fermented milk products in India (Behare et al., 2009). The product finds its name and well documented for its medicinal values in the ancient Hindu scriptures. Dahi is having a semisolid texture that develops from fermentation of cow or buffalo milk or from their combinations by inoculating single or mixed strains of LAB and/ or lactose fermenting yeasts such as *Lc. lactis* subsp. *lactis, Lc. lactis* subsp. *diacetylactis, Lc. lactis* subsp. *cremoris, Stre. thermophilus, Leuconostocs* spp., *L. bulgaricus, L. plantarum,* etc. Based on the intensity of acidity required in dahi, the specific strains should employed during the manufacturing dahi, for instance, mesophilic streptococci are used to prepared dahi that have medium acidity, whereas lactobacillus spp. are used to produce sour dahi (Yadav et al., 1993).

A sweet aroma, clean acid taste with firm and even consistency is the desired character of a worth-quality dahi. It is regularly consumed with meals like chapatti, rice, and vegetable curries. It is also incorporated with pieces of fresh fruits and vegetables that is termed as *Raita.* Dahi finds wide applications to cure gastrointestinal disorders including diarrhea and dysentery attributed to low pH and antimicrobial compounds formed during fermentation. It is recommended as appetite and vitality improver. Dahi is referred for its health-improving and disease-treating possessions in the Ayurveda, a system of medicine, too (Aneja et al., 2002).

In mishti dahi, one of the popular varieties of dahi, milk, is added with 10–15% cane sugar or jaggary that leads to development of a smooth body and texture with a cream to light-brown color appearance and pleasant aroma. In addition to this, incorporation of caramel (1–2%) is also practiced to enhance flavor. In eastern India and Bangladesh, mishti dahi is consumed as a dessert item during meal.

1.4.1.2.2   Shrikhand

Shrikhand is a dahi-derived sweetened fermented milk product famous in western and southern regions of India. After fermentation, dahi is kept in muslin cloth to drain off whey to obtain semisolid mass called chakka that is further mixed with sugar, color, flavor and cardamom, saffron, or dry fruits to accomplish a desired level of consistency (Yadav et al., 1993). By making use of a basket or a desludging centrifuge, or quark separator, to prepare shrikhand, whey is removed from dahi to obtain chakka in which 55–60% sugar is mixed in a planetary mixer at industrial level (Chandan, 2013). If required, a specific quantity of plastic cream is mixed with the chakka to get a minimum of 8.5% fat in shrikhand. In modern plants, the products is heat treated (thermization) to prevent further increase in acidity during storage by the starters. Shrikhand is consumed primarily as a dessert and having higher shelf life than dahi.

1.4.1.2.3 Lassi

Lassi or chhas/chhachh is a refreshing beverage of North India which is derived by churning dahi for development of butter granules by frequent addition of cold water. Lassi is obtained by manual separation of butter granules from the churned dahi. The product is consumed with or without addition of salt, sugar, fresh fruits, dry fruits, spices (cumin powder, curri leaves powder, and chili flakes), or flavors depending upon the consumer's preference. Through standardization of processes and application of modern UHT (ultra high temperature), treatment of milk a considerable progress has been made toward the industrial manufacturing of lassi in last few years (Aneja et al., 2002).

1.4.1.3 SCANDINAVIAN FERMENTED MILKS

From different regions of the world, Scandinavian countries (Sweden, Denmark, Norway, Finland, and Iceland) showed very high per capita consumption for fermented milk products. As reported by Schultz (2011), per person, 28.48 kg of yoghurt is consumed in Sweden only. Though most of the fermented milks belonging to Scandinavian countries possess a ropy and viscous body, they differ significantly in flavor and texture. Ymer, taette, skyr, viili, langfil, and keldermilk are several examples of popular fermented milks of these regions.

1.4.1.3.1 Ymer

Ymer is Denmark-oriented fermented milk. It has a pleasant acidic flavor, buttery taste and is characterized with very high protein content (5–6%).

In the traditional process of ymer making, after fermentation, the whey is removed from curd either by draining or heating. In modern methods, in milk, protein enrichment is usually achieved through ultrafiltration. After that, the milk is fermented with mesophilic ropy starters *Lc. lactis* subsp. *lactis* biovar. *diacetylactis* and *Leuc. mesenteroides* subsp. *cremoris* and incubated at 20°C for 18–24 h. Once fermentation finishes, the fermented curd is cooled and packaged in plastic or glass containers.

1.4.1.3.2 Skyr

Skyr is traditional fermented milk which belonged to Iceland. It differs with other products in a manner that during fermentation, a small amount of rennet is added along with yoghurt cultures and lactose-fermenting yeast in skim milk. It results into development of adequate amount of lactate, acetate, alcohol, and CO_2 with a heavier body in the finished product. Manufacturing technology involves fermentation of milk at 40 °C for 4–6 h, until a pH of 4.6 is achieved followed by further incubation at 18–20 °C for 16–18 h for acidification up to pH 4.0. Modern processes involved centrifugation following pasteurization to obtain around 13% protein content and concentrate other solids by making use of a clarifier-type separator at 35–40 °C.

1.4.1.3.3 Viili

Viili is traditional fermented milk of Scandinavian country Finland. It is fermented using mesophilic cultures *Lc. lactis* subsp. *lactis*, and *Leuc. mesenteroides* subsp. *cremoris* together with a mold *Geotrichum candidum*. After seeding cultures, milk is distributed in individual cups and incubated for 24 h at 20 °C. Once lactic starter develops 0.85–0.9% acidity, the fungus grows on surface cream layer to give rise velvety texture with musty odor. The product is having high viscosity and ropiness due to liberation of mucopolysaccharides. It is available in plain and fruit-flavored form in market.

1.4.1.3.4 Taette

Taette is moderately viscous/ropy sour milk of Scandinavian origin, especially popular in Finland. Taet-majolk is similar product commercially available in market. The fermenting microflora involves *Lc. lactis* subsp. *hollandicus* (which accounts to develop slimy milk), *L. delbrueckii* subsp. *bulgaricus*, *B. aciductislogus*, and yeast *Saccharomyces taette* that may develop 1% lactic acid, 0.3–0.5% alcohol, and traces of CO_2 in the final product.

1.4.1.4 FERMENTED MILKS OF RUSSIA AND EASTERN EUROPEAN

1.4.1.4.1 Kefir

Kefir is typical Russian fermented milk, which is also popular in certain European and Asian countries. It is believed to be originated in the Caucasian mountains of the former USSR. It is characterized by mildly acidic, alcoholic, and distinctly effervescent product. Traditional method of kefir manufacturing involves natural fermentation of milk obtained from goat, sheep, or cow with kefir grains that are of irregular sized white-yellow colored grains with cauliflower like appearance. During kefir manufacturing, these grains are added to milk; they swell and turn white forming slimy jelly like products. At the end of fermentation, grains are filtered off and reinoculated in the subsequent lot.

Kefir grains are made up of polysaccharide gum called "kefiran" and milk residue embedded with lactobacilli, streptococci, micrococci, and yeasts in specific synbiotic relationship. The major strains lactobacilli involve *L. kefir*, *L. kefirogranum*, *L. casei*, *L. helveticus*, *L. fermentum*, and *L. brevis* with mesophilic lactococci, and leuconostocs as well as acetic acid bacteria, whereas *Saccharomyces cerevisiae*, *S. kefir*, *Candida kefir*, *C. maris*, *Zygosaccharomyces* spp. and *Torula* spp. are the major yeasts involved with kefir fermentation (Hutkins, 2006). Kefir grains contain both lactose-fermenting as well as not lactose-fermenting yeasts either on the external surface or interior part of the kefir grain. Kefir contains lactic acid, ethanol, and CO_2 with typical flavor compounds acetaldehyde, diacetyl, and acetone. Kefir contains a large number of bioactive peptides because of the native flora with high proteinase activities. Kefir is also marketed in the other parts of the world like the United States and Europe, which varies from traditional product in that milk, is fermented with a mixed species of lactobacilli, lactococci, leuconostocs, and yeasts to confer small amount of alcohol and CO_2 production for mouth feel. To improve the sensory and organoleptic properties, the commercially available kefir is blended with either sugar and/or fruit juices/flavors.

1.4.1.4.2 *Kumiss*

Kumiss is also effervescent, mildly acidic alcoholic fermented milk. It is known as *kimiz* in Turkey and *airag* in Mongolia. Traditionally, it is prepared from mare or cow milk or combination of thereof that is inoculated with *L. delbrueckii* subsp. *bulgaricus*, *L. acidophilus*, and lactose-fermenting yeast *Torula kumiss* or *Saccharomyces lactis* (Yadav et al., 1993). Usually, no curdling is visible in the finished product as mare's milk has very low protein content (only 2%). Kumiss contains 0.6–1.0% lactic acid, 0.7–2.5% alcohol and adequate amount of CO_2 to give fizziness with tingling mouth feel. Kumiss is considered as refreshing therapeutic drink and is traditionally used for the treatment of pulmonary tuberculosis. It has perceived many health benefits, especially to cure GI disturbances, hypertension, or allergies.

1.4.1.5   MIDDLE EASTERN FERMENTED MILKS

Historically, fermented milks have been associated with the people of Middle East countries. It is believed that yoghurt was also originated from Bulgaria, and its regular consumption was correlated with longevity of Bulgarians. Traditional products of Middle East regions include Bulgarian sour milk, leben, laban rayeb, matzoon, zabady, etc.

1.4.1.5.1 *Leben*

Leben is the fermented milks originated in the Middle East countries. It resembles to concentrated yoghurt or kefir-like product. The microflora found in Leben mainly includes *L. lactis*, *Stre. thermophlilus*, *L. bulgaricus*, and lactose fermenting yeast. After fermentation of milk, the fermented curd is hanged in cloth bag to remove whey. In Turkey, a goat or sheep's skin bag is used, whereas in Egypt porous, earthenware is used to remove moisture. In modern methods, fermented curd is concentrated using a quarg separator. The related fermented milks are known under different names such as "*Lebneh*"/ "*Labneh*" from Labanon, Jordan or Arabian countries; "*Tan*" from Armenia; "*Torba*" from Turkey; "*Gioddu*" from Italy; and "*Matzum*" from Russia (Prajapati and Nair, 2008).

1.4.1.5.2 Laban Rayeb

Traditional process of manufacturing laban rayeb involves pouring of raw milk into deep earthenware pots followed with spontaneous fermentation for 3–4 days. The fat rise on the surface is used to make butter and the curdle milk is called laban rayeb which smells like butter milk and have slight acidic taste. The fermenting microflora of product may vary due to seasonal factors in which mesophilic lactococci may predominate during winter time, whereas thermophilic lactobacilli may prevail during summer (Mistry, 2001).

Production method of laban rayeb varies in some regions and thus, the same product termed differently. For example, when fermentation is carried out in a goat's skin bag, the product is termed as laban khad while if fermentation is done in earthenware pots laban zeer. In Sudan, laben rayeb is prepared from buffalo milk, mixed with cream, and diluted with water. Such preparation is consumed as traditional drink namely Goubasha. Another product, Karish cheese, is manufactured by pouring laban rayeb onto a reed mat to drain off water. Intermittently, it is spread and squeezed on mat and hanged for 2–3 days. Finally, the cheese is hung on mat for a few hours after sprinkling dry salt.

1.4.1.5.3 Mazun

Mazun or matzoon, originated from Armenia, milk is fermented with *L. mazun*, *L. bulgaricus*, *Stre. thermophilus*, spore-forming bacillus, and lactose-fermenting yeast. It is having many synonyms like matsoon, matsoun, matsoni, matzoun, madzoon, etc.

1.4.1.5.4 Zabady

An Egyptian fermented milk zabady is prepared by fermenting milk with yoghurt culture followed by heating–boiling the curd to concentrate the finished product. Further concentration of milk solids is achieved by heating it and separating the whey (Yadav et al., 1993).

Other products of Middle East countries differ a little in appearance as well as organoleptic properties from above-mentioned popular fermented milks and yoghurt. For instance, Jameed is yoghurt which is salted and dried to preserve in Jordan. Doogh is yoghurt-based beverage which is sometimes carbonated and seasoned with mint. Ayran is liquid yoghurt added with salt.

1.4.2 FERMENTED FRUITS AND VEGETABLES

It is difficult to store fresh fruits and vegetables at ambient temperature as they are highly perishable. Various fermented vegetable and fruit products should have arisen from the time people began to collect and store it. Fruits are slightly acidic and naturally rich in sugars and watery portion containing soluble nutrients termed juice. Such conditions favor proliferation of yeasts and molds, and thus, they are used to make alcoholic beverages. It was known since long before that salt helps to preserve food and also to improve the organoleptic properties, and based on this principle, people began to add salt or sea water which resulted into shelf-life extension of vegetables (Prajapati and Nair, 2008). Later on, may be natural fermentations of individual vegetables or mixed vegetables resulted in development of diverse fermented products namely sauerkraut, kimchi, pickles, etc. which may contain carrot, cabbage, radish, turnip, cucumber, beet, pepper, and/or olive. In Table 1.3, major fermented vegetables and food products have been listed.

1.4.2.1 SAUERKRAUT

Sauerkraut is a German term which means "sour cabbage." This is a popular fermented product of several European countries, China, Korea, and the United States. It is prepared by natural fermentation in which cabbage is shredded and mixed with in 2–2.5% salt (brine). The natural microflora associated with cabbage begins the fermentation process and reduces the pH by production of organic acids. The organic acids together with salt contribute to enhance the shelf life for longer period. In addition to improve organoleptic qualities of sauerkraut, salt prevents softening of the tissues and contributes to generate a crisp texture (Prajapati and Nair, 2008; Swain et al., 2014). The manufacturing of fermented vegetables such as sauerkraut depends on a progression of various microorganisms which are inherently found in the raw material. During the fermentation, some of these microbes emerge out early, perform a particular function and then disappear from the product. Other microorganisms, on the contrary, may appear later during the process of fermentation and stay behind throughout the duration of

TABLE 1.3 Traditional Fermented Fruits and Vegetables of Various Parts of the World.

Fermented product	Fruit and vegetables	Microorganisms	Country
Sauerkraut	Cabbage	Leuc. mesenteroides, L. brevis, L. plantarum, Pd. pentosaceous	US, Europe, Asian countries
Kimchi	Cabbage, radish, various vegetables	Leuc. mesenteroides, L. brevis, L. curvatus, L. plantarum, L. sakei, Stre. lactis, Pd. pentosaceus, Weissella confusa, W. koreensis	Korea
Pickles	Vegetables and Fruits	Lactic acid bacteria	US, Asian countries
Olive	Olive	L. brevis, L. pentosus, L. plantarum, L. casei, Leuc. mesenteroides	Spain, Italy
Gundruk	Cabbage, Radish, Mustard, Cauliflower	Pediococcus and Lactobacillus spp.	Nepal, India
Inziangsang	Mustard leaves	L. plantarum, L. bevis	India
Soibum/Soidon	Bamboo shoot	Lactobacillus spp., Leuc. mesentroides, Enterococcus durans, Lc. lactis, Bacillus subtilis, B. licheniformis, B. coagulans, and the yeasts Candida spp., Saccharomyces spp., Torulopsis spp.	India
Mesu	Bamboo shoot	L. plantarum, L. brevis and L. pentosaceus	India
Lung-seij	Bamboo shoot	Lactic acid bacteria	India
Ziang-sang	Leaves of hangam (Brassica sp.)	L. plantarum, L. brevis and Pd acidilactici	North-East India
Goyang	Green vegetables	L. plantarum, L. brevis, Lc. lactis, Enterococcus faecium, Pd. pentosaceus	North-East India
Anishi	Leaves of edible yam (Colocasia sp.)	Unknown	North-East India
Khalpi	Cucumber	L. plantarum, L. brevis, Leuc. fallax	North-East India, Nepal
Sinki	Radish	L. plantarum, L. brevis, L. fallax, L. fermentum, Pd. pentosaceus	India, Nepal, and Bhutan

TABLE 1.3 *(Continued)*

Fermented product	Fruit and vegetables	Microorganisms	Country
Sayurasin	Mustard, Cabbage	*L. confuses, L. plantarum, Leuc. mesenteroides, Pd. pentosaceus*	Indonesia
Tempoyak	Duriyan (*Duriozibethinus*)	*Bacillus coagulans, Leuc. mesenteroides, L. brevis, L. mali*	Malaysia
Nozawana-Zuke	Turnip	*L. curvatus, L. sakei*	Japan
Sunki	Leaves of otaki-turnip	*L. brevis, L. plantarum, P. pentosaceus*	Japan
Paocai	Cabbage, Cucumber, Celery and Radish	*Leu. mesenteroides, L. plantarum, L. brevis*	China
Yan-taozih	Peaches	*S. lactis, W. paramesenteroides, W. minor, W. cibaria, E. faecalis*	China and Taiwan
Yan-jiang	Ginger	*S. lactis, W. cibaria*	Taiwan
Yan-dong-gua	Wax gourd	*W. cibaria, W. paramesenteroides, L. sakei*	Taiwan
Yan-tsai-shin	Broccoli	*W. cibaria, W. minor, W. paramesenteroides, Leuc. mesenteroides, L. plantarum*	Taiwan
Suan-tsai	Chinese Cabbage, Cabbage, Mustard leaves	*Pd. pentosaceus, Tetragenococcus halophilus*	Taiwan
Jiang-gua	Cucumber	*Weissella cibaria, W. hellenica, Pd. acidilactici, L. plantarum*	Taiwan
Pobuzihi	Cummingcordia	*S. lactis, L. fermentum, L. plantarum, W cibaria*	Taiwan
Dhamuoi	Cabbage, Various vegetables	*Leu. mesenteroides, L. fermentum, L. plantarum*	Vietnam
Duamuoi	Mustard or Beet	*L. pentosus, L. plantarum, P. pentosaceus*	Vietnam
Camuoi	Eggplant	*L. fermentum, L. pentosus*	Vietnam

TABLE 1.3 *(Continued)*

Fermented product	Fruit and vegetables	Microorganisms	Country
Burongmustala	Mustard leaf	*L. brevis, P. cerevisiae*	Philippines
Dakguadong	Mustard leaf	*L. brevis, L. plantarum*	Thailand
Pak-Gard-Dong	Mustard leaf	*L. mesenteroides, P. cerevisiae, L. brevis*	Thailand
Pak-sian-dong	Leaves of Pak-sian (*Gynadropsispen-taphylla*)	*P. cerevisiae, L. brevis, L. plantarum, L. pentosus,*	Thailand
Fufu	Cassava roots	*Lactobacillus* spp., *Leuconostoc* spp., *S. cerevisiae*	Africa
Gari	Cassava roots	*Corynebacterium, Geotrichumcandidum, L. plantarum, Leuconostocs, Alcaligenes* spp., *Candida* sp.	West Africa
Lafun	Cassava roots	*Leuconostocs, Corynebacteria Candida,*	West Africa and Nigeria
Peujeum	Cassava roots	Yeasts, molds	Java
Chickwa-ngue	Cassava roots	Bacteria	Congo

Source: Modified from Das and Deka (2012) and Swain et al. (2014).

fermentation process as well as in postfermentation from moderate to high amounts. *Leuc. mesenteroides* is the primary organism associated during the initial phase in sauerkraut fermentation followed by *L. brevis*, *L. plantarum*, and *Pd. pentosaceous* (Plengvidhya et al., 2007). The probable reason could be the ability of *Leuconostocs* spp. to grow more promptly over a broad range of temperature and salt concentrations in compared to other LAB. Ratio of acetic:lactic acid is about 1:4 in the final product.

1.4.2.2   KIMCHI

Kimchi is the example of Korean-fermented food which contains acid fermented vegetables like carrot, cabbage, ginger, green onions, green chilies, and garlic. Depending on the processing methods, type of raw material, season, and region more specific terms are applied for these pickled vege-tables. The vegetables are salted after cutting into pieces, prebrined, mixed together with different spices like red pepper, green onion, ginger, garlic, and seasonings agents then followed by fermentation at low temperature (25 °C). Kimchi fermentation is a temperature-dependent process. It ripens in 1 week at 15 °C and takes 3 days at 25°C. Together with rice and other dishes, kimchi is consumed at every meal. Kimchi has characterized by its little salty-sweet, sour (acidic), and carbonated taste. Kimchi conquered its optimum taste when the pH reaches to about 4.0–4.5 and acidity to 0.5–0.6, and at this point, concentration of vitamin C is also maximum (Swain et al., 2014). The major species associated with kimchi fermentation are *Leuc. mesenteroides, Leuc. pseudomesenteroides, Leuc. citreum, Leuc. gasicomitatum, L. plantarum, L. curvatus, L. brevis, L. sakei, Lc. lactis, Pd. pentosaceus, Weissella confusa*, and *W. koreensis* (Surh et al., 2008, Swain et al., 2014).

1.4.2.3   PICKLED VEGETABLES

From centuries, pickled vegetables either made in houses or small factories are very much popular in Asian and Middle East countries. The vegetables pickled include cucumbers, carrots, onions, green turmeric, green chilies, radish/turnips, beets, cauliflower, and peppers. Pickled vegetables are served as appetizers with cooked rice or chapatti in every meal (Prajapati and Nair, 2008). In Malaysia, homemade pickles are termed as *jeruk* that are prepared from fruits and vegetables. In western states of India, chilies are mixed with dahi like fermented milks which given rise to enhanced taste and flavor along with long-lasting quality.

1.4.2.4   OLIVES

Originally, an olive belongs to the Mediterranean area, but now fermented olives are popularly consumed in Europe and United Kingdom. According to Vaughn (1985), there are four kinds of fermented olives available such as California-ripe, brined Greek-type, Siciliano-type green, and Spanish-type green. All these varieties differ in the level of salting, lye treatment, and fermentation period. The major microflora associated with all type of olive fermentations are *L. plantarum*, *L. casei*, *Leuc. Mesenteroides*, and yeasts like *Saccharomyces cerevisiae* and *Candida boidinii*. Olives are given lye treatment to remove bitterness.

1.4.2.5   GUNDRUK

Gundruk is an indigenous vegetable-based fermented product of the Himalayas and northern parts of India. It is characteristically nonsalted and somewhat resembles to other fermented vegetables like kimchi, sauerkraut, and sunki. For fermentation of gundruk, fresh leaves of local vegetables including rayosag (*Brassica rapa* subsp. *campestris* var. *cuneifolia*), mustard (*Brassica juncea* (*L.*) *Czern*), cabbage (*Brassica* spp.), and cauliflower (*Brassica oleraceaL.* var. *botrytis L.*) are kept for wilting for a day or two (Das and Deka, 2012). Once the leaves get wilted, they are mildly crushed, pressed, and packed into airtight container for natural fermentation for about 2–3 weeks. The preparation is removed and allowed to sun-dried for 2–4 days once the desired texture is obtained through fermentation. The predominant microflora of Gundruk includes various LAB such as *L. plantarum*, *L. casei*, *L. casei* subsp. *pseudoplantarum*, *L. fermentum*, and *P. pentosaceous* (Tamang et al., 2005; Karki et al., 1983). Gundruk is served as pickle or used to make soup.

1.4.2.6   KHALPI

A traditional fermented cucumber (*Cucumis sativus* L.) based product khalpi is commonly consumed in eastern India and Nepal. During preparation, ripened cucumber is cut into suitable pieces, sun-dried for a day or two, and kept into a bamboo vessel. It is covered with dried leaves to make airtight and allowed to ferment naturally for 3–5 days at ambient temperature that makes the preparation sour in taste. Khalpi is consumed as pickle by adding salt, powdered chilies, and mustard oil. As reported by Tamang et al. (2005), the major microorganisms isolated from khalpi include *L. plantarum*, *L. brevis*, and *Leuc. fallax*.

1.4.3 FERMENTED CEREAL AND LEGUME-BASED PRODUCTS

Legume and cereal-based products play vital role in the diet of vegetarian population of the world. Preparation of major traditional fermented food products at different regions utilizes common types of cereals (rice, ragi, wheat, corn, or sorghum) and legumes (black gram, bengal gram, green beans, or soya beans). A number of major fermented cereal–legume-based foods originated from different regions of the world are listed in Table 1.4 with their details of origin and microflora involved. Though the traditional methods used to ferment cereals and legumes are quite simple, rapid, and economical, use of modern microbial biotechnology has changed them hastily. Many of these products have quite complex microbiology, for few foods, it is not even known yet. Nevertheless, most of these products involve natural fermentation aid by mixed population of bacteria, yeasts, and molds. During the fermentation process, some of these microflora may act parallel and some in a sequential manner (Blandino et al., 2003; De Vuyst et al., 2009).

Cereal grains and legumes are rich in proteins, dietary fiber, carbohydrates, minerals, and vitamins. However, in compared to milk or vegetable based products, the nutritional values and organoleptic properties of cereal grains and legumes are often inferior or poor. The probable reasons could be the occurrence of antinutritional compounds like polyphenols, phytic acid, and tannins; low starch availability; deficiency of some essential amino acids (like lysine); and the coarse texture (Chavan and Kadam, 1989). During natural fermentation of cereals and legumes, a decrease in the carbohydrate content and oligosaccharides is generally observed; some amino acids may be produced while some may be utilized by the associated microflora; and usually improved availability of certain B group vitamins (folic acid, riboflavin, and thiamine) is observed.

TABLE 1.4 Some Cereal and Legume-Based Fermented Foods of Different Parts of the World.

Product	Substrate	Microorganisms involved	Nature of product	Country/region	Product use
Adai	Cereal/legume	*Pediococcus* spp., *Streptococcus* spp., *Leuconostocs* spp.	Solid	India	Breakfast or snack food
Anarshe	Rice	Lactic acid bacteria	Solid	India	Breakfast or sweetened snack food
Idli	Black gram and rice	*Leuconostocs, Lactobacilli, Streptococci,* yeasts	Solid/spongy	India	Steam cooked breakfast food
Dhokla	Rice, black gram, Bengal gram, pigeon pea and/or wheat	*Leuconostocs, Lactobacilli, Streptococci,* yeasts	Solid/spongy	India	Steam cooked breakfast food
Khaman	Bengal gram	*Leuconostocs, Lactobacilli,* yeasts	Solid/spongy	India	Steam cooked breakfast food
Dosa/dosai	Black gram and rice	*Leuconostocs, L. fermentum, Saccharomyces* spp.	Solid	India	Spongy fried breakfast food
Handvo	Rice, black gram, Bengal gram, pigeon pea	Lactic acid bacteria, yeasts	Solid/spongy	India	Steam cooked breakfast food
Ambali	*Ragi* flour and rice	*Leu. mesenteroides, L. fermentum, Streptococcus faecalis*	Semisolid	India	Breakfast
Koozhu	*Ragi* flour, boiled rice, yoghurt/buttermilk	Lactic acid bacteria	Semisolid	India	Breakfast
Fermented rice	Rice	*Streptococcus faecalis, Pd. acidilactici* and other lactic acid bacteria	Semisolid	India	Breakfast
Jalebies/jeri/jilawii/ zoolbia	Wheat flour	Yeasts, lactobacilli	Solid	India, Nepal, Pakistan, Middle East	Syrup-filled confectionery

TABLE 1.4 (Continued)

Product	Substrate	Microorganisms involved	Nature of product	Country/region	Product use
Puda/Pudla	Bengal gram, mung dal, wheat	Lactic acid bacteria, yeasts	Solid	India	Pancake snack food
Siddhu/Khobli	Wheat flour	Yeast	Semisolid	India	Snack/staple food
Chilra/Lwar	Wheat/barley and buckwheat flour	Yeast	Solid	India	Snack/staple food
Marchu	Wheat flour	Yeast	Solid	India	Snack/breakfast
Chakuli	Par-boiled rice, black gram		Solid	India	Snack or breakfast
Bhallae	Black gram	Lactic acid bacteria, yeasts	Deep fried patties	India	Snack after soaking in curd or water
Bhatura	White wheat flour	Lactic acid bacteria, yeasts	Deep fried bread	India	Breakfast
Kulcha	White wheat flour	Lactic acid bacteria, yeasts	Flat bread	North India and Pakistan	Staple food
Nan	Unbleached wheat flour	Yeasts	Solid	India, Pakistan, Afghanistan, Iran	Snack food
Vadai	Black gram, millet flour	Leuconostocs, H. anomala, Saccharomyces	Deep fried patties	India	Snack
Waries	Black gram flour, cow pea flour	Candida spp., Saccharomyces spp.	Solid	India	Spongy, spicy condiment
Kurdi	Wheat	Unknown	Solid	India	Noodles like snacks
Kanji	Rice and carrots	Hansenulaanomala	Liquid	India	Sour liquid added to vegetables

TABLE 1.4 (Continued)

Product	Substrate	Microorganisms involved	Nature of product	Country/region	Product use
Kinema/Kenima	Soybeans	*Bacillus subtilis, Enterococcus faecium.*	Solid	Nepal, North East India	Snack food
Hawaijar	Soybeans	*Bacillus subtilis, B. cereus, B. licheniformis, Staphylococcus aureus, S. sciuri, Alkaligenes* spp.	Semisolid	North East India	Paste and condiments
Tungrumbai	Soybeans	*Bacillus subtilis, Enterococcus faecium, Candida parapsilosis, Saccharomyces bayanus, Saccharomycopsisfibuligera, Geotrichum candidum*	Semisolid	North East India North East India	Condiments
Aakhone/Bekang/ Peruyyan	Soybeans	Lactic acid bacteria, *Bacillus* spp.	Semisolid	North East India	Condiments
Hopper (Appa)	Rice or wheat flour and coconut water	Baker's yeast, acid-producing bacteria	Semisolid	Sri Lanka	Breakfast
Lao-chao	Rice	*R. oryzae, R. chinensis, Chlamydomucor oryzae, Saccharomyces* spp.	Soft, glutinous	China, Indonesia	Eaten with vegetables
Ang-kak	Red rice	*Monascus purpureus*	Powder	China	Dry red powder used as colorant
Chee-fan	Soybean wheat curd	*Mucor* spp., *A. glaucus*	Solid	China	Eaten fresh, like cheese
Minchin	Wheat gluten	*Paecilomyces* spp., *Aspsesrgillus* spp., *Claadosporium* spp., *Fusarium* spp.	Solid	China	Condiment

TABLE 1.4 (Continued)

Product	Substrate	Microorganisms involved	Nature of product	Country/region	Product use
Mantou	Wheat flour	*Saccharomyces* spp.	Spongy	China	Breakfast, snacks
Meitauza	Soybean cake	*Actinomucor elegans*	Solid	China, Taiwan	Oil fried or cooked with vegetables
Sufu	Soybean whey curd	*A. elegans, Mucorhiemalis, M. silvaticus, M. Subtilissimus*	Solid	China, Taiwan	Soybean cake, condiment
Meju	Soybeans	*A. oryzae, Rhizopus* spp.	Paste	Korea	Seasoning agent
Kichudok	Rice, Takju	*Saccharomyces*	Spongy solid	Korea	Breakfast, snack food
Tape	Cassava roots or Rice	*S. cerevisiae, H. anomala, Rhizopusoryzae, Mucor* spp., *Endomycopsisfibuliger*	Solid/paste	Indonesia and nearby regions	Soft, solid eaten as a staple
Tempeh	Soybeans	*Rhizopus* spp.	Solid	Indonesia and nearby regions	Roasted, oil fried as meat substitute
Kecap	Soybeans, wheat	*Asp. oryzae, Lactobacillus* sp., *Hansenula* spp., *Saccharomyces* spp.	Liquid	Indonesia and nearby regions	Condiment and seasoning agent
Ketjap	Black soybeans	*Asp. oryzae*	Syrup	Indonesia	Seasoning agent
Ontjom	Peanut cake	*Neurospora intermedia, R.*	Solid	Indonesia	Roasted or fried food used as meat
Khanomjeen	Rice	*Lactobacillus, Streptococcus*	Solid	Thiland	Noodles like product
Puto	Rice	*Leuconostocs, Streptococcus faecalis, S. cerevisiae*	Solid paste	Philippines	Snack food
Tao-si	Wheat flour, Soybeans	Unknown	Semisolid	Philippines	Seasoning agent

TABLE 1.4 (Continued)

Product	Substrate	Microorganisms involved	Nature of product	Country/region	Product use
Balaobalao	Rice/shrimp mixture	Unknown	Solid	Philippines	Staple food
Hama natto	Soybeans, wheat flour	*Aspsergillus oryzae, Streptococcus spp., Pediococcus spp.*	Soft	Japan	Seasoning agent for meat fish or eaten as a snack
Natto	Soybeans	*Bacillus natto*	Solid	Japan	Cake used as a meat substitute
Miso	Rice and soybeans	*Aspergillus spp., Torulopsis etchellsii, Lactobacillus spp., Saccharomyces rouxii*	Paste	Japan, China	Paste, soup base
Soya sauce	Soybeans and wheat	*A. oryzae, A. sojae, Lactobacillus spp., Saccharomyces rouxii*	Liquid	Japan, China, Philippines and Oriental countries	Seasoning agent for meat, fish and cereals
Dawadawa	African locust bean	Spore forming bacteria, LAB, yeasts	Solid	West Africa, Nigeria	Eaten fresh or in stews
Mahewu	Maize	Lactic acid bacteria	Solid	South Africa	Staple food
Burukutu	Sorghum and cassava	Lactic acid bacteria, *Candida* spp., *S. cerevisiae*	Liquid	Savannah region of Nigeria	Liquid creamy drink
Dalaki	Millet	Unknown	Spongy solid	Nigeria	Thick porridge
Kwunu-Zaki	Millet	LAB, yeasts	Paste–semisolid	Nigeria	breakfast dish
Ogi	Maize	Lactic acid bacteria, *Cephalosporium spp., Fusarium spp., Aspsergillus spp., Penicillium* spp., *S. cerevisiae*	Paste	Nigeria, West Africa	Staple breakfast food

TABLE 1.4 (Continued)

Product	Substrate	Microorganisms involved	Nature of product	Country/region	Product use
Kenkey	Maize	Corynebacteria, Saccharomyces, molds	Solid	Ghana	Steamed, eaten with vegetables
Koko	Maize	Enterobacterclocae, Acinetobacter., L. plantarum, L. brevis, Candida mycoderma, Saccharomyces cerevisiae	Spongy solid	Ghana	Porridge as staple
Injera	Wheat, barley, teff, maize	Candida guiliermondii	Solid/spongy	Ethiopia	Bread substitute
Taotjo	Roasted wheat meal or glutinous rice, soybeans	Unknown	Semisolid	East Indies	Condiment
Banku	Maize and cassava	Lactic acid bacteria, yeasts	Solid	Ghana	Used as a staple food
Uji	Maize, sorghum or millet flour	Lactobacilli, Pediococci, Leuconostocs	Semisolid	Kenya, Uganda and Tanzania	Breakfast and lunch
Kisra	Sorghum flour	Yeasts, lactobacilli, acetobacter	Spongy bread	Sudan, Arabian Gulf, and Iraq	Staple food
Nasha	Sorghum	Streptococcus, Lactobacillus, Candida, Saccharomyces cerevisiae	Spongy	Sudan	Porridge as a snack
Bongkrek	Coconut press cake	Rhizopus oligosporus	Solid	Central Java	Roasted and fried in oil, meat substitute
Shamsy bread	Wheat flour	Yeasts	Spongy bread	Egypt	Staple food
Kishk	Wheat flour, fermented milk	L. plantarum, L. brevis, L. casei, Bacillus subtilis, yeast	Solid	Egypt, Syria and Arabian countries	Breakfast meal

TABLE 1.4 *(Continued)*

Product	Substrate	Microorganisms involved	Nature of product	Country/region	Product use
Tarhana/Trahanas	Wheat flour, sheep milk yoghurt, cooked vegetables and spices	*Lactic acid bacteria, yeasts*	Solid	Greece and Turkey	Breakfast meal and soup
Bagni	Millet	*Unknown*	Liquid	Caucasus	Drink
Poi	Taro corms	*Lactobacillus spp., Candida vini, Geotrichum candidum*	Semisolid	Hawaii	Taken with fish or meat
Pozol	White maize	*Molds, yeasts*	Solid	Mexico	Beverage or porridge
Atole	Maize	Lactic acid bacteria	Solid	Southern Mexico	Porridge based on maize dough
Bogobe	Sorghum	Unknown	Spongy	Botswana	Soft porridge staple
Chicha	Maize	Aspergillus, penicillium, yeasts, bacteria	Spongy solid	Peru	Eaten with vegetables
Jamin-bang	Maize	Yeasts, bacteria	Spongy solid	Brazil	Bread, cake-like
Ilambazilokubilisa	Maize	LAB, yeasts and molds	Spongy solid	Zimbabwe	Porridge as weaning food
Mutwiwa	Maize	LAB, bacteria and molds	Spongy solid	Zimbabwe	Porridge
Kaanga-Kopuwai	Maize	Bacteria, Yeasts	Spongy solid	New Zealand	Soft, slimy eaten as vegetable

Source: Modified from Soni and Sandhu (1999), Prajapati and Nair (2008), Das and Deka (2012).

1.4.3.1 TRADITIONAL RICE AND LEGUME-BASED FERMENTED FOODS

1.4.3.1.1 Idli and Dosa

Idlis are naturally fermented steam cooked products made from the mixture of rice (*Oryza sativa*) and black gram (*Phaseolus mungo*) in usually 3:1 ratio. The ingredients are soaked separately and ground before fermentation, and the batter is steamed to produce soft, spongy pancakes often eaten for breakfast (Soni and Sandhu, 1989; Thakur et al., 1995). The microflora found included *Leuc. mesenteroides*, *L. fermentii*, *Stre. faecalis*, *P. cerevisiae*, *Weissella cibaria*, Bacillus spp., and many a time yeasts like *Candida* spp. and/or *Saccharomyces* spp. The microflora usually involved with biochemical changes and biosynthesis of vitamins like thiamine, riboflavin, and vitamin B_{12}.

Dosa is a thin, crisp, and fried pancake made from overnight fermented batter of black gram and rice mixture that is widely popular in South India and parts of Western India. Dosa are also prepared from green beans. Either the batter is autofermented or inoculated with starter of previous batch. It is cooked in form of pancakes. It is reported that the bacterial population ranged from 10 cells per gram, and the yeast population ranged from 0 to 10 cells per gram in the batter (Beuchat, 1978; Evans and McAthey, 1991). Yeasts contribute to confer typical flavor to dosa batter. Dosa is also served by filling stuffing of vegetables (like potato, onion) inside or with toppings on the surface and name differently like masala dosa, onion dosa, uttapm, etc. The microflora of dosa includes *Leuconostocs* spp., *L. fermentum*, *Candida* spp., and *Saccharomyces* spp.

1.4.3.1.2 Dhokla and Handvo

Dhokla is popular fermented steam cooked product of Western India, particularly of Gujarat and Maharashtra state. They are made in a similar way as like idli however bengal gram (*Cicer arietinum*), wheat, and/or curd are also added to the mixture (Rati Rao et al., 2006). The steamed dhokla is seasoned with oil and spices and garnished with coriander leaves before consumption. The microflora presents mainly included *Leuc. mesenteroides*, *L. fermentii*, *Stre. faecalis*, and *Bacillus* species like *B. subtilis*, *B. licheniformis* in combination with several yeasts. Changes in the batter properties are similar to that for idli. Handvo is made by mixing rice, Bengal gram, black gram, and other legumes in various proportions. After 7–8 h of natural fermentation, the dough is added with salt, spices, oil, seasame seeds, and/or shredded vegetables and steam cooked. The fermenting microflora is similar to that of dhokla.

1.4.3.1.3 Koozhu

Koozhu is fermented cereal-based product popular in South India. In the preparation, a flour of *Eleusine coracana* (ragi, sorghum, or millet) is mixed with water and left overnight to develop acidity. Then, this sour dough is cooked with moderate heat to evaporate most of the water. It requires constant stirring to obtain right consistency. After that a small part of cooked rice is added, the mixture is allowed to cool. Prior to consumption, yoghurt and salt are also mixed into kozhu.

1.4.3.1.4 Ambali

It is fermented ragi and rice-based product. For the preparation, a thick batter is made from mixing of ragi (millet) flour that is left for fermentation for 14–16 h. Later, the fermented batter is moderately heated together with almost three-fourth part of partially cooked rice and stirred continuously till completion. The finished product is having decreased pH (from 6.4 to 4.0) and up to 20% increase in the volume due to CO_2 production. The major microflora involved in the fermentation is *Leuc. mesenteroides, L. fermentum*, and *Stre. faecalis* (Ramakrishnan, 1993).

1.4.3.2   TRADITIONAL WHEAT-BASED FERMENTED FOODS

1.4.3.2.1 Bread and Sourdough

Bread is considered as one of the widely consumed acceptable staple foods in almost all regions of the world and is a good source carbohydrates, protein, and fat and other micronutrients (including vitamins and minerals) that are essential for human health (Oluwajoba et al., 2012). The whole bread is fermented by using yeast cultures, and *Saccharomyces cerevisiae* is the usually used to leaven which ferments the sugar and other carbohydrates from the flour produce CO_2. To produce rapid, uniform, and reliable quality of bread each time, majority of bakers leaven dough with commercially available baker's yeast. Alternatively, several artisan bakers develop their own yeast culture to make breads at commercial level.

In many European countries, sourdough which is similar to bread is popular. It is obtained through fermentation of dough using inherent cultures of lactobacilli and yeasts for longer period in compared to bread. Sourdough also differs from bread as because of lactic fermentation, it tastes mildly sour. It is important during production of rye-based breads because alone yeasts do not produce good-quality product through dough fermentation. Generally, sourdough is defined as a microbial ecosystem involving mainly LAB and yeasts together in a cereal flour and water matrix. Even though the type and quality of sourdough and the flour are the main source of autochthonous bacteria and yeasts during natural fermentation processes, it is quite difficult to establish linkage among the species diversity of the sourdough microflora with the fermentation latter (De Vuyst et al., 2009, 2014). Among LAB, *L. sanfranciscensis*, *L. brevis*, *L. alimentarius*, *L. delbrueckii* subsp. *delbrueckii*, *L. plantarum*, *L. fermentum*, *L. acidophilus*, *W. confusa*, *Leuc. citreum*, and *Lc. lactis* subsp. *lactis* are the commonly found microflora of different sourdough, whereas major yeast species associated with sourdough are *Saccharomyces cerevisiae*, *Saccharomyces exiguous* (*Kazanchastania exigua*), *Candida milleri*, and *C. humilis* (Corsetti et al.,, 2001; Zhou and Therdthai, 2012).

1.4.3.2.2 Jalebi

Jalebi is a sweetened fermented product of Arabic or Persian region which reached to India in 1450 A.D. and popularized as dessert. In the manufacturing of jalebi, a batter of wheat flour is inoculated with baker's yeasts and fermented for 8–10 h. Afterwards, the batter is squeezed through a fine cotton cloth having a hole of 3–5 mm diameter and prepared as rings-spirals in hot oil or ghee and fried up to yellow to light brown color and immediately dipped in hot sugar syrup (70–75%). The finished crispy product is called Jalebies, which are served as dessert after removing from hot sugar syrup.

1.4.3.2.3 Kishk (Fugush)

A mixture of wheat and fermented milk, popularly called kishk, is an important food of the people of Middle East. During the kishk manufacturing, wheat grains are washed, soaked, and boiled; sun-dried and grinded to make powder. This powder and traditional fermented milk known as laban zeer mixed and fermented for 24 h. After fermentation, the mixture is sun-dried in form of small balls-lumps. Kishk is having little sour taste, usually taken into breakfast or used to make curry. The completely dried balls can be kept for years (Blandino et al., 2003).

1.4.3.2.4 Tarhana (Trahanas)

Tarhana is consumed popularly in Greece and Turkey and very similar to kishk. It is prepared by from wheat flour, fermented sheep milk, cooked vegetables, salt, pepper, and other spices together with yeast, and the mixture is fermented for a day to week. After that the fermented dough is sun-dried. Such dried balls or biscuits can be preserved for years (Blandino et al., 2003). Tarhana has a typical yeasty aroma and flavor and acidic-sour taste. It serves good amount of protein, dietary fibers, and vitamins. Tarhana is consumed as breakfast snacks and used to make soup or curry with meal.

Kishk and tarhana have gained nutritional importance in diet because of the augmentation of the cereal protein together with milk proteins and also in acceptable amount. Presence of very low moisture with acidic pH (about 3.8–4.2) provides poor conditions for the growth of contaminating and spoilage microorganisms (Haard et al., 1999).

1.4.3.3 TRADITIONAL MAIZE-BASED FERMENTED FOODS

1.4.3.3.1 Ogi

Ogi is a fermented cereal product from West Africa which is important weaning food for infants. It is manufactured by fermentation of maize, sorghum, or millet. Depending on the type of substrate and form in which it is eaten, ogi is given different names such as *akamu, eko, koko, agidi, furah*, or *kamu* in different regions of West Africa (Blandino et al., 2003). To prepare ogi, the grains are steeped into earthen pot or plastic pot for 1 to 3 days. LAB, predominantly *L. plantarum*, yeasts, *Saccharomyces, Candida* spp., and molds are the main fermenting microflora of ogi. Other bacteria such as *Corynebacterium* are responsible to hydrolyse starch, whereas the yeasts contribute to develop flavor in the product (Caplice and Fitzgerald, 1999). The fermented grains are then after milled with water to make ogi slurry. Ogi possess slight acidic flavor and taste comparable to that of yoghurt. It has a typical distinct aroma which makes it diverse from other fermented cereal-based products. Depending on the cereal grain, the color of finished product varies for instance cream-white, reddish brown, or dark gray color for maize, sorghum, and millet (Onyekwere et al., 1993).

1.4.3.3.2 Kenkey

Kenkey is traditional maize-based product in form of fermented dough of Ghana. During the preparation, maize grains are soaked into water for a day or two, and then, water is drained off from hydrolyzed grains which are further wet-milled, and stiff dough is prepared by adding water and naturally fermented. LAB, particularly *L. fermentum* and *L. reuteri*, fungi involving *Saccharomyces*, *Candida*, *Aspergillus*, *Penicillium*, and *Fusarium* spp. were found to predominate and contribute to flavor development in kenkey (Blandino et al., 2003; Jespersen et al., 1994).

1.4.3.3.3   Pozol

Pozol is popular fermented maize-based dough in form of small balls of different shapes and size of South-Eastern Mexico. Initially, the maize grains are boiled in limewater, wet-grinded coarsely. Small compact balls are made from this mixture and kept for fermentation from few hours to several days at ambient temperature after wrapping in banana leaves. A diverse complex microflora which gets incorporated during the process is responsible for the dough fermentation (Nanson and Field, 1984). Various LAB including *L. delbrueckii* subsp. *bulgaricus*, *L. casei*, *L. plantarum*, *L. alimentarium*, *Lc. lactis*, *Lc. suis*, *Bacillus* spp., and *Clostridium* spp. have been isolated from pozol (Escalante et al., 2001).

1.4.3.4   TRADITIONAL SORGHUM-BASED FERMENTED FOODS

1.4.3.4.1   Injera/Enjera

Injera is the traditional fermented foods of Ethiopia. Tef (*Eragrostis* tef) is the main cereal to make it; however, different cereals like maize, sorghum, millet, and barley are used to make injera. Injera is nutritionally rich in calcium and iron like minerals (Zegeye, 1997). Initially, the cereal grains are dehulled and milled to make flour that is blended with water to prepare dough using a fluid from previous batch of fermented dough as starter. It is allowed to ferment for 2–3 days. Subsequently, the dough converted in to batter, poured onto a pan, and covered with a tight lid to retain the steam during heating. After 2–3 min, the product is removed from pan and placed on a basket. Various types of molds and yeasts species like *Penicillium*, *Pullaria*, *Aspergillus*, *Hormodendrum*, *Rhodotorula*, *Candida*, and a few of unidentified bacteria are main microflora involved in fermentation of injera (Ashenafi, 1993). Injera is a round, spongy, and flexible product that has uniform holes on the surface and has a slight sour taste and flavor.

1.4.3.4.2 Kisra

Kisra is fermented food similar to that of injera and popularly consumed in Arabian countries like Sudan and Iraq (Blandino et al., 2003). It is prepared from sorghum or pearl millet flour that is fermented as dough and baked in form of thin sheets. The major microflora in kisra is *Lactobacillus* spp., *Acetobacter* spp., and yeast *S. cerevisiae*. Kisra has higher amount of vitamins like riboflavin and minerals because of fermentation (Mahgoub et al., 1999).

1.4.3.4 TRADITIONAL SOYA BEAN-BASED FERMENTED FOODS

Soya beans provide cheap source of most essential proteins. Soya protein is nutritionally comparable to milk protein casein and animal-derived products like meat and egg. The whole grains possess considerable amounts of flavonoids, terpenoids, alpha-linolenic acid, isoflavones, and other natural antioxidants like carotene, ascorbic acid, and tocipherol (Anderson et al., 1995; Das and Deka, 2012). Hence, the therapeutic values of soya bean products are also noteworthy. Fermented soya-based food includes soya sauce, miso, natto, sufu, tempeh, kinema, douche, and doenjang.

1.4.3.4.1 Soya Sauce

Soya sauce, a dark brown colored liquid condiment, is obtained from fermented rice and/or wheat with soya bean. It is popularly called *chau yau* or *pak yau*, *kenjang*, *shoyu*, *ketjap*, *kecap*, *see-ieu*, and *toyo* in China, Korea, Japan, Malaysia, Indonesia, Thailand, and Philippines, respectively (Wang and Hesseltine, 1970; Prajapati and Nair, 2008). While preparing the sauce, coarse wheat flour is mixed with cooked soya beans, keeping 55% (w/w) initial moisture of the mixture, inoculated by molds, and left for 3 days of fermentation at 25–35 °C. This mixture is known as koji, which is further immersed in a brine solution (1:3 w/v ratios). Now, this koji with brine is known as moromi having about 18–21% salt, and it is put for fermentation for few months. Afterwards, the liquid portion (soya sauce) is separated, sieved, pasteurized, and packed (Franta et al., 1993; Rowan et al., 1998). It is believed that longer the fermentation period, better the soya sauce quality. The major groups reported in the literature include molds like *Aspergillus oryzae* and *A. soyae*, yeasts—*Zygosaccharomyces rouxii* and *Candida* spp., and halotolerant LAB-like pediococci (Blandino et al., 2003).

1.4.3.4.2   Miso

Miso is a fermented paste obtained from soya beans originated in China. It is popularly famous as *miso* in Japan, whereas in China, Korea, Indonesia, Thailand and Philippines, it is known as *chiang, jang* or *deoenjang, tauco, taochieo*, and *tao-si*, respectively, over thousand years (Prajapati and Nair, 2008). In such preparations, mainly soya beans are mixed with rice or barley and fermented for few months by a fungi *Aspergillus oryzae* (Steinkraus et al., 1993). Miso finds its use as soup base or as sauce that can be served with vegetable dishes, meat, seafood, or poultry products.

1.4.3.4.3   Tempeh

Tempeh is popular soya-based fermented product of Indonesia and produced in Malaysia, Singapore, the United States, and Canada. The fermentation process of tempeh with retention of the entire bean confers tempeh very high proportion of proteins, dietary fibers, vitamins, and minerals (Beuchat, 1983). It is used as a meat analog worldwide in vegetarian cuisine. During tempeh manufacturing, soya beans sometimes together with wheat are soaked to get soften, dehulled, and cooked partially. The fermenting starter *Rhizopus oligosporus* or *R. oryzae* in form of spores is mixed into this mixture, and to lower the pH as well as to make selective environment that may favors the growth of the tempeh mold over competitors, vinegar may be added too. After this, the beans are fermented for 24–36 h at ambient temperature after as a thin layer (Keuth and Bisping, 1994). Tempeh has a firm body texture, and as ages, its characteristic earthy flavor becomes more pronounced. Frequently, it is cut into pieces, soaked in brine solution or salty sauce, and fried. Because of the fermentation procedure, the soy carbohydrates get more digestible particularly the oligosaccharides that are linked with gas and indigestion problems reduced significantly. In Indonesia, many traditional tempeh-manufacturing shops utilizes starter cultures offering biosynthesis of vitamins such as B_{12} such as strains of *Citrobacter freundii* and *Klebsiella pneumonia* (Keuth and Bisping, 1994).

1.4.3.4.4 Natto

Natto is a famous Japanese-fermented product made from fermentation of whole soya beans with *Bacillus subtilis* var. *natto*. *Natto* means "contributed beans" products which have a strong pleasant smell and flavor, slimy texture, and dark brown color. Natto serve as a substitute for fermented fish and meat-based foods due to high nutritional significance at reasonably low price. It is also known as *thua-nao* in Thailand, whereas in China, Philippines, and Eastern India, it is popularly termed as *tu-si, tao-si tao-tjo* (Prajapati and Nair, 2008). Usually, it finds its use as seasoning to vegetables or cooked meat and seafood and is also served with rice.

1.4.3.4.5 Sufu

Soya curd is popularly known as *sufu* which means "molded milk", and *tosufu* means "molded bean milk." It is a popular product of China and is referred as "Chinese cheese." It has many synonyms in China like *tosufu, fu-su, fu-ru, foo-yue*, etc. because of the numerous dialects used, whereas in other countries, it is termed differently like *tao-hu-yi* in Thailand and Taiwan, *tahuri* in Philippines, *chao* in Vietnam, and *takaoan* in Indonesia. The main microorganisms involved in fermentation process include molds such as *Aspergillus elegans, Mucor hiemalis, M. silvaticus, M. subtilissimus, Rhizopus* spp., and yeasts (Prajapati and Nair, 2008).

1.4.3.4.6 Kinema

Kinema is a soya bean-based fermented food of North-east India. It is produced by the natural fermentation of bacterial species namely *B. subtilis* and *Enterococcus faecium*; however, in market, samples of *Candida parapsilosis* contaminating yeast and a mold *Geotrichum candidum* were detected sometimes (Sarkar et al., 1994; Das and Deka, 2012). For preparation of kinema, soya beans are required to soak for overnight, 10 boiled and grinded lightly through a wooden mortar and pestle to increase the surface area during prompt fermentation. Then to maintain an alkaline condition, a pinch of fire-wood ash (1%) is mixed and placed on a bamboo basket that is previously lined with *Glaphylopteriolopsis erubescens*, a type of fresh fern. A jute bag is used to cover such basket and placed for natural fermentation in ambient conditions. Afterwards, for 1–2 days, the full thing is put over an earthen oven. Appearance of white viscous mass on the beans with ammonical odor is indicative of completion of fermentation process. In winter, kinema has as shelf life of 5–7 days, whereas in summer, it is only 2–3 days (Tamang et al., 2009).

1.4.3.4.7   Doenjang

Doenjang is fermented soya-based product from Korea. While making, soya beans are boiled, stone-ground to produce coarse paste which is arranged into blocks termed *meju* (Surh et al., 2008). Such blocks are exposed to sunlight, and on the surface of the blocks, dried rice plants are attached which served as a rich source of bacteria, especially *B. subtilis*. From the meju, these bacteria utilize soya protein and water and increase their population and develop unique odor of ammonia in the meju (Kim et al., 2009). On the basis of block size, the meju are placed into big opaque pottery jars containing brine solution after 1–3 months. Diverse beneficial bacteria enhance the nutritional significance of the mixture further during storage. At the end of fermentation, the liquids and solids are separated; the liquid is Korean soy sauce while the solid part is doenjang that have very salty and quite thick consistency. Traditional household doenjang utilizes only soya beans and brine solution; however, the products contain wheat flour in addition to soya beans in many commercial variants (Kim et al., 2009). To heighten the savory flavor of doenjang, some people also adjoin dried fermented ground anchovies. Doenjang find its use as seasoning agent, dipping condiments, or as a paste with vegetables.

1.4.3.4.8   Douchi

Douchi is made by fermentation of whole soya beans followed by salting. The black type soya bean is most commonly used, and the processes turn the beans soft, and mostly dry (Batt, 1999). If white beans are used, they become brown after fermentation. Douchi is used only as a seasoning and to flavor fish or vegetables.

1.4.4 FERMENTED FISH PRODUCTS

Development of fermented fish based products like sauce, paste, and salted fish have been practiced from old periods in the different countries particularly in coastal regions. Production of fermented fish products could be correlated well with the geographical distances due to poor roads and transportation facilities, the availability of fresh fish was not possible to inland consumers. In Europe, fermented fish products find use as condiments only, whereas Southeast countries like Thailand, Philippines, Malaysia, Indonesia, and Japan consider such products as staple foods (Chojnacka, 2009; Prajapati and Nair, 2008).

In the world, fermented fish pastes are served as condiments with vegetables, breads, and other foods, so they are more popular in compare to fish sauces, though in market a number of fish sauces exist. In general, the fermentation time for paste is shorter than for sauces. Fish is rich in protein, thus make an important contribution in the daily diet. Fresh fish are highly perishable and, hence, if they are not preserved by salting, drying, boiling, smoking, or any other means, it may get spoiled in short duration. Preservation of fish by means of fermentation is the very common mode in Southeast Asian countries, and compared to salted or dried fish, fermented fish in form of sauce or paste is more popular. Fish and crustaceans mainly used in the fermentation from South-East Asia are shown in Table 1.5.

TABLE 1.5 Fish and Crustaceans Mainly Used in the Fermentation from South-East Asia.

Product group	Species
Saltwater fish	Anchovies, herring, deep-bodied herring, Fimbriated herring, mackerel, round scad, slipmouth
Freshwater fish	Carp, catfish, climbing perch, gourami, mudfish
Shellfish and crustaceans	Shrimp, mussels, oysters, ctopus

Source: van Berkel et al. (2004).

For many years, fermented fish was considered as a Southeast Asian food only. These products are salted prior to fermentation during which fish tissue gets transformed into simpler easily digestible form, it may take several months (van Berkel et al., 2004). Fish contains its own microbiota in different parts of the body like in gut and gills inherently which carry out putrefactive changes in the fish fleshes when it dies. Moreover, the inherent microorganisms of the salt may also contribute in this process. Most of the LAB are associated with normal fermentation, whereas *Lactobacilliaceae*, *Enterobacteriaceae*, *Bacilliaceae*, and *Micrococcaceae* are spoilage type microorganisms. The common pathogens associated with fish are *Bacillus cereus*, *Staphylococcus* spp., *Cl. perfringens*, *Cl. botulinum*, *Salmonella* spp., *Vibrio parahaemolyticus*, and xerolytic molds. Fermented fish products of different countries are comprised in Table 1.6.

During fermentation process, because of higher concentration of salt fish, protein is broken down through enzymatic actions. It is believed that microorganisms do not have any significance in this protein breaking process; even so, they do contribute to enhance the specific taste and flavor of the finished product (van Berkel et al., 2004).

TABLE 1.6   Fermented Fish Products of Different Countries.

Fish sauce	Kind of fermentation	Country/region
Nuoc-mam, Nampla	Enzymatic	South-East Asia—Vietnam
Patis	Enzymatic	Philippines
Shottsuru	Enzymatic + Microbial	Japan
Fish Paste		
Bagoong	Microbial	Philippines
Balao-balao	Microbial	Philippines
Belachan	Enzymatic + Microbial	Philippines
Ngapi	Enzymatic + Microbial	Philippines
Prahoc	Enzymatic	Kampuchea
Trassi	Enzymatic + Microbial	Indonesia
Colombo cure	Enzymatic	Indonesia
Pedah-siam	Enzymatic	Indonesia
Sushi	Microbial	Japan
Anchoa	Enzymatic	South America
Momone	Enzymatic	Africa-Ghana

Source: van Berkel et al. (2004).

A few traditional fermentation processes also involve addition of some simple source of fermentable carbohydrates like boiled rice, for instance, during the production of fermented fish product called sushi (van Berkel et al., 2004). Such technique enhances the growth of fermenting microflora like LAB. The fermented fish products have higher shelf life and are much safer due to production of lactic acid and subsequent reduction of the pH. There are mainly two major groups of traditional fermented fish products: (1) products in which fermentation is done by the enzymes present in the fish tissues and gut in the presence of high salt and (2) products that utilize either boiled or roasted rice during fermentation.

1.4.5 FERMENTED MEAT PRODUCTS

Fermented meat products or sour meat offer a different organoleptic taste to consumers taking meat in their regular diet and thus, its popularity is growing day by day. During the early stages of storage added salt, nitrite, and/or nitrate together with reduced pH by means of lactic acid biosynthesis through fermenting microflora like LAB, and in the later stage, low water activity (a_w) due to drying contributes to extend the keeping quality of fermented meat sausages more often (Hutkins, 2006; van Berkel et al., 2004). Most starter cultures selected for their metabolic activity also enhance the flavor and aroma of these products. Meat sausages are known as *salami*, *cervelatwurst*, *nham*, *plockwurst*, or *boulogna* in different countries and are made of beef, pork, chicken, or lamb.

Fermented meat products are highly popular in European countries as compared to other regions, though they are manufactured and consumed globally (Table 1.7). According to survey, about more than 6 billion kg of meat products are consumed in only in Scandinavia, Germany, France, Italy, and Spain, and to that also around 3–5% of all meat is mainly utilized in form of sour sausages. The most commonly used starters are *L. plantarum*, *Pd. Pentosaceus*, and *Pd. acidilactici* in America, whereas *Staphylococcus xylosu*, *S. carnosuss*, and to a lesser extent *Micrococcus* spp. are the common microflora found in the fermentation of meat products (Hutkins, 2006).

Selected strains of LAB can effectively inhibit the spoilage as well as some pathogenic bacteria without affecting the organoleptic properties of the fermented meat sausages (Metaxopoulos et al., 2002). Thus, LAB finds exclusive use to enhanced shelf life, safety, and quality aspects for fermented meats.

TABLE 1.7   Fermented Meat Products of the World.

Product	Meat type	Country
Moist sausages	Beef pork	North America
Lebanon bologna		Italy, France, USA
Mortadella		
Semidry sausages	Beef/pork	USA
Summer sausage	Pork	Italy, France, USA
Thuringercervelat		
Dry sausages	Beef/pork	Italy, North America
Pepperoni	Pork	Europe, USA, Mexico, Middle East countries
Salami	Beef/pork	
Chorizo		Italy
Whole meats		
Country-cured hams	Pork	North America
Parma hams	Pork	Italy

Source: Modified from Hutkins (2006).

1.5   BENEFITS OF FERMENTED FOODS

In general, fermentation helps to improve the shelf life of the raw substrate for extended period of time.

1.5.1   HIGHER SHELF LIFE AND ORGANOLEPTIC TASTE

The key purpose behind fermentation of any food is to enhance the keeping quality without much effect on the original properties. For example, fermented milks like yoghurt will last longer than milk; sauerkraut and pickles can be preserved for a long time than fresh vegetables. Apart from this, fermentation serves to enrich the organoleptic qualities of the original raw substrate, increase in the aroma and flavor of end product along with improved body and texture. Many fermented foods such as smelly cheese, fermented milks, sauerkraut, and idli are particularly enjoyed because of their strong flavors. More than this, fermentation enables to create verities of food.

1.5.2 ENHANCED DIGESTIBILITY AND NUTRITIONAL VALUE

The fermentation process actually makes the food more digestible by breaking down hard-to-digest carbohydrates (starch, cellulose, and pectin) and proteins into simpler form in fermented product. Fermentation results in enhanced concentrations of protein, dietary fibers, vitamins, and minerals.

During fermentation of sourdough and other cereals, LAB synthesizes many metabolites that have been shown to confer a valuable effect on the body, texture, and appearance of fermented product. Fermentation also improves staling of bread because of production of organic acids, biopolymers and/or some enzymes by fermenting microbiota. A kind of biopolymers like exopolysaccharides produced by food grade bacteria have been shown to have a potential to substitute expensive hydrocolloids which find application as viscosity enhancer or texture improvers for the products like bread, idli, and dhokla (Nisha et al., 2005).

The outcome of fermentation on the concentration of protein and amino acids is always a debatable topic. As an instance, the amount of available amino acids like lysine, tryptophan, and methionine increases during maize fermentation (Nanson and Field, 1984). During fermentation, cereal-based fermented product, idli batter a noteworthy raise in all essential amino acids, has appeared. Likewise, fermentation considerably improves the protein quality and the level of lysine in cereals like millet, maize, and sorghum (Hamad and Fields, 1979). In contrast, though tyrosine and methionine contents increased, it did not elevate lysine content in sorghum *kisra* bread (McKay and Baldwin, 1990). Authors also reported an increase in the tryptophan level during *uji* fermentation with a significant decline in the level of lysine (McKay and Baldwin, 1990). Such outcomes suggest that fermentation process variably influences the nutritive value of different substrates and food products.

In terms of vitamins, it is observed that fermented foods like sauerkraut and kimchi have increased vitamin C content. Other traditional fermented food products show a big jump in B complex vitamin like thiamine, riboflavin, and biotin content. Recently, Spano (2011) employed vitamin B_2 biosynthesizing strains of LAB to enrich fermented sourdough as well as pasta naturally with riboflavin. Tempeh, a traditional Indonesian soya bean-based fermented food, has shown elevated levels of B group vitamin from *Streptococcus* and *Enterococcus* spp., as reported by Keuth and Bisping (1994). A few species of Propionibacteria such as *P. shermani* is also responsible for producing vitamin B_{12} in certain cheese varieties (Burgess et al., 2004). It is also documented that fermentation improves bioavailability of zinc, iron, and calcium like minerals in the fermented food (Sasikumar, 2014).

1.5.3 DESTRUCTION OR DETOXIFICATION OF ANTINUTRITIONAL COMPOUNDS AND TOXINS FROM RAW FOOD MATERIALS

The raw vegetables, fruits, legumes, or cereals may have spoilage causing or pathogenic microorganisms on it. Once the fermentation process starts, organic acids and other metabolites formed by fermenting microflora found to inhibit growth of unwanted microorganisms; hence, fermented foods are safer than eating raw ingredients. Fermentation may help to destruct or detoxify some undesirable substances inherent in raw food materials such as toxins, phytic acid, nonstarch polysaccharides (NSPs), oligosaccharides, tannins, lectins, saponins, polyphenols, and enzyme inhibitors. These compounds have been reported to increase disease susceptibility, cause gastric disturbances, intestinal damage, and reduced growth performance (Anderson et al., 1995; Steinkraus et al., 1993; Rose, 2015). Fermentation processes carried away by LAB and certain molds have shown a considerable effect on destruction and/or drop of these antinutritional substances and the flatulence problem. Fermenting microflora have the ability to synthesize phytase enzyme. Phytic acid primarily occurs in cereals and legumes in the form of complexes with polyvalent cations like Ca, Mg, Zn, Ir, and some proteins, and fermentation gives optimal pH for enzymatic degradation. Hence, such diminution in phytic acid may also raise the levels of soluble minerals (Chavan and Kadam, 1989; Haard et al., 1999).

During the *idli* fermentation, a marked reduction (up to 35–40%) in phytate content was acknowledged. Similarly, appreciable amount of reduction up to 33.8% in the concentration of nondigestible oligosaccharides like raffinose, verbascose, and stachyose due to fermentation and steaming of *idli* was also reported. During the preparation of *idli*, fermentation and steaming assist to eliminate chymotrypsin-inhibiting activity, though trypsin-inhibiting activity remains unaffected (Sasikumar, 2014). Polyphenol oxidase enzyme naturally present in the food grain or fermenting microflora may reduce concentration of polyphenols during fermentation process.

Most often food products are contaminated with several toxins either inherently or by invasion through various microorganisms. Secondary toxic metabolites called mycotoxins like aflatoxin, ocratoxin A, and fumonisins are frequently reported from molds. Certain techniques are available to decontaminate food products from such toxins, for instance, alkaline ammonia treatment can destruct mycotoxins; however, these methods make use of harmful chemicals that can impair/reduce the nutritional value of foodstuff. As majority of toxins are heat-stable, cooking or heat treatment can not eliminate toxins either. However, in last few years, elimination of mycotoxins from foods has been reported during fermentation process (Mokoena et al., 2005; Chelule et al., 2010; Dalie et al., 2010). LAB destruct mold toxins without leaving any toxic residues or effect on nutritional quality and flavor of contaminated food, and thus, employing LAB in fermentation is considered as a vital alternative (Bata and Lásztit, 1999). The detoxifying activity is assumed to be through toxin-binding ability or possible enzymatic interaction of LAB (Zinedine et al., 2005; El-Nezami et al., 2002; Haskard et al., 2001). Similarly, subsequent LAB fermentation of cassava has also revealed positive results through detoxification of cyanogens, cassava toxin (Caplice and Fitzgerald, 1999). Fermentation is imperative to ensure safety to consume cassava, and thus, it is crucial to carry out fermentation to eliminate the toxins from the leaves of *Cassia obtusifolia* prior to the production of Sudanese product *Kawal.*

1.5.4   HEALTH-PROMOTING EFFECTS

Apart from improved digestibility, enhanced nutrient content as well as organoleptic characteristics, traditionally fermented foods, have been connected with many health beneficiary activities such as longevity, gastric disturbances, GI tract infections, tuberculosis, leishmaniasis, and other illnesses in different regions. Many scientific studies have proven health-promoting effects of different fermented foods through in vitro and in vivo investigations (Choi et al., 1997; Surh et al., 2008), and it is shown in Table 1.8. For instance, yoghurt, dahi, and kumiss like fermented milks have shown antimicrobial activity and effectiveness against gastric disturbances. Kimchi contains various health-promoting components, including vitamin C, β-carotene, chlorophyll, and dietary fiber. These compounds are assumed to prevent cancer (La Vecchia et al., 2001; Song, 2004). In addition, kimchi also believed to defend against disease due to its antimutagenic, antioxidative, and angiotensin-converting enzyme inhibition activities (Park et al., 1998; Kim et al., 1991). Similarly, sauerkraut has been shown specifically to be cancer fighting, and a digestive aid. Soya-based fermented food *Doenjang* possess flavonoids, vitamins, and plant hormones like phytoestrogens that are often claimed to own anticarcinogenic activities (Surh et al., 2008).

TABLE 1.8   Health Promoting Activity of Several Fermented Food Products.

Fermented food product	Bioactive compounds	Reported health benefits
Kimchi	Beta carotene, chlorophyll	Antimutagenic activities, angiotensin converting activity
Sauerkraut	Vitamin C, dietary fibers	Anticancer property, antioxidant activity
Olives	Vitamin C, flavonoids	Antioxidant activity
Soya bean fermented products	Isoflavones, alpha linolenic acid	Antioxidant activity
Sourdough	Riboflavin	Cofactor in various biochemical reactions
Doenjang	Flavonoids, vitamins, minerals, plant hormones (phytoestrogens)	Anticarcinogenic properties
Yoghurt	Folic acid	Antioxidant activity, antimicrobial activities

1.5.5 EXCELLENT SOURCE OF PROBIOTICS AND PREBIOTICS WITH ENHANCED FOOD SAFETY AND QUALITY

With more and more research showing the importance of a healthy gut, the significance of foods that can aid our gut health is becoming more prominent. Fermented foods are an excellent source of live microorganisms, especially health beneficiary bacteria known as probiotics. Probiotics have been defined as the viable microbial supplements ingestion which provides benefit to the host (FAO/WHO, 2001). It comprises specific strains of *Lactobacillus* spp., *Bifidobacterium* spp., *Bacillus* spp., and some yeast. However, chiefly the members of the genera *Lactobacillus* and *Bifidobacterium* are employed in commercial products as probiotic bacteria. The major probiotic *Lactobacillus* species are *L. casei, L. acidophilus, L. reuteri, L. johnsonii, L. rhamnosus,* and *L. gasseri, whereas among Bifidobacteria, B. longum, B. bifidum,* and *B. infantis are the main probiotic strains.* The potential health-promoting activities of such beneficial microbes are mainly caused by their ability to maintain the balance of intestinal microflora, prevent GI tract infections through inhibition of spoilage and/or pathogenic microorganisms and stimulate the immune system by activation of immunestimulatory cells (like macrophages, WBCs) of the host's body. In general, the antimicrobial activities of such bacteria is associated with their ability to produce organic acids, hydrogen peroxide, CO_2, antimicrobial peptides, and/or bacteriocins that may create unfavorable environment for other spoilage causing or harmful microorganisms in food system (Patel and Shah, 2014).

Fermented plant origin products also offer prebiotic activity in addition to their probiotic significance. Prebiotics are generally defined as "a non-digestible food component (mainly oligosaccharides) that beneficially affects the host by stimulating growth and/or activity of certain bacterial components of the intestinal microflora" (Gibson and Roberfroid, 1995). Prebiotics are not metabolized or absorbed in the small intestine; however, few beneficial gut microflora can assimilate it. The gluco- and fructo-oligosaccharides, galacto-oligosaccharides, raffinose, and stachyose are an example of well-known prebiotics. Furthermore, extracellular polysaccharides produced by several bacteria as well as hydrolyzed plant derived products like cell wall polysaccharides or xylo-oligosaccharides are recently evolved novel prebiotics.

In last few years, many new probiotic fermented products have been developed by incorporation of probiotics into milk, fruit juices, or cereals. For instance, yakult, yoghurt, kefir, cheese etc. are example of commercial available probiotic fermented milk products (Table 1.9). Yosa, an oat-based fermented product, contains probiotic bacteria along with prebiotic beta glucan fibers naturally present in oat bran. Velli is also a similar product from Finland. A lactic starter having an antagonistic activity against several diarrheagenic bacteria have been employed to produce Dogik, an improved variety of fermented food ogi (Olukoya et al., 1994).

1.5.6 FERMENTATION PROCESSES: SOURCE OF INCOME AND HELP TO MAINTAIN SOCIAL WELL-BEING

The production of fermented food products is a low-input small-scale venture that can offer revenue and employment to many people around the world. Fermentation provides access to safe, economical, and nutritious foods particularly to individuals with limited purchasing power. Fermented products play a significant role by contributing to the livelihoods of village people as well as peri-urban dwellers equally, by means of improved food safety and income making through valuable small-scale business preference. Though substrates may be seasonal, the availability of variety of fermentable substrate throughout a year may serve as a regular source of income. Moreover, fermentation does not depend of climate and mostly all by-products can be recycled as feed for farm animals. Fermentation techniques

TABLE 1.9 Commercially Available Milk and Cereal Based Probiotic Fermented Products.

Product	Raw material/ substrate	Bacteria used and levels	Purported physiological benefit(s)	Country/Company
Activia yoghurt	Milk	*Stre. thermophilus, L. bulgaricus, L. casei* DN-114 001, *L. rhamnosus* LGG, *Bifidobacteriumlactis* DN-173 010	Maintenance of healthy gut flora	Danone, France
Yakult	Skim milk	*L. casei*Shirota	Maintenance of gut flora, modulation of the immune system, regulation of bowel habits and constipation, effects on some gastrointestinal infections	Yakult Honsha Co., Ltd., Japan
Probiotic cottage cheese	Milk	*L. acidophilus* and *Bifidobacterium* spp.	—	Horizon Organic, USA
Kefir	Milk	Probiotic lactic acid bacteria and yeasts	—	Lifeway, USA
Dogik	Maize	Probiotic Lactobacilli strain	Antimicrobial activity against diarrheagenic bacteria	
Togwa (Tansania)	Sorghum or maize	*Lactobacillus plantarum, L. brevis, L. fermentum, Pd. pentosaceus*	Reduces enteropathogen occurrence in rectal swabs of children, improves intestinal mucosa barrier function in children with acute diarrhea	
Yosa	Oat bran and flavored with fruit, berries	Probiotic Lactobacilli strain	Improves intestinal mucosa barrier function in children with acute diarrhea	BiofermeOy (Ltd.), Finland

TABLE 1.9 *(Continued)*

Product	Raw material/ substrate	Bacteria used and levels	Purported physiological benefit(s)	Country/Company
Velli	Oat bran and flavored with fruit, berries.	*Lactobacillus acidophilus* La-5, *Bifidobacterium bifidum* Bb-12, and other strains	Potential synbiotic effects with probiotic and prebiotic components, beta-glucan soluble fiber, which may contribute to cholesterol lowering and blood glucose attenuation	Velle Oats, St. Petersburg, Russia
Fermented Oatmeal soup for enteral feeding	Oatmeal, malted barley flour	Probiotic *Lactobacillus* strains, development made with *L. plantarum* 299v, *L. reuteri*, and other strains	Maintains intestinal function and structure after surgical operations, diarrhea, etc., strains able to colonize human intestinal mucosa, reduction of sulphite reducing clostridia, Enterobacteria and other undesirable bacteria	Sweden

are greatly linked with a variety of traditional cultural, religious, as well as domestic activities. It can become the source of revenue or business for women, the disabled, and landless poor people of the developing nations with appropriate training and access to inputs.

1.6 SAFETY ASPECTS OF FERMENTED FOODS

Majority of the food products are manufactured in unhealthy surroundings. Such foods are often contaminated with harmful microorganisms and play a key role in child mortality through diarrhea or related gastric illnesses, toxic food infections, nutrient malabsorption, and malnutrition. According to one survey, in the tropical regions of the world, above 13 million infants and children who are less than 5 years of age die annually (Anonymous, 2013). In that, respiratory infections are the major cause of death followed by diarrheal diseases and have the maximum negative impact upon the development of newborns and children. Polluted water and food along with unhealthy hygiene practices are the major causes for food infections including diarrhea and other gastric disorders (Motarjemi, 2002).

LAB and related food grade microorganisms have been regarded as GRAS. Moreover, LAB and related microorganisms have been frequently reported to possess antagonistic effect against many spoilage causing and pathogenic microorganisms. This significant character is ascribed with their ability to synthesize antimicrobial compounds like organic acids, hydrogen peroxide, carbon dioxide, and bacteriocins or antimicrobial peptides. Production of organic acids leads to the decrease in pH of the food environment and to develop unfavorable environment for the growth of unwanted microorganism. Bacteriocin and antimicrobial peptides synthesize by different LAB have broad as well as narrow spectrum of activity against specific group of microorganism (Servin, 2004; Patel et al., 2013).

Despite the fact that fermented foods usually have the benefit of a rational status for safety and quality, several noteworthy outbreaks of foodborne infection have been recorded with their consumptions. A diverse type of microorganism, including both harmful and harmless, are always associated with any foodstuffs in different amounts and among them which microorganisms will dominate depends on a number of factors. Occasionally, some microbes initially occur at low numbers in food, such as LAB may dominate other microorganisms through inhibiting or suppressing their growth or vice-versa. As mentioned in the report of workshop (FAO/WHO, 2006), small-scaled fermentation processes were reevaluated and the related nutritional as well as safety aspects of such technologies together with fermented end products were analyzed. According to this report, depending on the fermentation method together with sanitary practices followed during manufacturing of several fermented food products indicated to pose a safety risk during Hazard Analysis and Critical Control Point (HACCP) studies. Fermented fish, meat, milk, and cereal-legume products have evidenced as a considerable source of microbial food infections or food poisoning than fermented vegetables (Motarjemi, 2002). Therefore, fermented food products should be investigated by following good manufacturing practices and HACCP principles at small food industries and cottage levels and must be guided on the critical control points to maintain the safety and security of fermentation.

1.7 FUTURE PROSPECTIVE OF THE FERMENTED FOODS

Fermented foods are generally safe and consumed regularly in diet in different parts of the world. However, lack of standardization in the fermentation processes, sanitary conditions of the environment, and personnel hygiene of the persons involved ultimately determine the safety as well as quality of end product. Food safety and hygienic both are of vital importance, and thus, much emphasis should be given upon personal hygiene to complement the nutritional and economic benefits of fermented products. Many fermented food products are manufactured in primitive conditions and unhygienic environments which results low yield, poor quality, and less shelf life; this is the biggest disadvantage associated in the development of fermented foods of the developing countries Most importantly, because the processes are often laborious and time consuming as well as lack of appeal in presentation plus marketing strategies at commercial level retard their popularity.

The spreading out of traditional fermented products both at national as well as international level should not get suppressed by imported food products. Fermentation technology required improving safety level and nutritional value of finished products, and thus, focus should be given to research and investigation in this direction. Innovations in the technologies should bring about value addition to the fermented products such as extended shelf life, enhanced nutritional, and health improving properties as well as appealing packaging and labeling. Microencapsulation is one of the recent techniques which are employed to conserve and propagate food grade starters for mass production of fermented food products. The implementation of selected strains of starters, such as probiotics and/or those have an ability to synthesize vitamins, antimicrobials (bacteriocins or antimicrobial peptides), or enzymes in fermented foods, is of fundamentally important for the industrialized manufacturing of various products to improve body-texture and organoleptic characteristics, or health-improving properties. On the other hand, to maintain traditional indigenous methods, it is also indispensable to keep the record and document it for the prospect generations. It will build a reference database for making guidelines for HACCP, taking decisive steps and corrective actions and to improve the fermentation parameters, and thus, ultimately assist food regulatory bodies, policy makers, as well as prospective researchers.

KEYWORDS

- **antinutritional factors**
- **detoxification**
- **fermented milks**
- **food safety**
- **functional foods**
- *kimchi*
- *khalpi*
- *kinema*
- **dhokla**
- **probiotics**

REFERENCES

Anderson, J. W.; Johnstone, B. M.; Cook-Newell, M. E. Meta-Analysis of the Effects of Soy Protein Intake on Serum Lipids. *N. Engl. J. Med.* **1995,** *333,* 276–282.

Aneja R. P.; Mathur, B. N.; Chandan, R. C.; Banerjee A. K. *Technology of Indian Milk Products.* Dairy India Year Book: New Delhi, India. 2002; pp 158–182.

Anonyms. *Levels & Trends in Child Mortality. A Report was Prepared at UNICEF on Behalf of the United Nations Inter-Agency Group for Child Mortality Estimation.* 2013. http//www.childinfo.org/files/Child_Mortality_Report_2013.pdf.

Ashenafi, M. Ethiopian Enjera. In *Handbook of Indigenous Fermented Foods;* Steinkraus, K. H., Ed.; Marcel Dekker: New York. 1993; pp 182–194.

Lactic Acid Bacteria, Classification & Physiology. Doctoral Thesis, Matforsk, Norwegian Food Research Institute, Norway, 2004.

Bata, A.; Lásztity, R. Detoxification of Mycotoxin-Contaminated Food and Feed by Microorganisms. *Trends Food Sci. Technol.* **1999**, *10*(6–7), 223–228.

Batt, C. A. *Encyclopaedia of Food Microbiology*. Academic Press: USA, 1999; pp 848–849.

Behare, P.; Singh, R.; Singh, R. P. Exopolysaccharide-Producing Mesophilic Lactic Cultures for Preparation of Fat-Free Dahi–An Indian Fermented Milk. *J. Dairy Res.* **2009**, *76*(1), 90–97.

Beheshtipour, H.; Mortazavian, A. M.; Mohammadi, R.; Sohrabvandi S.; Khosravi-Darani, K. Supplementation of *Spirulina platensis* and *Chlorella vulgaris* algae into Probiotic Fermented Milks. *Compr. Rev. Food Sci. Food Saf.* **2013**, *12*(2), 144–154.

Bernardeau, M.; Vernoux, J. P.; Henri-Dubernet, S.; Gueguen, M. Safety Assessment of Dairy Microorganisms, the *Lactobacillus* genus. *Int. J. Food Microbiol.* **2008**, *126*, 278–285.

Beuchat, L. R. *Food and Beverage Mycology*. AVI Publishing Co. Inc.: USA, 1978.

Beuchat, L. R. Indigenous Fermented Foods. In. *Biotechnology. Food and Feed Production With Microorganisms*; Reed G., Ed.; Verlag Chemie: Weinheim, 1983.

Blandino, A.; Al-Aseeri, M. E.; Pandiella, S. S.; Cantero, D.; Webb, C. Cereal-Based Fermented Foods and Beverages. *Food Res. Int.* **2003**, *36*, 527–543.

Burgess, C.; O'Connell-Motherway, M.; Sybesma, W.; Hugenholtz, J.; van Sinderen, D. Riboflavin Production in *Lactococcus lactis*, Potential for in situ Production of Vitamin-Enriched Foods. *Appl. Environ. Microbiol.* **2004**, *70*(10), 5769–5777.

Caplice, E.; Fitzgerald, G. F. Food Fermentations, Role of Microorganisms in Food Production and Preservation. *Int. J. Food Microbiol.* **1999**, 50, 131–149.

Chandan, R. C.; Nauth, K. R. Yogurt. In *Handbook of Animal-based Fermented Food and Beverage Technology*, Second Edition; Hui, Y. H., Chandan R. C. Eds.; CRC Press: Boca Raton, FL. Chapter 12. 2012.

Chandan, R. C. History and Consumption Trends In. *Manufacturing Yogurt and Fermented Milks*, Second Edition; Chandan R. C., Kilara, A. Eds.; John Wiley and Sons, Inc.: USA, Chapter 1. 2013.

Chavan, J. K.; Kadam, S. S. Critical Reviews in Food Science and Nutrition. *Food Sci.* **1989**, *28*, 348–400.

Chelule, P. K.; Mbongwa, H. P.; Carries, S; Gqaleni, N. Lactic Acid Fermentation Improves the Quality of Amahewu, A Traditional South African Maize-Based Porridge. *Food Chem.* **2010**, *122*(3), 656–661.

Choi, M. W.; Kim, K. H.; Park, K. Y. Effects of Kimchi Extracts on the Growth of Sarcoma-180 Cells and Phagocytic Activity of Mice. *J Korean Soc. Food Sci. Nutr.* **1997**, *26*, 254–260.

Chojnacka, K. *Chemical Engineering and Chemical Process Technology–Vol. V–Fermentation Products*; 2009; pp 1–31. http://www.eolss.net/sample-chapters/c06/e6-34-09-09.pdf, Accessed on 14.12.14.

Corsetti, A.; Lavermicocca, P.; Morea, M.; Baruzzi, F.; Tosti, N.; Gobbetti, M. Phenotypic and Molecular Identification and Clustering of Lactic acid Bacteria and Yeasts from Wheat (Species *Triticum durum* and *Triticum aestivum*) Sourdoughs of Southern Italy. *Int. J. Food Microbiol.,* **2001**, *64*(1–2), 95–104.

Dalie, D. K. D.; Deschamps, A. M.; Richard-Forget, F. *Lactic acid Bacteria-Potential for Control of Mould Growth and Mycotoxins, A Review*. Elsevier: Kidlington, ROYAUME-UNI, 2010.

Das, A. J.; Deka, S. C. Fermented Foods and Beverages of the North-East India. *Int. Food Res. J.* **2012,** *19*(2), 377–392.

De Vuyst, L.; Van Kerrebroeck, S.; Harth, H.; Huys, G.; Daniel, H. M.; Weckx, S. Microbial Ecology of Sourdough Fermentations, Diverse or Uniform? *Food Microbiol.* **2014,** *37,* 11–29.

De Vuyst, L.; Vrancken, G.; Ravyts, F.; Rimaux, T.; Weckx, S. Biodiversity, Ecological Determinants, and Metabolic Exploitation of Sourdough Microbiota. *Food Microbiol.* **2009,** *26*(7), 666–675.

El-Nezami, H.; Polychronaki, N.; Salminen, S.; Mykkanen, H. Binding Rather than Metabolism May Explain The Interaction of Two Food-Grade Lactobacillus Strains with Zearalenone and its Derivative {Alphaacute}-Zearalenol. *Appl. Environ. Microbiol.* **2002,** *68*(7), 3545–3549.

Escalante, A.; Wacher, C.; Farres, A. Lactic Acid Bacteria Diversity in the Traditional Mexican Fermented Dough Pozol as Determined by 16S rDNA Sequence Analysis. *Int. J. Food Microbiol.* **2001,** *64,* 21–31.

Evans, I. H.; McAthey, P. Comparative Genetics of Important Yeasts. In *Genetically Engineered Proteins and Enzymes from Yeasts, Production Control*; Weiseman A. Ed.; Ellis Horwood: New York, 1991.

FAO/WHO. Health & Nutritional Properties of Probiotics in Food Including Powder Milk With Live Lactic Acid Bacteria. Cordoba, Argentina, 1-4 October 2001, 2001. ftp,//ftp.fao.org/es/esn/food/probio_report_en.pdf.

FAO/WHO. Fermentation, Assessment and Research. A Report of a Joint FAO/WHO Workshop on Fermentation as Household Technology to Improve Food Safety. WHO Document WHO/FNU/FOS/96.1, World Health Organization, Geneva, 2006.

Fox, P. F.; Fox, P. F. Fundamentals of Cheese Science. Springer: USA, 2000; pp.388. Accessed on 21.12.14.

Franta, B.; Steinkraus, K. H.; Mattick, L. R.; Olek, A.; Farr, D. Biochemistry of Saccharomyces. In. *Handbook of Indigenous Fermented Foods*; Steinkraus K. H. Ed.; Marcel Dekker: New York, 1993; pp 517–519.

Galle, S.; Schwab, C.; Arendt, E.; Ganzle, M. Exopolysaccharide-Forming *Weissella* Strains as Starter Cultures for Sorghum and Wheat Sourdoughs. *J. Agric. Food Chem.* **2010,** *58,* 5834–5841.

Gibson, G. R.; Roberfroid, M. B. Dietary Modulation of the Human Colonic Microbial, Introducing the Concept of Prebiotics. *J. Nutr.* **1995,** *125,* 1401–1412.

Goddik, L. Sour Cream and Crème Fraiche. In *Handbook of Animal-Based Fermented Food and Beverage Technology*, Second Edition; Hui Y. H. Ed.; CRC Press: Boca Raton, FL, 2012; pp 235–246.

Haard, N. F.; Odunfa, S. A.; Lee, C. H.; Quintero-Ramı́rez, R.; Lorence-Quinones, A.; Wacher-Radarte, C. Fermented Cereals. *A Global Perspective. FAO Agric. Serv. Bull.* 1999, *138*.

Hamad, A. M.; Fields M. L. Evaluation of the Protein Quality and Available Lysine of Germinated and Fermented Cereal. *J. Food Sci.* **1979,** *44,* 456.

Haskard, C. A.; El-Nezami, H. S.; Kankaanpaa, P. E.; Salminen, S.; Ahokas, J. T. Surface Binding of Aflatoxin B1 by Lactic Acid Bacteria. *Appl. Environ. Microbiol.* **2001,** *67*(7), 3086–3091.

Heller, K. J.; Wilhelm Bockelmann, W.; Schrezenmeir, J.; deVrese, M. Cheese and its Potential as a Probiotic Food. In *Handbook of Functional Fermented Foods;* Farnworth, E. D. Ed.; CRC Press: USA, 2008; pp 243–250.

Holzapfel, W. Use of Starter Cultures in Fermentation on a Household Scale. *Food Contr.* **2002**, *8*(5–6), 241–258.

Hutkins, R. W. *Meat Fermentation. Microbiology and Technology of Fermented Foods.* Blackwell Publishing: USA, 2006; pp. 207–229.

Jespersen, L.; Halm, M.; Kpodo, K.; Jacobson, M. Significance of Yeasts and Moulds Occurring in Maise Dough Fermentation for Kenkey Production. *Int. J. Food Microbiol.* **1994**, *24*, 239–248.

Karki, T.; Okada, S.; Baba, T.; Itoh, H.; Kozaki, M. Studies on the Microflora of Nepalese Pickles Gundruk. *Nippon Shokuhin Kogyo Gakkaishi,* 1983, *30*, 357–367.

Keuth, S.; Bisping, B. Vitamin B12 Production by *Citrobacter freundii* or *Klebsiella pneumoniae* During Tempeh Fermentation and Proof of Enterotoxin Absence by PCR. *Appl. Environ. Microbiol.* 1994, *60*(5), 1495–1499.

Kim, S. H.; Park, K. Y.; Suh, M. J. Inhibitory Effect of Aflatoxin B1 Mediated Mutagenicity by Red Pepper Powder in the *Salmonella assay* System. *J. Korean Soc. Food Sci. Nutr.* **1991**, *20*, 156–161.

Kim, T. W.; Lee, J. H.; Kim, S. E.; Park, M. H.; Chang, H. C.; Kim, H. Y. Analysis of Microbial Communities in Doenjang, a Korean Fermented Soybean Paste, Using Nested PCR-Denaturing Gradient Gel Electrophoresis. *Int. J. Microbiol.* **2009**, *131*, 265–271.

La Vecchia, C.; Altieri, A.; Tavani, A. Vegetables, Fruit, Antioxidants and Cancer, A Review of Italian Studies. *Eur. J. Nutr.* **2001**, *40*, 261–267.

Mahgoub, S. E. O.; Ahmed, B. M.; Ahmed, M. M. O.; El Agib El Nazeer, A. A. Effect of Traditional Sudanese Processing of Kisra Bread and Hulu-Mur Drink on Their Thiamine, Riboflavin and Mineral Contents. *Food Chem.* **1999**, *67*, 129–133.

McKay, L. L.; Baldwin, K. A. Applications for Biotechnology, Present and Future Improvements in Lactic Acid Bacteria. *FEMS Microbiol. Rev.* **1990**, *87*, 3–14.

Metaxopoulos, J.; Mataragas, M.; Drosinos, E. H. Microbial Interaction in Cooked Cured Meat Products Under Vacuum or Modified Atmosphere At 4 °C. *J. Appl. Microbiol.* **2002**, *93*(3), 363–373.

Mistry, V. V. Fermented Milks and Cream. In *Applied Dairy Microbiology*, Second Editionl; Marth, E. H., Steele, J. L., Eds.; Marcel Dekker: New York, NY, Chapter 9, 2001.

Mokoena, M. P.; Chelule, P. K.; Gqaleni, N. Reduction of Fumonisin B1 & Zearalenone by Lactic Acid Bacteria in Fermented Maize Meal. *J. Food Protect.* **2005**, *68*, 2095–2099.

Molnár, N.; Sipos-Kozma, Z.; Tóth, Á.; Ásványi, B.; Varga, L. Development of a Functional Dairy Food Enriched with Spirulina *(Arthrospira platensis). Tejgazdaság.* **2009**, *69*(2), 15–22.

Motarjemi, Y. Impact of Small Scale Fermentation Technology on Food Safety in Developing Countries. *Int. J. Food Microbiol.* **2002**, *75*(3), 213–229.

Nanson, N. J.; Field, M. L. Influence of Temperature on the Nutritive Value of Lactic Acid Fermented Cornmeal. *J. Food Sci.* **1984**, *49*, 958–959.

Nisha, P.; Ananthanarayan, L.; Singhal, R. S. Effect of Stabilizers on Stabilization of Idli (Traditional south Indian Food) Batter During Storage. *Food Hydrocoll.* **2005**, *19*, 179–186.

Olukoya, D. K.; Ebigwei, S. I.; Olasupo, N. A.; Ogunjimi, A. A. Production of DogiK, an Improved Ogi (Nigerian fermented Weaning Food) with Potentials for Use in Diarrhoea Control. *J. Trop. Pediatr.* **1994**, *40*(2), 108–113.

Oluwajoba S. O.; Malomo, O.; Ogunmoyela, O. A. B.; Dudu, O. E. O.; Odeyemi, A. Microbiological and Nutritional Quality of Warankashi Enriched Bread. *J. Microbiol. Biotechnol. Food Sci.* **2012**, *2*, 42–68.

Onyekwere, O. O.; Akinrele, I. A.; Koleoso, O. A. O. In *Handbook of Indigenous Fermented Foods;* Steinkraus, K. H. Ed.; Marcel Dekker: New York, 1993; pp. 212–222.

Park, K. Y.; Cho, E. J.; Rhee, S. H. Increased Antimutagenic and Anticancer Activities of Chinese Cabbage Kimchi by Changing Kinds and Levels of Sub-Ingredients. *J. Korean Soc. Food Sci. Nutr.* **1998**, *27*, 169–182.

Patel, A.; Shah, N. Recent Advances in Antimicrobial Compounds Produced by Food Grade Bacteria in Relation to Enhance Food Safety and Quality. *J. Innov. Biol.* **2014**, *1*(4), 189–194.

Patel, A.; Shah, N.; Ambalam, P.; Prajapati, J. B.; Holst, O.; Ljungh, A. Antimicrobial Profile of Lactic Acid Bacteria Isolated from Vegetables and Indigenous Fermented Foods of India Against Clinical Pathogens using Microdilution Method. *Biomed. Environ. Sci.* **2013**, *26*(9), 759–764.

Plengvidhya, V.; Breidt, F.; Lu, Z.; Fleming, H. P. DNA Fingerprinting of Lactic acid Bacteria in Sauerkraut Fermentations. *Appl. Environ. Microbiol.* **2007**, *73*, 7697–7702.

Prajapati, J. B.; Nair, B. The History of Fermented Foods. In *Handbook of Functional Fermented Foods;* Farnworth E. D., Ed.; CRC Press: USA, 2008; pp 1–24.

Ramakrishnan, C. V. Indian Idli, Dosa, Dhokla, Khaman, and Related Fermentations. In *Handbook of Indigenous Fermented Foods*; Steinkraus K. H., Ed.; Marcel Dekker: New York, 1993; pp 149–165.

Rati Rao, E.; Vijayendra, S. V. N.; Varadaraj. Fermentation Biotechnology of Traditional Foods of the Indian Subcontinent. In *Food Biotechnology;* Shetty, K., Paliyath, G., Pometto, A., Levin, R. E., Eds.; CRC Press: USA, 2006; pp 1759–1794.

Rose, A. Soy and Phytic Acid, Stick with Fermented Tempeh and Miso. *Reducing Phytic Acid in Your Food, A Visual Analysis of the Research on Home Kitchen Remedies for Phytic acid*. Rebuild Market. 2015. Accessed on 20.01.15.

Rowan, N. J.; Anderson, J. G.; Smith, J. E. Potential Infective and Toxic Microbiological Hazards Associated with the Consumption of Fermented Foods. In *Microbiology of Fermented Foods;* Wood J. B., Ed.; Blackie Academic and Professional: London, 1998; pp 812–837.

Sarkar, P. K.; Tamang, J. P.; Cook, P. E.; Owens, J. D. Kinema–A Traditional Soybean Fermented Food, Proximate Composition and Microflora. *Food Microbiol.* **1994**, *11*, 47–55.

Sasikumar, R. Fermentation Technologies in Food Production. In *Progress in Biotechnology for Food Applications*; Wing-Fu, L. Ed.; OMICS Group eBook: USA, 2014.

Schultz, M. Dairy Products Profile. Agricultural Marketing Resource Center-Dairy Products Profile, 2011. http://www.agmrc.org/commodities_products/livestock/dairy/dairy products profile.cfm. Accessed on 1/11/2014.

Servin, A. L. Antagonistic Activities of Lactobacilli and Bifidobacteria Against Microbial Pathogens. *FEMS Microbiol. Rev.* **2004**, *28*, 405–440.

Song, Y. O. The Functional Properties of Kimchi for the Health Benefits. *J. Food Sci. Nutr.* **2004**, *9*, 27–33.

Soni, S. K.; Sandhu, D. K. Indian Fermented Foods, Microbiological and Biochemical Aspects. *Indian J. Microbiol.* **1990**, *30*, 135–157.

Soni, S. K.; Sandhu, D. K. Fermented Cereal Products. In *Biotechnology, Food Fermentations*, Vol. 2; Joshi V. K., Pandey A., Eds.; Educational Publishers and Distributors: New Delhi, 1999; pp 895–950.

Soni, S. K.; Sandhu, D. K. Fermentation of Idli, Effects of Changes in Raw Material and Physico-Chemical Conditions. *J. Cereal Sci.* **1989**, *10*, 227–238.

Spano, G. Biotechnological Production of Vitamin B2-Enriched Bread and Pasta. *J. Agric. Food Chem.* **2011,** *59,* 8013–8020.

Steinkraus, K. H. Classification of Fermented Foods, Worldwide Review of Household Fermentation Techniques. *Food Control.* **1997,** *8,* 311–317.

Steinkraus, K. H.; Ayres, R.; Olek, A.; Farr, D. Biochemistry of Saccharomyces. In *Handbook of Indigenous Fermented Foods;* Steinkraus K. H., Ed.; Marcel Dekker: New York, 1993; pp 517–519.

Surh, J.; Kim, Y. L.; Kwon, H. Korean Fermented Foods, Kimchi and Doenjang. In *Handbook of Functional Fermented Foods;* Farnworth E. D., Ed.; CRC Press: USA, 2008; pp 333–350.

Swain, M. R.; Anandharaj, M.; Ray, R. C.; Rani, R. P. Fermented Fruits and Vegetables of Asia, A Potential Source of Probiotics. *Biotechnol. Res. Int.* Article ID 250424, 19 pages, 2014, http://dx.doi.org/10.1155/2014/250424.

Tamang, J. P.; Chettri, R.; Sharma, R. M. Indigenous Knowledge on North-East Women on Production of Ethnic Fermented Soybean Foods. *Indian J. Tradit. Knowl.* **2009,** *8*(1), 122–126.

Tamang, J. P.; Tamang, B.; Schillinger, U.; Franz, C. M. A. P.; Gores, M.; Holzapfel W. H. Identification of Predominant Lactic Acid Bacteria Isolated from Traditionally Fermented Vegetable Products of the Eastern Himalayas. *Int. J. Food Microbiol.* **2005,** *105*(3), 347–356.

Tamime, A. Y.; Robinson, R. K. Yogurt Science and Technology, Third Edition; Woodhead Publishing Limited, Cambridge, England/CRC Press: Boca Raton, FL, 2007.

Thakur, S.; Prasad, M. S.; Rastogi, N. K. Effect of Xanthan on Textural Properties of Idli (Traditional South Indian Food). *Food Hydrocoll.* **1995,** *9,* 141–145.

van Berkel, B. M.; Boogaard, B. V.; Heijnen, C. Agrodok 12, Preservation of Fish and Meat. Edited by Marja de Goffau-Markusse, Agromisa Foundation, Wageningen: The Netherlands, 2004; pp 1–86.

Vaughn, R. H. The Microbiology of Vegetable Fermentation. In *Microbiology of Fermented Foods*, volume 1, Wood, B. J. B., Ed.; Elsevier Applied Sci. Publ.: London, 1985; pp. 49–110.

Wang, H. L.; Hesseltine, C. W. Sufu and Lao-chao. *J. Agric. Food Chem.* **1970,** *18,* 572–575.

Yadav, J. S.; Grover, S.; Batish, Y. K. *A Comprehensive Dairy Microbiology*. Metropolitan: New Delhi, India, 1993; pp 350–453.

Zegeye, A. Acceptability if Injera with Stewed Chicken. *Food Qual. Pref.* **1997,** *8,* 293–295.

Zhou, W.; Therdthai, N. Fermented Bread. In. *Handbook of Plant-Based Fermented Food and Beverage Technology,* Second Edition; Hui, Y. H., Özgül Evranuz E., Eds.; CRC Press: USA, 2012; pp 477–526.

Zinedine, A.; Faid, M.; Benlemlih, M. In Vitro Reduction of Aflatoxin B1 by Strains of Lactic Acid Bacteria Isolated from Moroccan Sourdough Bread. *Int. J. Agric. Biol.* **2005,** *7*(1), 67–70.

CHAPTER 2

FERMENTED FOODS: RECENT ADVANCES AND TRENDS

AMRITA POONIA*

Centre of Food Science & Technology, Institute of Agricultural Science, Banaras Hindu University, Varanasi 221005, Uttar Pradesh, India

E-mail: dramritapoonia@gmail.com; amritapoonia@yahoo.co.in

CONTENTS

ABSTRACT

Foods are preserved by fermentation since ancient times. Fermentation helps to enhance the shelf life, value addition, microbiological safety and also increases the digestibility of the foods. Fermented foods are still among the most popular types of foods consumed. A variety of food products can be developed using fermentation process. These products are healthy and suit to the consumer's demands and their lifestyle. These foods are cheaper, improve heath, and provide a variety of flavors and textures to the consumers. Every civilization has certain traditional foods which are based on fermentation process. In India, every region consumes a considerable amount of fermented food in their daily diet, be it pickles and *Idli* in South or curds in North.

2.1 INTRODUCTION

Fermentation changes the raw materials into products that possess a particular aroma, mouth feel, and appearance which enriches the human diet. Preservation of foods by use of fermentation is a very old way of preserving foods in the world. It not only helps in preserving the food but also helps in the development of distinct new food products with improved nutritional quality. In fermented food products, components such as alcohols, organic acids, ketones, and aldehydes are the main end products. The proteins are partially hydrolyzed, are more soluble, and available nutritionally. Vitamins like riboflavin, Vitamin B_{12}, and the precursors of vitamin C increase during fermentation. Fermentation in foods is caused by the activity of microorganisms or their enzymes. Different types of fermentations involved in the production of fermented food include the alcoholic fermentation, acetic acid fermentation, lactic acid fermentation, alkaline fermentation, or their combinations.

2.2 PROPERTIES OF FERMENTED FOODS

Fermented foods are actually more nutritious than unfermented one which can come about in many different ways. They are popular since ancient times because of the many reasons, which are explained in the following sections.

2.2.1 PRESERVATION

Preservation of foods is a well-known technique since long ago. It is a need-based approach because the perishable food products get spoiled soon. Preservation of food and retention of nutrition are through all the stages of processing. Fermentation is one of the primary concerns to the food scientists. The various nutrients like vitamins, proteins, amino acids, minerals, etc. are sensitive to methods used in preservation.

2.2.2 NUTRITIVE VALUE

Fermented foods are reported to be some more nutritious and with reduced mutagenicity of the foods by degrading the mutagenic substances during fermentation. Many people are lactose intolerance due to deficiency of *β-galactosidase*, but the lactose intolerant can consume yoghurt and *dahi*. Lactic acid bacteria (LAB), present in cultured dairy products, get colonized in the gut of the host and prevent the entry of the pathogens and thus plays a vital role in the immune system. Components such as alcohols, organic acids, ketones, and aldehydes are the main-end products. The proteins are partially hydrolyzed, made more soluble, and available nutritionally. Vitamins such as riboflavin, vitamin B_{12} and the precursors of vitamin C have increased during fermentation. It is also found that the selected probiotic strains may provide health benefits that extend beyond these macronutrients by contributing to gastrointestinal health, enhanced immunity and antioxidant activity, or antihypertensive effect for even broader overall well-being.

2.2.3 SENSORY AND FUNCTIONAL ATTRIBUTES

Fermentation changes the raw material into products that possess typical aroma, mouth feel, and appearance which is unthinkable for the raw material. Fermented foods are totally different from the raw materials used for fermentation. Alcoholic fermentation is the important food fermentation and used in the production of alcoholic beverages. The alcoholic beverages are important in relation to nutrition, indirectly by providing a pleasurable aspect to the intake of food, fulfilling the subjective desire of man, and quenching one's thirst. Earlier, wheat flour was used for making simple breads. But by using fermentation, many other products can be prepared. Similarly, barley has been transformed into one of the most popular drinks, that is, beer with unique organoleptic and functional attributes is one of the most common example.

2.2.4 EXCEPTIONAL FLAVOR

Some fermented food products could not be prepared without the use of microorganisms to develop a particular flavor during fermentation. For example, beer, wine, different types of cheese, meat products, sauerkraut, and many more products are difficult to prepare. Fermented foods have a particular flavor which could not be developed by using any other methods of food preservation. Fermented foods are prepared by direct addition of acids and enzymes to enhance the activities of fermentative microorganisms. These food products do not have that particular flavor and other unique properties of their traditional fermented foods.

2.2.5 ECONOMIC VALUE

Fermentation is an age-old process of value addition. By using microorganisms at the right stage of fermentation and proper ageing of milk results in wonderful fermented milk products (FMPs) with delicate aroma and texture. Fermented products can earn more money as compared with the raw materials used for their preparation. Barley is very cheap, but if fermentation is done in a particular environment and precise conditions, the value of the premium beer is very high. Similarly, some wines are very costly and have been sold for more than 20,000 US dollars per bottle. Most interesting thing about fermented foods is that they are made from cheap raw materials like barely, wheat, milk, etc., and mostly fermented products have very high profit from margins.

2.3 FERMENTED FOODS: PAST, PRESENT, AND FUTURE

Fermentation is one of the oldest methods of food preservation (Chavan and Kadam, 1989) in the world. *Vada, idli, dosa*, bread, beer, *dahi*, toddy, cheese, and wine are the Indian fermented foods which are prepared and consumed from thousands of years and strongly linked to the culture and tradition. Methods of fermentation of foods have been described since dawn of civilization. Earlier, peoples were not aware about the role of microorganisms in fermentation. But in the middle of the century, two main events changed the vision of the peoples for food fermentations. First revolution in this area was industrialization, due to which large number of peoples started to run toward town and cities for employment which resulted in increased demand of food. In 1850s, the concept of microbiology as a science came to existence and formed the biological basis of fermentation. This was the second important change in food fermentations. Now the science and role of microorganisms involved in food fermentation was clear for the first time (Caplice and Fitzgerald, 1999). After these revolutions, the technologies were developed for the industrial production of fermented products from various types of raw materials. Different aspects of fermentation and research work are actively carried out all over the world.

Fermented dairy products are more popular all over the globe. Names of different fermented products vary from region to region. The cultured milks can be classified on the basis of type of fermentation (Table 2.1). Fermented dairy products, wine, and beer are examples of fermented foods which have been researched for application of most advanced scientific techniques to produce on industrial scale in highly automated plants. The consumption of these products is going up, whereas on the other hand, many traditional fermented foods are vanishing due to urbanization, modernization, and introduction of western foods.

TABLE 2.1   Classification of Cultured Milks Based on Type of Fermentation.

A) Lactic fermentation

  i) Mesophilic type:

    Example: Cultured buttermilk, filmjolk, langofil, and tatmjolk

  ii) Thermophilic type:

    Example: Bulgarian buttermilk, dahi, yoghurt, and zabadi

  iii) Therapeutic or probiotic type: This group of food products comprises definitely maximum number of foods which are popular in world

    Example: ABT, Acidophilus milk, onka, vifit, and yakult

B) Yeast—lactic fermentations: *Example:* Acidophilus yeast milk, kefir, and koumiss

C) Mould—lactic fermentations: *Example:* Villi

D) Lactic fermentation with draining of whey: *Example:* Labneh, Ymer, *shrikhand*, and Skyr

---

Source: Modified from Robinson and Tamine (1995).

2.4 FERMENTED FOODS: GLOBAL TRENDS AND CONSUMPTION PATTERNS

There is an increasing demand of functional dairy products from consumers. Due to this, the probiotic foods, low fat, and low calorie dairy drinks become more popular. Health conscious people prefer those foods that are low in calorie, high fiber which promote good health and can prevent from various life style diseases. Consumers prefer the foods which are easy to use, fit into the lifestyle, good flavor, and less price. Due to this type of consumer's demands, the current and future demand for new product development occurred. These kinds of new products are the need of the future.

Probiotic drinks are booming in the market of different countries like South Korea, Brazil, and Western European markets due to their health claims, portability, and snack appeal. These health claims are similar to that of the cultured dairy products which are beneficial to the digestive system as well as to the immunity system (Chandan, 1999). Western Europe is the second largest country for fermented dairy drinks after Latin America. In United States, Danone's Actimel was launched first time as a functional dairy drink. Worldwide, Danone's Actimel was the second largest brand of fermented dairy drinks. Danone launched its creamy stirred yoghurt in 2011 to target the health-conscious population in India.

FMGs are becoming more popular these days. In India, about 10% of the total milk produced is used for fermented dairy products. The main fermented products in India are *dahi, lassi, chhach,* and *shrikhand.* These products are consumed by most of the Indian as an important part of diet. *Dahi* is an age-old indigenous. Presently, 80% of total packaged *dahi* is supplied by Amul, Mother Dairy, Nestle, and Britannia. Some international companies such as Danone, Fonterra, etc. are also in competition. In future, there will be more competition in the market because of the entry of the main business houses such as Reliance, ITC, and Cavin Care. According to the reports of Research and Market (2014), India's probiotic marker is projected to grow around 19% till 2019.

2.5 FERMENTED FOODS WITH BIOACTIVE PEPTIDES

Bioactive peptides are the specific protein fragments that have a positive impact on body functions and influence the health. They contain 3–20 amino acid residues per molecules. These peptides are inactive within the sequence of the parent protein but can be released by the digestive enzymes during gastrointestinal transit or by fermentation. Bioactive peptides can be produced either by hydrolysis with commercial enzymes or by fermentation with LAB. Use of bioactive peptides in fermented foods contributes an added health benefit to the consumers over and above the nutrient content of the product. So, by use of these products, the risk of chronic diseases can be reduced. Now, the market of these products has high growth rates due to consumers demand.

Europe, Japan, and Australia are the dominating countries for fermented dairy products. Use of bioactive peptides in functional foods is a new concept. These types of fermented foods are also available in the market. During fermentation, the angiotensin converting enzyme (ACE)-Inhibitory peptides are produced which provide health benefits (Hernandez-Ledesma et al., 2005). Some examples of these foods are: Calpis, which is sour milk and certified by the Food for Specified Heath Use (FOSHU) in Japan. Evolus is another product which is launched by Valio in Finland. The blood pressure lowering effect of Evolus has been studied in many clinical trials.

Japanese Ministry of Health and Welfare has enough scientific proof toward health claims under FOSHU system. FOSHU has been established in 1993, and a list of 400 foods has been approved and bear FOSHU label. For successful future of these products, new technologies of separation and enrichment are the new area of research. More research should be focused to find the association of bioactive peptides in food. Recent techniques to stabilize the bioactive peptides in food are part of the new area to be focused. Chromatography and membrane separation are new techniques to add the active peptides from the hydrolysates of many food proteins. There is a need to find out the mechanisms by which bioactive peptides exert these activities. This area of research is a challenge and can revolutionize the research on bioactive peptides. Milk proteins as a source of bioactive peptides offer a good approach for health promotion by means of well-planned diet. It also provides opportunities to the dairy industry to touch new horizons.

2.6   PROBIOTICS, PREBIOTICS, AND SYNBIOTICS

Probiotics are the live microorganisms, which when administered in adequate amounts, confer a health benefit on the host. In the last 2–3 decades, a number of microorganisms have been reported with scientifically claimed health benefits. India has a rich diversity of microorganisms which can be tested and used as probiotoics in traditional as well as novel food systems. Prebiotics are nondigestive substances that support growth of probiotic organisms in the gut. Synbiotics are the products containing a combination of probiotic cultures and prebiotic ingredients. Probiotics, prebiotics, and synbiotics are the new concept in the functional foods. Probiotic dairy products are the important segment and have tremendous scope for growth and development.

The organisms that are best studied and generally regarded as probiotics species of bacteria are *Bifidobacterium*, *Lactobacillus*, and *Streptococcus*, and they also include yeasts such as *Saccharomyces boulardii*. These probiotics inhibits the growth of bad bacteria in intestine by colonizing the gut. They also have antimicrobial activity and help in increasing our immunity by making our body more resistant to diseases and infections. The world famous Amul has introduced several probiotic dairy products in market which include Pro-Life Ice-cream, Probiotic *lassi*, probiotic buttermilk, and probiotic fruit yoghurt. Yakult Danone India has launched one of the first probiotic drink—Yakult. Nestle India entered probiotic milk market by launching its new low-fat product Nesvita *dahi*. Mother dairy has also launched Nutrifit, a fermented probiotic drink, in Delhi and NCR.

2.7 EXOPOLYSACCHARIDES FROM FERMENTED DAIRY PRODUCTS AND HEALTH PROMOTION

Exopolysaccharides (EPSs) are polymers that are produced by plants, fungi, and bacteria. EPSs are gaining popularity and have drawn attention of the researchers all over the world. An area that has not received much attention is that of the prebiotic potential of EPSs from LAB or yeasts, particularly those found in fermented dairy products. Can EPSs in fermented dairy products act in a similar way to conventional prebiotics? Prebiotic may be defined as "a nondigestible selectively fermented ingredient that allows specific changes, both in the composition and/or activity in the gastrointestinal microbiota that confers benefits upon host well-being and health" (Gibson and Roberfroid, 1995; Roberfroid, 2007).

Antipathogen effect that has been reported for EPSs from LAB is the potential of an EPS to act as a type of interspecies signaling molecule. Kim et al. (2009) demonstrated that *Lactobacillus acidophilus* A4 polysaccharide has caused down-regulation of a number of *Escherichia coli* O157:H7 genes involved with the formation of biofilms. This ability to reduce the levels of biofilm could have a significant effect on the treatment of infections resistant to antibiotic therapy because of biofilms; it could be used as an adjuvant to antibiotic therapy, in this instance. In brief, although there is some emerging evidence that EPSs have an antiinfective role to play in dairy foods, more specific evidence is needed to evaluate the benefits. Another interesting aspect lies in the transfer of genetic characteristics—the ability of EPS to aggregate bacteria will decrease the distances between cells, and thereby facilitate the transfer of genetic characters (Badel et al., 2011). This "cross-talk" may afford the sharing of a number of characteristics that would be beneficial not only to both strains or species, but to the host as well.

Oxidative stress is the main reason in the initiation and development of inflammatory bowel disease, and by manipulating this pathway may soothe the disease progress. Sengul et al. (2011) studied the effect of EPS from two (probiotic) strains of *Lactobacillus delbrueckii* sub sp. *bulgaricus* on experimentally induced colitis in a rat model. With the high-EPS group, a significant improvement in the oxidative stress parameters was found than in the low-EPS group (Li et al., 2013); using a rat pheochromocytoma line (PC 12) demonstrated that a purified EPS (EPS-3) from the fermentation of *Lactobacillus plantarum* LP6 could protect PC12 cells from oxidative damage induced by H_2O_2. EPS isolated from *Lactococcus lactis* sub sp. *Lactis* showed both in vitro and in vivo superoxide anion, hydroxyl radical, and 2,2-diphenyl-1-picrylhydrazyl scavenging activities (Guo et al., 2013). Similarly, EPS from *Lactobacillus paracasei* sub sp. *Paracasei* (strain NTU 101) and *L. plantarum* (strain NTU 102) has also demonstrated antioxidant activity (Liu et al., 2011).

2.8 CONJUGATED LINOLEIC ACID PRODUCTION IN FERMENTED FOODS

Conjugated linoleic acid (CLA) may be defined as a group of positional and geometric isomers of conjugated dienoic derivatives of linoleic acid having double bonds in a conjugated position ranging from 2, 4 to 15, and 17. Each positional isomer has four geometric isomers, those being *cis*, *trans*; *trans*, *cis*; *cis*, *cis*; or *trans*, *trans*. CLA may refer to a total of 56 possible isomers (Buccioni et al., 2012). CLAs are found in food derived from ruminants. CLAs are a group of fatty acids shown to have beneficial effects in animals. The presence of CLA isomers in ruminant fat is related to the biohydro-genation of polyunsaturated fatty acids (PUFAs) in the rumen. PUFAs of natural origin have double bonds, mostly in isolated positions. From these, numerous CLA isomers have been shown to be present in food.

In milk fat, one of the most important sources of CLA is the positions of double bonds range from 6, 8 to 13, 15; therefore, the term "total CLA content" includes a total of 32 isomers (Bessa et al., 2000; Kramer et al.,

2004). CLAs may be added to the diet either, by manipulating the diet of animals by directly adding CLA in food products. Food products from ruminant animals are the rich source of CLAs. Dairy products are the major source in the human diet. The concentration of CLA in milk fat is 3–6 mg/g of fat, but the level of CLA in milk vary among herds.

A possible means of production of fermented foods rich in CLA is the selection of raw materials with high CLA content and the preservation of their elevated CLA levels during fermentation. The main natural sources of CLA in human nutrition are, undisputedly, the products of ruminants, including their milk, meat, and adipose tissue. The enzymatic processes responsible for the formation of CLA proceed much more intensively in the tissues and organs of ruminants than those of monogastric animals. The predominant source of CLA in human diets is ruminant-derived food products. Vegetable oils usually contain only negligible amounts of CLA. The total CLA value of raw milk samples has been shown to extend over a wide range, from 2 to 37 mg/g CLA in fat (Jiang et al., 1996). Meats of ruminant origin have markedly higher CLA concentrations than those from monogastric animals (Chin et al., 1992). The highest amount of CLA was found in lamb and beef (19.0 and 13.0 mg/g fat, respectively). Corn oil, olive oil, and coconut fat were reported to contain approximately 0.2 mg/g CLA. In samples obtained from stores, the *cis*-9, *trans*-11 CLA isomer accounted 45% of the total amount of CLA, whereas corn oil processed in the laboratory did not have this isomer in detectable amounts.

A number of strains and probiotic cultures, that is, *Lactococcus, Sreptococcus, Enterococcus, Lactococcus, Bifidobacterium*, and *Propionibacterium* can biosynthesize CLA. Through diet manipulation, CLA content of ruminant fat can be changed. CLA producing bacteria (*L. acidophilus, Lactobacillus casei*) are the new hope for development of CLA enriched dairy products like ultra-high temperature milk, butter, yoghurt, and cheese. Encapsulation is a new technique which is used to transform liquids into stable and free flowing powders and protect them from oxidation. It also helps to reduce off-flavors, enhance stability of the microcapsules under fluctuations of temperature and moisture, and make it suitable for incorporation into various food products.

CLA present in the fats of ruminants has been reported to have many health benefits. Nowadays, in general, numerous health benefits are attributed to CLA, encompassing the range of common diseases in Westernized populations such as cancer, diabetes, and cardiovascular diseases (CVD). Positive effects on body composition, the immune system and bone health have also been observed (Jahreis et al., 2000; Khanal, 2004; Kraft and Jahreis, 2004; O'Shea et al., 2004). However, most experiments were carried out using animal trials and are not inevitably conclusive for humans; moreover, results obtained in many areas of health-effect investigations are contradictory. Observations in animal trials pointed out that differences exist between mammalian species regarding their response to CLA. It can be assumed that not all the impacts that were proved for animals also pertained to humans (Wahle et al., 2004). To clarify the effect of dietary CLA intake on the process of human carcinogenesis, more clinical studies are needed. The anti-diabetic effect of CLA may depend on both species and types of isomer. Rumenic acid seemed to be inactive, but the role of *trans*-10, *cis*-12 isomer is controversial: some studies verified decreasing glucose levels and increased insulin sensitivity with increased intake of *trans*-10, *cis*-12 CLA (Khanal, 2004), whereas others reported opposite effects, that is, the promotion of insulin resistance (Khanal, 2004). Human studies, however, did not prove such strong evidence of the weight loss and body fat reduction-inducing effects of CLA (Bhattacharya et al., 2006; Park and Pariza, 2007).

2.9   BIOPRESERVATION EFFECTS IN FERMENTED FOODS

These days' consumers are more health conscious and much aware about the adverse health effects from the use of chemical additives in foods. Due to this concept, there is an increase demand of minimally processed foods with biopreservatives. Bactreiocins from LAB is one of the biopreservative which attracted the attention of scientists from many years and can be used as a "biological" immunity to foodstuffs (Cotter et al., 2005).

LAB constitutes a vast group of gram positive bacteria, which have common morphological, metabolic, and physiological traits. They are anaerobic bacteria, nonsporulating, acid tolerant, and produce mainly lactic acid as an end product of their fermentation. The LAB, that is, *Lactobacillus*, *Lactococcus*, *Leuconostoc*, *Pediococcus*, and *Streptococcus* have been used since long in the processing of fermented foods. LAB are generally regarded as safe in food preservation and widely accepted. They produce lactic acid and other organic acids which lower the pH and helps in food preservation. Stiles and Hastings (1999) reported that preservation effect is enhanced by the production of other antimicrobial compounds, such as hydrogen peroxide, CO_2, diacetyl, acetaldehyde, and bacteriocins.

Pawar et al. (2000) studied the activity of nisin at concentration of 400 and 800 IU/g and in combination with 2% sodium chloride against *Listeria monocytogenes* in minced raw buffalo meat. Antilisterial activity of nisin A and pediocin AcH in decontamination of artificially contaminated pieces of raw pork was studied. It was reported that nisin A was considerably more efficient than pediocin AcH. During storage, after 2 days surviving rate of bacteria in meat treated with each bacteriocin resumed growth at a rate similar to that of the control. Nisin was reported to be more stable than pediocin AcH (Murray and Richard, 1997).

Foegeding and Stanley (1991) investigated the inactivation of a mix of five strains of *L. monocytogenes* during a meat fermentation challenge started with a pediocin-producing (Bac+) *Penicillium acidilactici* strain. By using an isogenic pediocin-negative (Bac-) derivative as negative control, the authors demonstrated that *L. monocytogenes* can survive the fermentation process when the pH was not less than 4.9, but in that case, pediocin released in situ contributed to inactivation of the pathogen during the drying process. McAuliffe et al. (1999) addressed the application of bacteriocinogenic strains of *L. lactis* as protective culture to improve the safety of cottage cheese. In experimental cheeses, started with *L. lactis* DPC4268-producing lacticin 3147 and with 4 log CFU/g of *L. monocytogenes* Scott A added, the authors measured bacteriocin activity in the curd of 2560 AU/ml throughout the first week of storage at 4 °C and a consequent reduction of the *L. monocytogenes* target strain up to 99.9% within 5 days. While the initial numbers of the pathogen remained unchanged in the control, cheese started with the nonproducer strain *L. lactis* DPC 4275. As the authors were unable to recover any *Listeria* in cheese through preenrichment over a week of storage at different temperatures, they could confirm the bactericidal nature of the lactic in 3147.

Dal Bello et al. (2012) applied four strains of bacteriocinogenic *L. lactis* (one nisin Z producer, one nisin A producer, and two lactic in 481 producers), isolated from cheese and previously evaluated for desired technological characteristics, as starter cultures in cottage cheese production, with the aim to control the growth of *L. monocytogenes*. In particular, the strain *L. lactis* subsp. *Cremoris* 40FEL3 (producer of nisin A), in combination with the high acidity reached during cheese manufacturing, was able to control and partially reduce the growth of the *L. monocytogenes* target strain inoculated in cheese at 3 log CFU/g. By considering that the in situ inhibitory effects were reduced in comparison with those observed in vitro, the authors considered the application of the above strain in cheese production as a possible additional measure to control *Listeria* spp. contamination. Lower efficacy of bacteriocins in food with respect to their inhibitory activity as evaluated with in vitro assays is a common problem. There are many factors in bacteriocins application in food biopreservation such as partitioning into polar or nonpolar food components and binding of the bacteriocins to food fat or protein particles and food additives (e.g., triglyceride oils), NaCl concentration, proteolytic degradation or inactivation by other inhibitors, changes in solubility and charge, changes in the cell envelope of the target bacteria, as well as the development of bacteriocin-resistant strains. A more detailed list of the factors limiting the bacteriocin efficacy in foods is given in Figure 2.1.

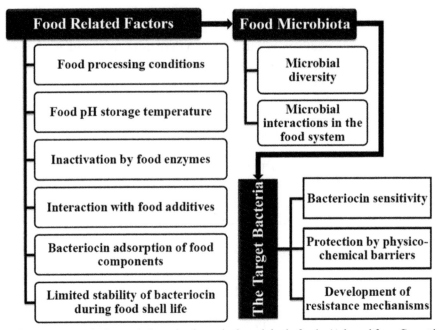

FIGURE 2.1   Main factors influencing bacteriocin activity in foods. (Adapted from Corsetti et al. 2015).

2.10   MODERN APPROACHES FOR ISOLATION, SELECTION, AND IMPROVEMENT OF BACTERIAL STRAINS

Chr. Hansen A/S (global company supplier of food cultures) has been collecting microorganisms for many decades and today is in possession of an extensive collection of bacteria, yeast, and molds that can be used in the development of industrial fermentation processes. This collection contains more than 20,000 deposits with the majority of these being pure single strains of bacteria. The determination and analysis of the complete sequence of the genetic material in an organism is commonly referred to as genomics. Due to a significant reduction in both the time required and the cost of genome sequencing, this has become a powerful method of strain characterization (Danielsen and Johansen, 2009). At present, there are more than 2900 bacterial genome sequences, including more than 100 genome sequences of LAB, available in public databases. Recombinant deoxyribonucleic acid (DNA) technology can be used to genetically improve bacterial strains for use in industrial processes including food applications. A working definition of "food-grade" genetically modified organisms has been elaborated (Johansen, 1999), and a large variety of food-grade genetic modifications can be made. Specific genes can be partially or totally eliminated from a strain or replaced with different alleles from other strains of the same genus. Likewise, new properties can be introduced into a strain by the transfer of a gene absent in the specific strain but present in other members of the genus. The presence of recombinant DNA can be used to isolate mutants with interesting properties.

2.11 CONTROLLING THE FORMATION OF BIOGENIC AMINES IN FERMENTED FOODS

Biogenic amines (BAs) are basic organic molecules of low molecular weight, generally resulting from the microbial decarboxylation of certain amino acids. The ability to form BAs is strain-specific yet quite widespread among species of LAB and Gram-negative bacteria and is based on the presence on amino acid decarboxylase enzymes. Thus, both the key microorganisms for the food fermentations and contaminating spoilage bacteria can contribute to BA formation in foods. Generally, histamine, tyramine, tryptamine, β-phenylethylamine, and the polyamines putrescine, cadaverine, spermine, and spermidine represent the most important BAs in foods (Shalaby, 1996).

Over the years, many analytical methods for BA detection have been developed. Onal et al. (2013) reported two methods of BA detection including: qualitative and semiquantitative methods. Semiquantitative methods include enzymatic and thin layer chromatography. Qualitative methods are chromatographic techniques such as high-performance liquid chromatography (HPLC), gas chromatography, capillary electrophoresis (CE), and ion-exchange chromatography. The most reliable and sensitive methods for detection of BA in foods are the HPLC-based techniques. Out of all these, ultra-high-pressure liquid chromatography (UHPLC) is a new emerging chromatographic technique that operates at higher pressures, up to 15,000 psi. It enhances the efficiency of HPLC in terms of increased resolution, sensitivity, and analysis speed. However, the expensive cost of laboratory equipment and the need for specialized operators make UHPLC a viable opportunity only for high-throughput applications. UHPLC analysis has been performed on fermented food samples to determine their BA content in wine, fish, cheese, and dry fermented sausages. Fiechter et al. (2013) have validated a UHPLC method for the simultaneous determination of amino acids and BAs in ripened acid-curd cheeses. In a single run, 23 amino acids and 15 BAs were separated in a retention time of only 9 min, showing very sensitive limits of detection (0.05–0.29 mg 100/g) and quantization (0.16–0.97 mg 100/g). Similarly, a derivatization treatment with diethyl ethoxymethylenemalonate followed by UHPLC allowed the simultaneous quantization of 22 amino acids, 7 Bas, and ammonium ions in cheese samples in 10 min (Redruello et al., 2013). Therefore, this approach will be valuable to monitor the content of BAs during the ripening of cheese and identify those foods potentially hazardous for their high levels of free amino acid precursors.

CE methods are quantitative methods becoming more popular due to their speed and high-resolving power. By using these techniques, separation and identification of highly polar compounds is possible that cannot be easily separated by traditional HPLC methods. The low sensitivity of CE can be overcome by coupling to mass spectrometric detection. Kvasnicka and Voldrich (2006) analyzed six BAs in different fermented foods (salami, cheese, wine, and beer), developing a CE method with conductometric sensitivity (µmol/l), speed (less than 15 min), and a detection procedure that does not require derivatization. Therefore, this method seems to be an interesting alternative to chromatographic techniques for routine analysis. Enzymatic sensors take less time for analysis and are more rapid and reliable for screening the quality of industrial foods. Hernandez-Cazares et al. (2012) reported that these sensors can detect the presence of BAs in dry fermented sausages. They may play a vital and useful tool for quality control in the meat industry. An overview of the developments and issues in the construction of biosensors for the detection of most common BAs found in food has been recently reported (Kivir and Rinken, 2011).

2.12 METABOLIC ENGINEERING OF LAB FOR IMPROVED FERMENTED FOODS

Industrial LAB has been used for more than a decade as powerhouses for metabolic engineering. In the absence of complete genomic information, most attention has initially been focused on the well-known conversion of sugars (mainly glucose or lactose) into lactic acid. The glycolysis and the pyruvate branch represent the major highway in LAB metabolism. Rerouting of this central metabolism has been used toward the food-grade production of chemicals (lactic acid, ethanol, and butanol), flavors and fragrances compounds (diacetyl, acetaldehyde, and 2,3-butanediol), food ingredients (alanine, polyols, and EPS), and production of nutraceuticals, such as vitamins.

Nutraceutical and other functional foods have received growing interest from consumers and the food industries as these foods offer health benefits. Apart from amino acids, sweeteners, and polyols, vitamin production by LAB has received considerable attention from different research groups. An increased level of folate (also known as vitamin B_{11}) has been realized in *L. lactis* by optimization of its biosynthesis pathway. Sybesma et al. (2003) reported increase by 10-fold the folate production in *L. lactis* by over expressing *folKE* with NICE system. The complex metabolic pathway for vitamin B_{12} synthesis has been characterized in *Lactobacillus reuteri* and used to increase its production via medium optimization (Santos et al., 2008). Moreover, by expressing the *L. plantarum* folate genes, a *L. reuteri* strain capable of producing vitamin B_{12} and folic acid has been generated (Santos et al., 2008). Similarly, a *L. lactis* strain simultaneously producing riboflavin (vitamin B_2) and folate has been engineered (Sybesma et al., 2004). This area of multivitamin production has significant potential, as has been recently reviewed (Leblanc et al., 2013).

2.13 ADVANCES IN STARTER CULTURE TECHNOLOGY

In the production of fermented food, starter cultures are used to prevent fermentation failure and to ensure high-product quality. Starter cultures are cultures with well-defined properties that ensure a fast, safe, and defined fermentation and lead to fermented food products with high and constant product quality. The use of defined starter cultures is state-of-the-art in the dairy industry and replaces traditional procedures in the production of meat and bakery products and other fermented commodities. To minimize damage during drying, we can use (a) protective agents that are added to the cell suspension before drying. They can be simple substances such as mono-, di-, oligo-, or polysaccharides, or more complex substances such as skim milk powder. Several mechanisms for the protective effect are proposed: (i) preferential hydration, (ii) water replacement, and (iii) glass transition (Santivarangkna et al., 2008).

Freeze drying, also called lyophilization, is a process where the concentrated cell suspension (with added protectants such as sugars and antioxidants) is first frozen under atmospheric pressure, and then the water is sublimed at reduced pressure (sublimation). At the end, the process residual water is desorbed at very low pressure (desorption). The advantage of freeze drying is that the drying method is suitable to maintain the quality of sensitive products, though it is a highly time-consuming and expensive drying technique because of the very low pressures and temperatures. There is no shrinkage during drying as only the water is removed from the frozen sample, and the solid substances build a solid and porous framework after drying (Oetjen and Haseley, 2004). Due to the high porosity, the freeze-dried cell suspension has a high specific surface area. This influences the sorption behavior as well as rehydration and storage characteristics (Haque and Roos, 2005). Freeze-dried samples are easy to rehydrate and therefore exhibit good instant properties. Freeze drying leads to high survival rates, high metabolic activity, and good rehydrate ability and therefore to a high product quality (Oetjen and Haseley, 2004; Kiviharju et al., 2005). Investment costs and energy cost per kg removed water are higher than for other drying processes (Table 2.2).

TABLE 2.2 Investment and Energy Costs of Different Drying Technologies Relative to the Freeze–Drying Process.

Drying technology	Relative investment costs at same throughput (%)	Specific energy costs (kW h/kg water)
Freeze drying	100	2.0
Vaccum	65.2	1.3
Spray	52.2	1.6
Air drying	43.5	1.9

Source: Foerst and Santivarangkna (2015).

For spray drying, the cell suspension is pumped as liquid concentrate to the dryer and separated in small droplets by means of a pressure nozzle (one or two fluid nozzles) or a rotary disc. The mode of particle formation in the drier also influences the viability of the bacteria after drying. Semyonov et al. (2011) developed an ultrasonic vacuum spray drier combined with fluidized bed drier to produce highly viable probiotics under reduced oxidative and thermal stresses. Chavez and Ledeboer (2007) combined spray drying with vacuum drying. The spray drying was carried out mainly in the constant rate period where the temperature is low, and the vacuum drying was carried out in the second drying stage to prevent thermal and oxidative stresses. An overview of the product characteristics with regard to handling and storage stability for the different drying technologies is given in Table 2.3.

TABLE 2.3 Product Viability After Drying, Storage Stability, and Handling for the Different Drying Technologies.

Description	Liquid culture (as reference)	Freeze dried	Spray dried	Vacuum dried
Initial viability	High	Intermediate	Low	Intermediate
Storage stability, room temperature	Low	High	Intermediate	High
Storage stability, refrigerated temperature	Intermediate	High	High	High
Handling	High	High	Intermediate	Intermediate

(Reprinted from Foerst, P.; Santivarangkna, C. *Advances in Starter Culture Technology: Focus on Drying Processes. In Advances in Fermented Foods and Beverages, Improving Quality, Technologies and Health Benefits, Seri. Food Science, Technology and Nutrition*; Holzapfel, W., Ed.; Woodhead Publishing: Sawston, Cambridge, 2015, Vol.; pp. 249–270. © 2015, with permission from Elsevier.)

2.14 QUALITY IMPROVEMENT AND FERMENTATION CONTROL IN VEGETABLES

For the preservation of vegetables, there are two methods in which the added salt (sodium chloride) is crucial: brining (or salting) and fermentation. Although both methods are closely related to each other, they are based on different principles of conservation. In the case of salting, with equilibrium salt contents above 10%, the stabilization is based solely on the nonspecific decrease of the water activity (a_w value) and the high ionic activity. In the case of lactic acid fermentation, the preserving effect, however, is primarily based on the formation of acids in the product leading to a decreasing pH value. Here, the effect of the added salt on the water activity (a_w) value is only of secondary importance. Brining is rarely practiced today owing to problems of salt removal and decreased consumer acceptance of high-salt foods.

The fermentation of plant substrates in itself constitutes a very complex network of independent and interactive microbiological, enzymatic, chemical, biochemical, and physical processes and reactions. Use of known cultures that can be grow fast and are highly competitive in the environmental conditions under which the product is kept. Fermentation is influenced by many exogenous factors, which affect the quality of the final products. These factors can be divided into four groups: (1) technological factors, (2) nature and amount of admixed ingredients and optional additives, (3) quality of the raw materials used, which in turn depends on numerous agricultural factors, and (4) nature of the microbiota introduced with the raw material.

2.15   CHALLENGES IN MAINTAINING VIABILITY OF PROBIOTICS

At present, the probiotic foods are only limited to the metropolitan cities. In small cities and towns, the electricity and cold storage are a big problem. In this area, research and development should be focused on improving the viability and stability of the product; it is the need of the hour to improve and develop cost-effective fermentation technology for maximizing the probiotic health properties of new and innovative fermented foods. Research should be focused on the genetic manipulation of bacteria for specific probiotic function.

Microencapsulation offers the potential for developing the innovative functional foods. Recent studies are highlighting the microencaplsulation of probiotic cell process complexity and diversity of wall materials of food systems used for the management of carrier of food matrices. By use of microencapsulation, bacterial cells become able to survive in the stressful conditions as well as in the aggressive environment. They become able to survive in the extremes of bile salts and gastric acid. LABs fermentation can also be regulated by the use of this technique. It can also increase the survival of probiotics during storage of the fermented dairy products. Natural polymers can be used in microencapsulation to reduce the cell losses during storage and processing of the product. Cryoprotectants may also be used to the fermented dairy products before fermentation which helps the microorganisms adapt to the environment.

2.16   FORTIFICATION OF ESSENTIAL INGREDIENTS

Foods are fortified with various ingredients since long time. Different types of essential ingredients are added to the cultured dairy products, and their effect on the quality of the products has been studied in various researches. Some of these ingredients incorporated into fermented foods which enhance consumers appeal are listed in the following sections

2.16.1 FORTIFICATION OF MINERALS AND VITAMINS

Calcium helps to prevent osteoporosis and cancer hypertension. Calcium (Ca) can be fortified in fermented foods and especially in the cultured milks. Pirkul et al. (1997) developed fortified yoghurt with calcium salts. They reported that calcium gluconate, calcium lactate, and combination of calcium lactate and calcium gluconate can enhance the calcium content by 37.6, 34.3, and 39.4%, respectively. Vitamins (such as C and E) and antioxidants can help to prevent cardiovascular rand cancer disease. Multivitamin–mineral mixes can be added to the fat-free fermented dairy products. It is a good meal for the consumers which are calorie conscious.

2.16.2 FORTIFICATION OF DIETARY FIBERS

Dietary fibers play important role in human nutrition. Foods with added dietary fibers are in great demand. Various fibers like guar gum, gum acacia, oat fiber, psyllium, and soy components can be fortified in cultured dairy products. The effect of seven types of insoluble dietary fibers available from different five sources (corn, oat, rice, soy, and sugar beet) were used to fortify sweetened plain yogurt. It was reported that addition of fiber caused acceleration in the acidification rate of the yogurts. Fortified yoghurts also showed increase in their apparent viscosity. Soy and sugar beet fibers caused a significant decrease in viscosity due to partial syneresis. Fiber addition may cause lower score of flavor and texture because fiber addition may lead to grainy and gritty texture. Similarly, yoghurt with low level of fat can be prepared by using glucan. As the level of glucan increases, the firmness and consistency of yoghurt also increases and synersis decreases. Palacio et al. (2005) fortified yogurt from whole milk, with 50 mg of calcium/100 ml of yogurt. Three levels of fiber from two wheat-bran sources were used. When this developed yoghurt was compared with the control, calcium and fiber were reported to increase the consistency of the product. Synersis was less but pH was increased in the product.

2.16.3 FORTIFICATION WITH FATTY ACIDS (ω-3)

Milk fat is the essential component of our daily diet. It contains saturated as well as unsaturated fatty acids. Milk fat found in various dairy products can be changed by reducing the level of saturated to unsaturated fatty acids. By altering the composition, it is possible to increase the contents of fatty acids that are more desirable for human nutrition such as the ω-3 PUFAs. The ω-3 fatty acids like α-linolenic are very important and play the role of precursors of long chain fatty acids such as docosahexaenoic acid (DHA) and eiocosapentaenoic acid (EPA). The EPA and DHA cannot be synthesized in the human body. They are vital for the functioning of brain and its proper development. Formation of plaque in the arteries can be reduced by their use. They are also believed to reduce cancer, allergy and improve the ability of learning. Some drinks fortified with DHA have been developed and are available in the market for the school children. There are two methods to increase the level of healthy fatty acids in dairy products. The use of particular bacterial culture is for production of FMGs. Milk fat can be substituted with oils having high levels of PUFAs. This kind of replacement with oil provides the yoghurts with higher synersis and with less firmness. Diana et al. (2004) developed a set-type fermented product in which milk fat has been replaced by oils enriched in PUFAs. It was found that their yoghurt flavor was not affected, but the texture of the product was adversely affected.

2.16.4 FORTIFICATION WITH ISOFLAVONES

Soy bean is the part of our diet since long time. It contains some functional ingredients which are of more interest commercially. Most of the leguminous plants contain isoflavones and phytochemicals, but soy bean is the richest source of isoflavones. Isoflavones present in it is the part of diphenol compounds which are known as phytoestrogens which are similar to the estradiol, the human estrogen. Soy isoflavones are believed to play vital role in reducing the risk of age-related and hormone-related diseases including cancer, menopausal symptoms, CVD, and osteoporosis (Setchell, 1998; Omoni and Aluko, 2005). Many studies showed that the inclusion of 1.7 g/day phytostrerols into the diet of hypercholesterolemic men had the effect of lowering blood cholesterol. Low-density lipoprotein (LDL) consumed from plant stanols (2–3 g) daily, reduced the cholesterol concentrations in adults, suffered with hyper and nonhyper cholesteromic. Consumption of fermented foods enriched with these components has a great potential for cholesterol management.

2.16.5 FERMENTED FOODS WITH PHYTOSTEROLS AND PHYTOSTANOLS

Phytosterols, phytostanols, and their esters are a group of steroid alcohols and esters that occur naturally in plants. Phytosterols can be extracted from tall oil which is a by-product of wood pulp industry. They can be purified by using distillation, extraction, crystallization, and washing. They are widely used to develop enriched food products that fulfill the health claims associated with sterol fortifications. Dietary intake of phytosterols should not exceed 3 g/day which is considered as safe limit. The quantities of these in natural diets vary from 167 to 437 mg/day. Vegetables oils, bread, and nuts are good source of it as compared to fruits and vegetables (Ostlund et al., 2002). Corn oil contains 8 g/kg and 0.5 g/kg in palm oil, with intermediate levels being found in commonly used oils (Phillips et al., 2002). Degree of variation of sterols in each plant species may be due to genetic factors, cultivation, and the different states of the plant. As part of normal healthy diet, most people eat 100–500 mg of phytosterol each day (Ostlund et al., 2002). Pytosterols or phytostanols currently incorporated into foods are esterifies to unsaturated sterols/stanol esters to increase lipid solubility. Functional fermented foods with phytosterol and phytostanol products reduce the serum concentration of total cholesterol by up to 15% and that of LDL cholesterol by up to 22%. Plant stanols plays vital role in reducing hypocholesteromic effect. Blood cholesterol level lowers, if an adult with hypercholestromic will intake 1.7 g of phytosterols in diet per day. It was revealed that inclusion of 2–3 g of plant stanols in daily diet lowers the LDL cholesterol concentration in adults with hyper and nonhypercholestermoic and 10–14% in children without causing any effect on high-density lipoprotein cholesterol.

Foods enriched with these components have therefore a great potential for cholesterol management. Oil-based products enriched with plant stanol esters can reduce LDL cholesterol concentrations by 10–14%. When these results were compared with oil-based product, it was found that the developed products also reduce LDL cholesterol in similar manner as oil-based products within a week. Awaisheh et al. (2005) prepared yoghurts with three nutraceuticals, that is, isoflavones, phytosterols and omega-3 fatty acids in a milk base. *Bifidobacterium infantis* or *Lactobacillus gasseri* were used as a single culture to prepare yoghurts. For rapid acidification and to reduce the time of production, a two-stage fermentation process was carried out with *Lactobacillus derlbrueckii* sub sp. *bulgaricus* and *Streptococcus thermophilus*. Added nutraceuticals do not have any adverse effect on the product. During storage at 50 °C, the probiotic cultures were viable more than 15 days.

2.16.6 FORTIFICATION WITH GAMMA (γ)-AMINOBUTYRIC ACID (GABA)

Gamma γ-aminobutyric acid (GABA) is a metabolic by-product of plant and microorganisms. It is not found in fresh foods. GABA is found in fruits and vegetables but not in dairy products. It can be produced during fermentation process. During lactobacillus fermentation, the GABA production by microorganisms has increased. This amino acid has been reported to reduce blood pressure in human beings due to its ability to block peripheral ganglia. Similar results have been reported by Manyam et al. (1981).

LAB and germinated soy bean extract were used to develop yoghurt with high levels of GABA. Fermented soy milk (GABA soy yoghurt) was prepared with starter and substrate and has the high GABA concentration of 424.67 µg/g DW as compared to the fermented milk produced by a conventional method; GABA had less than 1.5 µg/g DW (Park and Oh, 2007). Hayakava et al. (2004) studied the blood-pressure-lowering effects of GABA and a GABA-enriched FMG. Low-dose oral administration of GABA and a GABA-enriched FMG to hypertensive and normotensive Wister–Kyoto rats showed that the effect (hypotensive) of FMG was because of FMG.

2.17 USE OF EPS AND TEXTURAL IMPROVEMENT OF FERMENTED FOODS

Polysaccharides are polymers produced by plants, fungi, and bacteria which have drawn attention of the researchers all over the world. EPSs produced by the LAB plays a vital role in preparation of cultured dairy products such as cheese, yoghurt, fermented cream, drinking yoghurt, and milk-based desserts. The generally recognized as safe status of lactobacilli attracts a lot of industries, simplifying the regulation and application in food product. LABs are the food grade microorganisms which produces EPS. These days they are used to determine the rheological qualities of milk products. EPS which are produced by LAB are used as an alternative to develop the desired texture in foods. They act as thickeners, stabilizers, emulsifiers, gelling agents, and water binding agents when added to the foods. EPS can be used as texturizing agents to increase the product's viscosity. They may be also used as stabilizers which binds the hydrated water and interact with proteins and miscelles in the milk, and the casein network get strengthen and becomes more rigid. Synersis in the fermented dairy products can be reduced by using EPS and thus improves the stability of the product. EPS not only improves rheology, texture, stability, and mouth-feel-to-fermented dairy products but also have been found to provide health benefits to consumers. Gluco-oligo-saccharides or fructo-oligosaccharides present in EPS materials produced short chain fatty acids when the micro flora of the colon acts on these fatty acids in the intestinal tract. They play vital role in reducing tumor, decrease cholesterol, immune modulatory effects, and act as a prebiotic to the micro-flora. In this manner, they provide nutritional benefits and thus improve the health.

2.18   USE OF FAT REPLACERS AND ARTIFICIAL SWEETENERS IN FERMENTED FOODS

Today, consumers are more aware about their health and want to have low calorie foods. Low calorie and low fat are highly acceptable to the consumers. These products are very popular and earn a high place in the consumer's life-styles. Low fat products were also popular many years back.

Nowadays, the fat replacers are used for this purpose. While using low levels of fat, some physical and flavor-related problems originate in the final product. So, fat replacers are used to solve these kinds of problems. Two types of fat replacers are used in food products: (1) modified starches or protein and (2) modified fat/oil-based products with digestion resistant bonds. Glycerol ethers and complex carbohydrates of fatty acid esters are the few examples. Commercially available fat replacers like N-oil[R]II, Lyca-dex[R] 100, and 200-maltodextrin, Litnesse™-improved polydextrose and Paselli[R]SA2, and P-fiber150C and 285F are the various examples of starch which were added at 1.5% to the low calorie yoghurts prepared with recon-stituted skim milk powder. Anhydrous milk fat added yoghurts were used as control with 3% milk fat. It was concluded that the product with 1% of inulin showed similar in quality attributes to the control yoghurt with 3% milk fat. It was reported that increased use of inulin in fat free yogurt negatively influ-enced some physical properties of yogurt. Increased level of inulin lowers the sensory scores, whey separation, and consistency of the product.

Artificial sweeteners can also be used in fermented foods as a source of calorie reduction. They can be used individually of in combinations. A number of low calorie and sugar free dairy products like cultured milk, dairy beverages, and whey-based beverages and yoghurts have been prepared using Aspartame and Acesulfame-K. Pinheiro et al. (2005) studied the effect of different sweeteners in low-calorie yoghurts. Farooq and Haque (1992) used aspartame and sugar esters to develop a nonfat low-calorie yoghurt and reported that addition of sugar esters improve the overall acceptability of nonfat low calorie yoghurt. They found that yoghurt with sugar esters, mainly stearate-type yoghurt with a hydrophile-lipophile balance range of 5–9, had firmer body, texture, and mouth feel than yoghurts without sugar esters. They also reported that skim milk yoghurts sweetened with aspartame had 50% fewer calories per serving than regular yoghurt containing 3.25% fat and 4% sucrose. Keating and White (1990) had developed plain and fruit-flavored yoghurts using nine different alternative sweeteners including aspartame, sodium, calcium saccharins, and acesulfame-K. They reported that among all the plain and fruit-flavored yoghurts, yoghurts sweetened with sorbitol and aspartame received highest sensory flavor scores. Fellows et al. (1991) used aspartame to develop a sundae-style yoghurt. They reported that during the manufacture, aspartame has excellent stability in fruit preparation. Aspartame, calcium saccharine, sucrose, fructose, acessulfame-K, dihydrochalcone, sorbitol, sodium saccharine, and sucrose plus monoammonium glyctyrrhizinate were used to prepare different types of plain and fruit yoghurts. On the basis of the sensory attributes, the yoghurt prepared by using aspartame was most preferred by the consumers. From all of these studies, it is concluded that aspartame and acesulfame-K are commonly used nonnutritive sweeteners and also give good results.

2.19 FRUIT FORTIFIED FERMENTED FOODS

Fortification of fermented dairy products with fruit juices or pulps is a new concept in the product development. Fruits are rich source of minerals, dietary fibers, and vitamins. They also have phytonutrients and good antioxidant properties. Fortification of foods with fruit components helps in value addition, and variety of products can be prepared. It also reduces the postharvest losses and enhances the economy of the country. By fortification, the milk and fruit producers as well as processors can earn more money. Use of fruits and vegetables in diet, one can improve the health. They also helps in reducing CVD, reduce risk of cancer, hypertension, and also delay the ageing process. Fruits are processed in various forms like juices, syrups, fruit pieces, crushed fruits, fruit purees, fruit preserve, frozen/osmodehydrozen and many other forms fruits, and many other fruit products.

A wide range of fruit fortified dairy products are available in the market. Fruit yoghurt has prepared by using different fruits, that is, mango, sapota, papaya, pineapple, and kokum. Three different levels of these fruits were used at the level of 10, 15, and 20, respectively. It was concluded that 20%

level of mango pulp and pineapple juice could be used in yoghurt with good results. Fruit flavored *dahi* was prepared using mango, banana, pineapple, and strawberry pulp with addition 6, 8, 6, and 4% pulp, respectively. It was found that mango pulp fortified *dahi* was found to be most acceptable on basis of sensory evaluation (Pandya, 2002). In a similar study, fruit-based *shrikhand* has also been prepared using apple, papaya, and mango (Bardale et al., 1986). Four types of yoghurt were prepared by using coconut milk with cow milk with various levels. This type of product was prepared in the regions in high coconut production interesting alternative option (Imele and Atemnkeng, 2001).

The addition of fruit extracts is a good option for product development because addition of whole fruits and vegetables may cause the problem of stabilization, high viscosity, and undesirable components and poor shelf life. Many functional foods or neutraceuticals are in demand with natural antioxidant compounds in the market. Bioactive polyphenolic compounds may be added to dairy products to improve their antioxidant potential. Karaaslan et al. (2011) studied the effect of addition of acidified ethanol extracts of four different grape varieties (Cabernet sauvignon, Chardonnay, Shyrah, and Merlot). Grape callus were used into yoghurt as functional ingredients. Yoghurts inoculated with red grape and callus extracts displayed high phenolic-anthocyanin content. It was concluded that the callus extracts showed higher antioxidant power as compared to the yoghurts containing chardonnay extracts and control samples. During storage, yoghurts supplied with grape callus extract displayed the greatest antioxidant power on the first day of storage compared with all the assayed samples. Singh et al. (2013) studied the effect of strawberry polyphenol on physico-chemical properties, total phenolic, and antioxidant activity of stirred *dahi*. Strawberry polyphenol extract was used to fortify stirred *dahi* (0.5 mg/ml). It was concluded that addition of very less amount of extract resulted in a seven-fold increase in the antioxidant activity of polyphenol-enriched stirred *dahi*, whereas acidity, pH, viscosity, and water-holding capacity remained comparable with the control. During storage, the product was acceptable up to 2 weeks when stored at 7–8 °C with no significant difference ($P > 0.05$) in the antioxidant activity and total phenolic content.

2.20 FERMENTED MILKS WITH ADDED WHEY

Lots of whey is produced in dairy plants every year. It is a by-product of cheese/*paneer* dairy plants and good source of lactoferrin, lactoperoxidase, immunoglobulins, and growth factors. Liquid whey contains approximately 93% water and almost 50% of the total solids present in milk, of which lactose is the main constituent. Whey is also rich in vitamin B_6, thiamin, pantothenic acid, B_{12}, and riboflavin. Whey added products have good digestibility and high protein efficiency ratio index which is more than 3.0. Being a good source of calcium, milk proteins, lactose, and water soluble vitamins, these products are good source of essential amino acids. Due to all these properties, whey also comes in the category of functional foods. By using new technologies, whey and its fractions play a very important role in new food product development. It can be used in various products and helps to raise the economy of the country. Addition of whey fractions as ingredients in various food products enhances the shelf life of the developed products. Whey addition also improves calcium, texture, taste, stability, and emulsification, improved flow properties and also improves functionality of the product. These functional properties of whey proteins make them interesting food ingredients.

The functional properties of whey solids make them suitable to be used in conjugation with FMGs. In several studies, whey has been used as an ingredient of functional dairy products. Whey protein concentrates (WPC) can be used to replace skimmed milk and milk protein concentrates (MPC). In one study, it is reported that addition of MPC and skimmed milk protein in yoghurt improves the viscosity of the product. It was found that yoghurts prepared with WPC have more synersis than the plain yoghurt. So, WPC is an important and useful ingredient in production of drinking type yoghurt.

Hernandez-Mendoza et al. (2007) reconstituted whey-containing pectin and sucrose with *L. reuteri* and *Bifidobacterium* to develop whey-based probiotic product. During storage, the product was acceptable upto 30 days and showed probiotic bacteria counts higher than 10^6 CFU/ml. Similarly, Almeida et al. (2008) fermented probiotic beverage by using fresh whey from Minas Frescal cheese. Recently, a biofunctional strawberry probiotic whey drink was prepared using yoghurt culture and probiotic culture of *L. acidophilus* which has antimicrobial, antihypertensive, and antioxidant properties. Macedo et al. (1999) used skim milk (cow), buffalo cheese milk whey, and soy milk for preparation of probiotic whey milk beverage.

2.21   USE OF UNDERUTILIZED FOOD SOURCES

Due to the shortage of food in developing and underdeveloped countries, malnutrition is a big problem. So, it is necessary to find out some alternative food sources to fulfill the demand of the peoples. Soybean is the alternative food source which is rich in protein, inexpensive to other food sources. Many soy-based food products has been developed which provide additional benefits to the consumers due to their antiatherogenic, anticholesterolemic, and hypolipidemic and also to their reduced allergenicity properties. Use of different types of millets with fermented foods may help to increase the utilization of these nonconventional food sources. By use of all these food sources, many types of fermented foods are prepared which are cheap, nutritious, and also add variety to the diet.

Soy milk is prepared from soy beans and various types of soy milk-based products; fermented milks could be prepared for consumers. Soy milk based fermented products has low cholesterol level, saturated fats, and lactose. Soymilk contains low carbohydrate, fat, Ca & P, and riboflavin, but high thiamin, niacin and Fe as compare with cow's milk. If we compare soymilk with buffalo milk, soy milk contain high amount of protein than buffalo milk. Soy milk is deficient in sulfur-containing amino acids. The beany flavor of soy milk can be modified during lactic acid fermentation. Nutritive value of the product can be enhanced by addition one part of soy milk. Nsofor and Chukwu (1992) prepared soy-yoghurt by using corn starch (2%), sodium citrate (0.3%), soybean solids (22%), sucrose (4%), and water and fermented with 5% of active mixed starter culture (*Lactobacillus bulgaricus* and *Streptococcus thermophiles*). Sensory evaluation of cow milk yoghurt and improved soy yoghurt showed nonsignificance difference for overall acceptability.

2.22 CEREAL-BASED FERMENTED FOODS

Wheat, rice, corn, and other cereal crops are dense source of energy as compared to the most succulent fruits and vegetables. They also contain dietary proteins, carbohydrates, B-complex group of vitamins, vitamin E, iron, trace minerals, and fiber. Fermentation further enhances the nutritive value, palatability, and functionality of cereals by reducing the antinutritional factors. Coarse grains like corn, sorghum, millet, barley and so on contain appreciable amounts of crude fiber and lack gluten like properties of wheat. Many cereal-based fermented beverages such as pearl millet and sorghum *lassi* (*Raabadi*), Tarhana, Kishk, and so on are manufactured. If these products are prepared at industrial scale, they will not only provide consumers a healthy choice but also at affordable prices. A range of fermented foods, their country of origin, and microorganisms which involved in the fermentation is given in Table 2.4.

TABLE 2.4   Examples of Some Fermented Foods.

Product	Country	Microorganisms	Substrate
Adai	India	Pediococcus, Streptococcus, Leuconostoc	Cereal/legume
Anarshe	India	LAB	Rice
Dhokla	Northern India	Leuconostoc mesenteroides, Streptococcus faecalis, Torulopsis candida, T. pullulans	Rice or wheat and bengal gram
Dosa	India	Leuconostoc mesenteroides, Streptococcus faecalis, Torulopsis candida, T. pullulans	Rice and bengal gram
Idli	South India, Sri Lanka	Leuconostoc mesenteroides, Streptococcus faecalis, Torulopsis, Candida, Tricholsporon pullulans	Rice grits and black gram
Jaanr	India, Himalaya	Hansenula anomala, Mucorrouxianus	Millet
Bread	International	Saccharomyces cerevisiae, other yeasts, lactic acid bacteria	Wheat, rye, other grains
Jalebies	India, Nepal, Pakistan	Saccharomyces bayanus	Wheat flour
Kanji		Hansenula anomala	Rice and carrots
Bongkrek	Indonesia	Rhizopus oligosporus	Coconut press cake
Nan	India, Pakistan, Afghanistan, Iran	Saccharomyces cerevisiae, LAB	Unbleached wheat flour
Gari	West Africa	Corynebacterium manihot, other yeast, lactic acid bacteria (Lb. plantarum, Streptococcus spp.)	Cassava root
Rabdi	India	Penicillium acidilactici, Bacillus, Micrococcus	Maise and buttermilk
Idli	Southern India	Leuconostoc mesenteroides, Streptococcus faecalis, Torulopsis, Candida, Trichosporon pullulans	Rice and black gram dhal
Taotjo	East India	Aspergillusoryzae	Roasted wheat meal or glutinous rice and soybeans
Vada	India	Pediococcus, Streptococcus, Leuconostoc	Cereal/legume
Kenkey	Ghana	Unknown	Maize

TABLE 2.4 *(Continued)*

Product	Country	Microorganisms	Substrate
Kimchi	Korea	*Lactic acid bacteria*	Cabbage, vegetables, sometimes seafood, nuts
Mahewu	South Africa	*Lactic acid bacteria*	Maize
Ogi	Nigeria, West Africa	*Lactic bacteria, Cephalosporium, Fusarium, Aspergillus, Penicillim spp., Saccharomyces cerevisiae, Candida mycoderma, C. Valida, or C. Vini*	Maize
Oncom	Indonesia	*Neurospora intermedia, or Rhizopus oligosporus*	Peanut press cake
Soy Sauce	The orient (Japan, China, Philippines)	*Aspergillus oryzae or A. Soyae, Lactobacillus, Zygosaccharomyces rouxii*	Soybeans and wheat
Tempeh	Indonesia Surinam	*Rhizopus oligosporus*	Soybeans
Nan	India	*Saccharomyces cerevisiae, Lactic acid bacteria*	White wheat flour
Cheese	International	*Lactibacillus lactis, S. thermophilus, Lactobacillus shermanii, bulgaricus, Propionibacterium shermanii*	Milk
Yoghurt	International	*S. thermophilus, Lb. bulgaricus*	Milk, milk solids
Fermented sausages	International	*Lactic acid bacteria (lactobacilli, pediococci), Catalase positive cocci (S. carnosus, S. xylosus, M. Varians) sometimes yeasts and/or moulds*	
Sauerkraut	International	*Lactic acid bacteria, Lc. mesenteroides, Lb. brevis, Lb. plantarum, Lb. curvatus, Lb. sake*	Cabbage
Pickles	International	*P. cerevisiae, Lb. plantarum*	Cucumber
Olives	Mediterranean	*Lc. Mesenteroides, Lb. plantarum*	Green Olives

Source: Blandino et al. (2003); Caplice and Fitzgerald (1999).

(Modified from 1. Blandino, A.; Al-Aseeri, M. E.; Pandiell, S. S.; Cantero, D.; Webb. C. Cereal-based Fermented Foods and Beverages. *Food Res. Int.* **2003**, *36*, 527–543. © 2003, with permission from Elsevier, and 2. Caplice, E.; Fitzgerald, G. F. Food Fermentations: Role of Microorganisms in Food Production and Preservation. *Int. J. Food Microbiol.* **1999**, *50*, 131–149. © 1999, with permission from Elsevier.)

2.23 CONCLUSION AND FUTURE PERSPECTIVE

Probiotic foods are popular and available in the market. Consumers are aware for the health benefits of probiotics and demanding the foods which are beneficial to them and can be prevent them from diseases. Lots of new fermented food products with innovative packaging have been launched in the market. These foods fit to the requirements of the consumers with new flavors for all occasions.

The main factors responsible for the growth of fermented foods are development of new fermented products and their publicity. The area of starter cultures and their functionality will be the main focus of research area in the coming years. Many natural resources are available from which food grade microorganisms as a starter cultures can be used. Enzymatic pathways of these starter cultures are very important part of research. By studying the pathways, specific strains with specific properties and desired characteristics can be exploited. More research is required in some specific areas like relationship between food and intestinal microbiota, human health & disease, and clinical studies related to the health benefits of probiotics to the human health.

KEYWORDS

- fermentation
- lactic acid bacteria
- fermented foods/beverages
- recent trends
- health benefits
- yoghurt
- probiotics

REFERENCES

Almeida, K. E.; Tamine, A. Y.; Oliveria, M. N. Acidification Rates of Probiotic Bacteria in Minas Frescal Cheese Whey. *LWT Food Sci. Technol.* **2008,** *41*, 311–316.

Awaisheh, S. S.; Haddadin, M. S. Y.; Robinson, R. K. Incorporation of Selected Nutraceuticals and Probiotic Bacetria into Fermented Milk. *Int. Dairy J.* **2005,** *15*, 1184–1190.

Badel, S.; Bernardi, T.; Michaud, P. New Perspectives for *Lactobacilli exopolysaccharides*. *Biotechnol. Adv.* **2011**, *29*(1), 54–66.

Bardale, P. S.; Waghmare, P. S.; Zanzad, D. M.; Khedkar, D. M. Preparation of Shrikhand Like Product from Skim Milk Chakka by Fortifying with Fruit Pulps. *Indian J. Dairy Sci.* **1986**, *39*, 480–483.

Bessa, R. J. B.; Santos-Silva, J.; Ribeiro, J. M. R.; Portugal, A. V. Reticulo-Rumen Biohydrogenation and the Enrichment of Ruminant Edible Products with Linoleic Acid Conjugated Isomers. *Livestock Prod. Sci.* **2000**, *63*, 201–211.

Bhattacharya, A.; Banu, J.; Rahman, M.; Causey, J.; Fernandes, G. Biological Effects of Conjugated Linoleic Acids in Health and Disease. *J. Nutr. Biochem.* **2006**, *17*, 789–810.

Blandino, A.; Al-Aseeri, M. E.; Pandiell, S. S.; Cantero, D.; Webb. C. Cereal-based Fermented Foods and Beverages. *Food Res. Int.* **2003**, *36*, 527–543.

Buccioni, A.; Decandia, M.; Minieri, S.; Molle, G.; Cabiddu, A. Lipid Metabolism in the Rumen: New Insights on Lipolysis and Biohydrogenation with an Emphasis on the Role of Endogenous Plant Factors. *Anim. Feed Sci. Technol.* **2012**, *174*, 1–25.

Caplice, E.; Fitzgerald, G. F. Food Fermentations: Role of Microorganisms in Food Production and Preservation. *Int. J. Food Microbiol.* **1999**, *50*, 131–149.

Chandan, R. C. Enhancing Market Value of Milk by Adding Cultures. *J. Dairy Sci.* **1999**, *82*, 2245–2256.

Chavan, J. K.; Kadam, S. S. Critical Reviews in Food Science and Nutrition. *Food Sci.* **1989**, *28*, 348–400.

Chavez, B. E.; Ledeboer, A. Drying of Probiotics: Optimization of Formulation and Process to Enhance Storage Survival. *Drying Technol.* **2007**, 25(7–8), 1193–1201.

Chin, S. F.; Liu, W.; Storkson, J. M.; Ha, Y. L.; Pariza, M. W. Dietary Sources of Conjugated Dienoic Isomers of Linoleic Acid, a Newly Recognised Class of Anticarcinogens. *J. Food Compos. Anal.* **1992**, *5*, 185–197.

Corsetti, A.; Perpetuini, G.; Tofalo, R. Advances in Fermented Foods and Beverages: Biopresevation Effects in Fermented Foods, 2015 http://dx.doi.org/10.1016/B978-1-78242-015-6.00013-X.

Cotter, P. D.; Hill, C.; Ross, R.P. Bacteriocins: Developing Innate Immunity for Food. *Nat. Rev. Microbiol.* **2005**, *3*, 777–788.

Dal Bello, B.; Cocolin, L.; Zeppa, G.; Field, D.; Cotter, P. D.; Hill, C. Technological Characterization of Bacteriocin Producing *Lactococcus Lactis* Strains Employed to Control *Listeria monocytogenes* in Cottage Cheese. *Int. J. Food Microbiol.* **2012**, *153*, 58–65.

Danielsen, M.; Johansen, E. Functional Genomics of Dairy Micro-Organisms and Probiotics in The Era of Low-cost DNA sequencing. *Aust. J. Dairy Technol.* **2009**, *64*, 102–105.

Diana, A. B. M.; Carolina, J.; Carmen, P.; Teresa, R. Effect of Milk Fat Replacement by Polyunsaturated Fatty Acids on the Microbiological, Rheological and Sensorial Properties of Fermented Milks. *J. Sci. Food Agric.* **2004**, 1559–1605.

Farooq, H. Z.; Haque, U. Effect of Sugar Esters on the Textural Properties of Non-Fat Low Calorie Yoghurt. *J. Dairy Sci.* **1992**, *75*, 2676–2680.

Fellows, J. W., Chang, S. W. and Shazer, W. H. Stability of Aspartame in Fruit Preparations Used in Yogurt. *J. Food Sci.* **1991**, 56 (3), 689–691.

Fiechter, G., Sivec, G.; Mayer, H. K. Application of UHPLC for the Simultaneous Analysis of Free Amino Acids and Biogenic Amines in Ripened Acid-Curd Cheeses. *J. Chromatogr. B.* **2013**, *927*, 191–200.

Foegeding, P. M.; Stanley, N. W. *Listeria innocua* transformed with an Antibiotic Resistance Plasmid As a Thermal Resistance Indicator for *Listeria monocytogenes. J. Food Prot.* **1991,** *54,* 519–523.

Foerst, P.; Santivarangkna, C. Advances in Starter Culture Technology: Focus on Drying Processes. In *Advances in Fermented Foods and Beverages, Improving Quality, Technologies and Health Benefits, Seri. Food Science, Technology and Nutrition*; Holzapfel, W., Ed.; Woodhead Publishing: Sawston, Cambridge, 2015, Vol.; pp. 249–270.

Gibson, G. R.; Roberfroid, M. B. Dietary Modulation of the Human Colonic Microbiota: Introducing the Concept of Prebiotics. *J. Nutr.* **1995,** *6,* 1401–1412.

Guo, Y.; Pan, D.; Sun, Y.; Xin, L.; Li, H.; Zeng, X. Antioxidant Activity of Phosphorylated Exopolysaccharide Produced by *Lactococcus lactis* sub sp. *lactis. Carbohydr. Polym.* **2013,** 97(2), 849–854.

Haque, M. K.; Roos, Y. H. Crystallization and X-ray Diffraction of Crystals Formed in Water–Plasticized Amorphous Spray–Dried And Freeze–Dried Lactose/Protein Mixtures. *J. Food Sci.* **2005,** 70 (5), 359–366.

Hayakava, K.; Kimura, M.; Kasaha, K.; Masumoto, K.; Sansawa, H.; Yamori, Y. Effect of a Gaminobutyric Acid-Enriched Dairy Products on the Blood Pressure of Spontaneously Hypertensive and Normotensive Wistar-Kyoto rats. *Brit. J. Nutr.* **2004,** *92,* 411–417.

Hernandez-Cazares, A. S.; Aristoy, M.C.; Toldra, F. An Enzyme Sensor for the Determination of Total Amines in Dry-Fermented Sausages. *J. Food Eng.* **2012,** *110,* 324–327.

Hernandez-Ledesma, B.; Miralles, B.; Amigo, L.; Ramos, M.; Recio. Identification of Antioxidant and Ace-Inhibitory Peptides in Fermented Milks. *J. Sci. Food Agric.* **2005,** *85,* 1041–1048.

Hernandez-Mendoza, A.; Robles, V. J.; Angulo, J.O.; Cruz, J. D. L.; Garcia, H.S. Preparation of a Whey-Based Probiotic Product with *Lactobacillus reuteri* and *Bifidobacterium bifidum. Food Technol. Biotechnol.* **2007,** 45(1), 27–31.

Imele, H.; Atemnkeng, A. Preliminary Study of the Utilization of Coconut in Yoghurt Production. *J. Food Technol. Afr.* **2001,** *6,* 11–12.

Jahreis, G.; Kraft, J.; Tischendorf, F.; Schöne, F.; von Loeffelholz, C. Conjugated Linoleic Acids: Physiological Effects in Animal and Man with Special Regard to Body Composition. *Eur. J. Lipid Sci. Technol.* **2000,** *102,* 695–703.

Jiang, J.; Bjorck, L.; Fonden, R.; Emanuelson, M. Occurrence of Conjugated *cis*-9, *trans*-11-Octadecadienoic Acid in Bovine Milk: Effects of Feed and Dietary Regimen. *J. Dairy Sci.* **1996,** *79,* 438–445.

Johansen, E. Genetic Engineering. (b) Modification of Bacteria. In *Encyclopedia of Food Microbiology*; Robinson, R.; Batt C.; Patel P., Eds.; Academic Press: London, 1999, pp. 917–921.

Karaaslan, M.; Ozden, M.; Vardin, H.; Turkoglu, H. Phenolic Fortification of Yogurt Using Grape and Callus Extracts. *LWT Food Sci. Technol.* **2011,** *44,* 1065–1972.

Keating, K. R.; White, C. H. Effect of Alternative Sweetners in Plain and Fruit-Flavoured Yoghurts. *J. Dairy Sci.* **1990,** *73,* 54–62.

Khanal, R. C. Potential Health Benefits of Conjugated Linoleic Acid (CLA): A Review. *Asian-Australasian J. Anim. Sci.* **2004,** *17,* 1315–1328.

Kim, Y.; Oh, S.; Kim, S. H. Released Exopolysaccharide (r-EPS) Produced from Probiotic Bacteria Reduce Biofilm Formation of Enterohemorrhagic *Escherichia coli. Int. Immunopharmacol.* **2009,** *11*(12) 2246–2250.

Kiviharju, K.; Leisola, M.; Eerikainen, T. Optimization of a *Bifidobacterium longum* Production Process. *J. Biotechnol.* **2005,** *117*(3), 299–308.

Kivirand, K.; Rinken, T. Biosensors for Biogenic Amines: the Present State of Art Mini-Review. *Anal. Lett.* **2011**, *44*, 2821–2833.

Kraft, J.; Jahreis, G. P hysiologische Wirkungen von konjugierten Linolsäuren. In *Lipide in Fleisch, Milch und Ei-Herausforderungfürdie Tierernährung*; ETH: Zürich; Kreuzer, M.; Wenk, C.; Lanzini, T. Eds.; 2004, pp. 81–93.

Kramer, J. K. G.; Cruz-Hernandez, C.; Deng, Z.; Zhou, J.; Jahreis, G.; Dugan, M. E. R. Analysis of Conjugated Linoleic Acid and Trans 18:1 Isomers in Synthetic and Animal Products. *Am. J. Clin. Nutr.* **2004**, *79*, 11375–11445.

Kvasnicka, F.; Voldřich, M. Determination of Biogenic Amines by Capillary Zone Electrophoresis with Conductometric Detection. *J. Chromatogr. A.* **2006**, *1103*, 145–149.

Leblanc, J. G.; Milani, C.; De Giori, G. S.; Sesma, F.; Van Sinderen, D.; Ventura, M. Bacteria as Vitamin Suppliers to Their Host: a Gut Microbiota Perspective. *Curr. Opin. Biotechnol.* **2013**, *24*, 160–168.

Li, J. Y.; Jin, M. M.; Meng, J.; Gao, S. M.; Lu, R. R. Exopolysaccharide from *Lactobacillus plantarum* LP6: Antioxidation and the Effect on Oxidative Stress. *Carbohydr. Polym.* **2013**, *98*(1), 1147–1152.

Liu, C. F.; Tseng, K. C.; Chiang, S. S.; Lee, B. H.; Hsu, W. H.; Pan, T. M. Immunomodulatory and Antioxidant Potential of *Lactobacillus* exopolysaccharides. *J. Sci. Food Agric.* **2011**, *91*(12), 2284–2291.

Macedo, R. F.; Renato, J.; Freitas, S.; Pandey, A.; Soccol, C. R. Production and Shelf-Life Studies of Low Cost Beverage With Soymilk, Buffalo Cheese Whey and Cow Milk Fermented by Mixed Cultures of *Lactobacillus casei ssp. shirota and Bifidobacterium adolescentis*. *J. Basic Microbiol.* **1999**, *39*, 243–251.

Manyam, B. V.; Katz, L.; Hare, T. A.; Kaniefski, K.; Tremblay, R. D. Isoniazid-Induced Elevation of Cerebrospinal Fluid (CSF) GABA Levels and Effects on Chorea in Huntington's Disease. *Ann. Neurol.* **1981**, *10*, 35–37.

McAuliffe, O.; Hill, C.; Ross, R. P. Inhibition of *Listeria monocytogenes* in Cottage Cheese Manufactured with A Lacticin 3147 Producing Starter Culture. *J. Appl. Microbiol.* **1999**, *86*, 251–256.

Murray, M; Richard, J. A. Comparative Study of the Antilisterial Activity of Nisin A and Pediocin AcH in Fresh Ground Pork Stored Aerobically at 50c. *J. Food Prot.* **1997**, *60*, 1534–1540.

Nsofor, L. M.; Chukwa, E. U. Sensory Evaluation of Soy Milk-Based Yoghurt. *J. Food Sci. Technol. Mysore.* **1992**, 29, 301–304.

O'Shea, M.; Bassaganya-Riera, J.; Mohede, I. C. M. Immunomodulatory Properties of Conjugated Linoleic Acid. *Am. J. Clin. Nutr.* **2004**, *79*, 1199S–1206S.

Oetjen, G.; Haseley, P. *Freeze-drying*, 2nd Ed., Wiley-VCH: Weinheim, 2004.

Omoni, A. O.; Aluko, R. E. Soybean Foods and Their Benefits: Potential Mechanisms of Action. *J. Nutr.* **2005**, *63*, 272–283.

Onal, A.; Tekkeli, S. E.; Onal, C. A Review of the Liquid Chromatographic Methods for the Determination of Biogenic Amines in Foods. *Food Chem.* **2013**, *138*, 509–515.

Ostlund, R. E.; JrRacette, S. B.; Okeke, A.; Stenson, W. F. Phytosterols that are Naturally Present in Commercial Corn Oil Significantly Reduce Cholesterol Absorption in Humans. *Am. J. Clin. Nutr..* **2002**, *75*, 1000–1004.

Palacio, A. A.; Morales, M. E.; Velez, R. J. F. Rheological and Physicochemical Behavior of Fortified Yoghurt, With Fiber and Calcium. *J. Texture Stud.* **2005**, *36*, 333–349.

Pandya, C. N. Development of Technology for Fruit Dahi. M.Sc. Thesis, Indian Dairy Research Institute, Karnal, India, 2002.

Park, K. B.; Oh, S. H. Production of Yoghurt with Enhanced Levels of Gamma-Aminobutyric Acid and Valuable Nutrients Using Lactic Acid Bacteria and Germinated Soybean Extract. *Bioresour. Technol.* **2007,** *98,* 1675–1679.

Park, Y.; Pariza, M. W. Mechanisms of Body Fat Modulation by Conjugated Linoleic Acid (CLA). *Food Res. Int.* **2007,** *40*(3), 311–323.

Pawar, D. D.; Malik, S. V. S.; Bhilegaonkar, K. N.; Barbuddhe, S. B. Effect of Nisin and its Combinations with Sodium Chloride on the Survival of *Listeria monocytogenes* Added to Raw Buffalo Meat Mince. *Meat Sci.* **2000,** *56,* 215–219.

Phillips, K. M.; Ruggio, D. M.; Toivo, J. I.; Swank, M. A.; Simpkins, A. H. Free and Esterified Sterol Composition of Edible Oils and Fats. *J. Food Compos. Anal.* **2002,** *15,* 123–142.

Pinheiro, M. V. S.; Oliveira, M. N.; Penna, A. L. B.; Tamine, A. Y. The Effects of Different Sweeteners in Low-Calorie Yoghurts-A Review. *Int. J. Dairy Technol.* **2005,** *58*(4), 193–199.

Pirkul, T.; Temiz, A.; Erdem, Y. K. Fortification of Yoghurt with Calcium Salts and its Effect on Starter Microorganisms and Yoghurt Quality. *Int. Dairy J.* **1997,** *7*(8–9), 547–552.

Redruello, B.; Ladero, V.; Cuesta, I.; Alvarez-Buylla, J. R.; Martin, M. C.; Fernandez, M. A. Fast, Reliable, Ultra High Performance Liquid Chromatography Method for the Simultaneous Determination of Amino Acids, Biogenic Amines and Ammonium Ions in Cheese, Using Diethyl Ethoxymethylenemalonate as a Derivatising Agent. *Food Chem.* **2013,** *139,* 1029–1035.

Roberfroid, M. B. Prebiotics: The Concept Revisited. *J. Nutr.* **2007,** *137*(3 Suppl. 2), 830S–837S.

Robinson, R. K.; Tamime, A. Y. Microbiology of Fermented Milks. In *Dairy Microbiology*, 2nd Ed., Robinson, R. K., Ed.; Elsevier Applied Science: London, 1995, Vol. 2, pp. 291–343.

Santivarangkna, C.; Higl, B.; Foerst, P. Protection Mechanisms of Sugars During Different Stages of Preparation Process of Dried Lactic Acid Starter Cultures. *Food Microbiol.* **2008,** *25*(3), 429–441.

Santos, F.; Wegkamp, A.; De Vos, W. M.; Smid, E. J.; Hugenholtz, J. High-Level Folate Production in Fermented Foods By the B12 producer *Lactobacillus reuteri* JCM1112. *Appl. Environ. Microbiol.* **2008,** *74,* 3291–3294.

Semyonov, D.; Ramon, O.; Shimoni, E. Using Ultrasonic Vacuum Spray Dryer to Produce Highly Viable Probiotics. *LWT Food Sci. Technol.* **2011,** *44,* 1844–1852.

Sengul, N.; Isık, S.; Aslım, B.; Ucar, G.; Demirbag, A. E. The Effect of Exopolysaccharide-Producing Probiotic Strains on Gut Oxidative Damage in Experimental Colitis. *Digest. Dis. Sci.* **2011,** *56*(3), 707–714.

Setchell, K. D. R. Phytoestrogen: The Biochemistry, Physiology, and Implications for Human Health of Soy Isoflavones. *Am. J. Clin. Nutr.* **1998,** 68, 1333S–1346S.

Shalaby, A. R. Significance of Biogenic Amines in Food Safety and Human Health. *Food Res. Int.* **1996,** *29,* 675–690.

Singh, R.; Kumar, R.; Venkateshappa, R.; Mann, B.; Tomar, S. Studies on Physicochemical and Antioxidant Properties of Strawberry Polyphenol Extract-Fortified Stirred Dahi. *Int. J. Dairy Technol.* **2013,** *66,* 103–108.

Stiles, M. E.; Hastings, J. W. Bacteriocin Production by Lactic Acid Bacteria: Potential for Use in Meat Preservation. *Trends Food Sci. Technol.* **1999,** *2,* 247–251.

Sybesma, W.; Burgess, C.; Starrenburg, M.; Van Sinderen, D.; Hugenholtz, J. Multivitamin Production in *Lactococcus lactis* Using Metabolic Engineering. *Metab. Eng.* **2004,** *6,* 109–115.

Sybesma, W.; Starrenburg, M.; Kleerebezem, M.; Mierau, I.; De Vos, W. M.; Hugenholtz, J. Increased Production of Folate by Metabolic Engineering of *Lactococcus lactis*. *Appl. Environ. Microbiol.* **2003**, *69*, 3069–3076.

Wahle, K. W. J.; Heys, S. D.; Rotondo, D. Conjugated Linoleic Acids: Are They Beneficial or Detrimental to Health? *Prog. Lipid Res.* **2004**, *43*, 553–587.

CHAPTER 3

MICROBIAL APPROACHES IN FERMENTATIONS FOR PRODUCTION AND PRESERVATION OF DIFFERENT FOODS

DEEPAK KUMAR VERMA*, DIPENDRA KUMAR MAHATO, SUDHANSHI BILLORIA, MANDIRA KAPRI, P. K. PRABHAKAR, AJESH KUMAR V., and PREM PRAKASH SRIVASTAV

Department of Agricultural and Food Engineering, Indian Institute of Technology, Kharagpur 721302, West Bengal, India

Corresponding author. E-mail: deepak.verma@agfe.iitkgp.ernet.in; rajadkv@rediffmail.com

CONTENTS

ABSTRACT

Food preservation has been used for the past years for maintaining edible products for consumption. New infectious diseases are emerging and adapting to infect humans in its own unique way. Besides these emerging and reemerging pathogens adaptation to the human environment, the changes in human lifestyle, population, and globalization of food supply have created new outbreaks for pathogens to infect humans. Preservation processes are designed to slow down the growth of microorganisms and are affected by food factors both intrinsic and extrinsic factors like pH, oxidation-reduction potentials, water activity, and natural antimicrobials. Several factors like heat, irradiation, and nonthermal processing are involved in inactivation of microorganisms. A combination of several techniques and processes can be used to make the food safer and to extend the shelf life of foodstuffs. A widespread suppression or eradication of the foodborne agents is not feasible. Continuous research efforts and newer strategies are essential for better understanding to control foodborne pathogens, and consumer education should be consistent which will create a quick response to the new and reemerging foodborne pathogens.

3.1 INTRODUCTION

Food preservation by fermentation technology is widely practiced. However, the current food industry is facing lot of challenges due to increasing consumer demands for fresh and nutritional food products, thus minimizing the use of preservatives for a safe and healthy living. The perception of food safety is essential and is of greater importance. The characteristic of a food product which is universally accepted is that food which is free or nearly free of human pathogens. To control pathogenic microorganism, humans have formulated many interventional methods to inactivate or inhibit the growth of microorganisms. The efficiency of these methodologies depends not only on the methodology but also on the target microorganisms, environment, and characteristics of the nutrition in the foodstuffs. "Emerging infectious diseases" are rapidly increasing and spreading, and there are a number of factors which aid in the spread of foodborne pathogens (Ray, 1992). The World Health Organization associates the spread and the reemerging foodborne pathogens with factors that include changes in human population, globalization of food supply, lifestyle, and accidental introduction of pathogens in new geographic areas and exposure of unfamiliar food borne hazards.

Preservation of food is a process of storing or maintaining food products with its stable properties which can provide maximum benefits. The stability of the food products may vary depending on the method of its handling, processing, storage, and distribution. Thus, it is essential to understand the handling, processing, storage, and its distribution for safety (Campbell-Plat, 1987). Food preservation technology is a high interdisciplinary science which is developing new techniques to satisfy current demands of consumer satisfaction and economic preservation in nutritional aspects, convenience, low demand of energy, absence of preservatives, and environmental safety. Food risk assessment is an integrative approach to food safety. The appropriate simulation and design of risk assessment modeling is the foundation of risk assessment model and is significant to predict and control the risk. At present, fermented foods are of customer choice and commercially very successful, and hence, it's foreseen that consumption of fermented foods will increase worldwide in near future (Ray, 1992).

3.2   FERMENTED FOOD

The production and consumption of fermented food date back to many thousand years with the evidence of fermentation of barley in beer. It is an age-old process in food biotechnology. Food products undergo fermentation by enzymatic or microbial action, thus producing biochemical changes and modifications (Campbell-Platt, 1987). In fermentation, organic substrates, mostly carbohydrates, are oxidized and act as an electron acceptor (Adams, 1990). In food processing, carbohydrates are converted to alcohol, lactic acid, carbon dioxide (CO_2), or organic acids with the help of microbial organisms such as bacteria, yeast, etc. in anaerobic conditions. Processes involving the production of ethanol by yeast or lactic acid bacteria (LAB)-producing organic acids are included in food fermentation.

3.3   FERMENTATION: AN IMPORTANT APPROACH IN FOOD INDUSTRY

Fermentation is one of the oldest forms of food preservation which increases the shelf life of highly perishable foods like meat, fish, fruit, and vegetables (Steinkraus, 1994, 2002). Fermentation involves enzymatic and microbial process and is found to have physical and nutritional benefits to the food by enhancing protein and fiber digestion and biosynthesis of essential amino acids, proteins, and vitamins. It also improves food safety by reducing toxic substances and also degrades antinutritional factors. One of the key factors of fermentation is that it progresses the shelf life of the food (Holzapfel, 2002). The food of fermentation has significant characteristics and aids in promotion of health in addition to the nutritional and safety requirements. Studies have also proven that fermented foods have therapeutic properties (Hansen, 2002). The vital roles of fermentation in food processing are discussed further.

3.3.1   ALLEVIATION OF LACTOSE INTOLERANCE

Some individuals are unable to digest lactose present in milk thus causing diseases such as diarrhea, pain in the abdomen and flatulence (Panesar et al., 2006). This intolerance is due to the lack or insufficient amount of enzymes β-galactosidase and β-phosphogalctosidase which convert them into monosaccharide components. Fermentation convert milk lactose to lactic acid, milk having cells of *Lactobacillus acidophilus* helps in the breakdown of lactose. LAB are able to convert lactose to lactic acid by utilizing the enzyme lactase. Apart from these, fermented food helps in stimulation of immune system, decrease in cholesterol level, give texture to food by softening it. Fermentation enhances the food quality and safety by probiotics. LAB are supposed to produce compounds of interest as various antimicrobial compounds, and even reduce blood serum cholesterol, alleviation of lactose intolerance, stimulation of the immune system. For these reasons, LABs have become center of study for food industry as well as international researchers (Khedid et al., 2009).

3.3.2   ANTIBIOTIC PROPERTIES

Hydrogen peroxide (H_2O_2) and bacteriocins produced during fermentation are inhibitory for other bacterial growth (Oyewole, 1997). Some compounds produced after acid fermentation has antitumor properties (Hirayama and Rafter, 1999). Gastrointestinal infections such as diarrhea are caused due to changes in the local flora of gastrointestinal tract by the evading pathogen. In milk products produced by fermentation, the LAB interferes with the colonization and spreading of food borne pathogens, resulting in the prevention of the pathogen (Gandhi, 2000). Dairy products such as yogurt, butter milk, and cheese have been considered to protect against breast cancer.

Experimental studies have showed that LAB produce antitumor effect by suppressing cancer initiation. Different mechanisms by which these bacteria induce anticancer effect includes changes in fecal enzymes that lead to colon carcinogenesis, cellular uptake of mutagenic substances, and enhancing immune response and reducing mutagens and tumor suppressors (Hosono et al., 1986).

LAB exerts antimicrobial in the biopreservation of foods by the production of specific compounds like organic acids, H_2O_2, CO_2, diacetyl, bacteriocins, and broad-spectrum antimicrobials like reuterin (De Vuyst and Vandamme, 1994a). LAB is deficient in true catalase which is responsible for the breakdown of H_2O_2 generated in the presence of oxygen (O_2), and accumulated H_2O_2 is responsible for the inhibition of microorganisms (Condon, 1987). H_2O_2 is also supposed to activate the lactoperoxidase system of fresh milk with the formation of hypothiocyanate and other antimicrobials (Reiter and Harnulv, 1984; Pruitt et al., 1986; Condon, 1987; De Vuyst and Vandamme, 1994b). The development of antibiotic resistances is a complex phenomenon involving interactions among humans, animals, bacteria, drugs, and the environment (Barbosa and Levy, 2000; O'Brien, 2002; Coast and Smith, 2003). The evidences of antibiotic resistances in LAB of food origin have threatened the food safety and public health (Teuber et al., 1999; Ge et al., 2007; Egervarn et al., 2009; Marshall et al., 2009).

3.3.3 FLAVOR ENHANCEMENT

Fermentation enhances the aroma and flavor of food. The organoleptic properties of fermented food make them preferable as compared to unfermented food products. The flavor of fermented food products is greatly influenced by acid fermentation as it produces lactic acid, thus resulting in lowering of pH causing sourness in food. During fermentation, metabolism of sugar produces acid or alcohol thus decreasing sweetness (McFeeters, 2004). Carbohydrates are converted to organic acids and other compounds like ethanol, diacetly, and acetaldehyde, thus increasing flavor and taste to the food. These are flavoring compounds which aid in the taste of the food as follows:

(a) Diacetly: Provides sweet taste to butter milk and culture sour milk.
(b) Acetaldehyde: Aids in the aroma and flavor of yogurt.
(c) 2-Acetyl-1-pyrroline: Compound is a precursor of ornithine and is liberated during the baking of bread dough, which gives a popcorn odor (Schieberle, 1995).

LAB ferments carbohydrates into lactic acid by homofermentation or to a mixture of lactic acid, CO_2, and acetic acid and/or ethanol by hetero-fermentation along with the production of other compounds like diacetyl, acetaldehyde, and H_2O_2, which are responsible for the flavor and texture of fermented foods. These may even lead to inhibition of undesirable microbes. LAB in tape fermentation is responsible for the development of its flavor (Nuraida and Owens, 2014). The salt is added in many Indonesian fermented foods (fruits, vegetables, meat, and fish) which are produced by natural fermentation to control the environment and to promote the growth of LAB over spoilage bacteria (Nuraida et al., 2014; Swain et al., 2014) as well as inhibition of pectinolytic and proteolytic enzymes responsible for softening and putrefaction (Swain et al., 2014). Kinema which is an ethnic-fermented soybean food of the Nepali community in the Eastern Himalayas is produced by the monoculture fermentation of soybean using *Bacillus subtilis* MTCC 2756 which gives it a pleasant nutty flavor and highly sticky texture (Tamang, 2015). Recent researches of many workers have been document in Table 3.1, that state flavor enhancement in meat products through microorganisms of genus *Candida, Debaryomyces, Pichia*, and *Williopsis* (Rantsiou et al., 2005; Aquilanti et al., 2007; Villani et al., 2007; Andrade et al., 2009a,b; Asefa et al., 2009; Purrinos et al., 2013; Ozturk, 2015).

TABLE 3.1   Microorganisms Responsible for the Flavor Enhancement in Fermented Meat Products.

Products	Microorganisms	Reference
Dry cured ham (Lacon)	*Debaryomyces hansenii*	Purrinos et al. (2013)
Dry cured Iberian ham	*Candida zeylanoides* and *D. hansenii*	Andrade et al. (2009a,b)
Dry cured meat product (beef Pastirma)	C. zeylanoides	Ozturk (2015)
Dry cured meat products	*C. zeylanoides* and *D. hansenii*	Asefa et al. (2009)
Dry sausages	*D. hansenii*	Aquilanti et al. (2007)
Fermented sausages	*C. kruseiii, C. sake, D. hansenii,* and *Williopsis saturnus*	Rantsiou et al. (2005)
Traditional salamis	*C. zeylanoides, D. hansenii,* and *Pichia guilliermondii*	Villani et al. (2007)

3.3.4   NUTRITIONAL VALUE

The preservation is assisted through lactic acid, alcoholic, acetic acid, and high salt fermentations. Besides this, fermentation is also responsible organoleptic characteristics which include development of flavors, aromas, and textures. Moreover, fermentation even improves digestibility and nutritional quality through enhancement of food matrices with vitamins, proteins, essential amino acids, and essential fatty acids (Steinkraus, 1994, 2002). Many food products have low nutritional value such as cereals, but it has been shown that lactic acid fermentation (LAF) enhances the nutrient content and digestive property of different foods (Nout, 2009). The quantity and quality of proteins and vitamins increase on the other hand antinutritional agents decreases after fermentation (Paredes-López and Harry, 1988). During the process of fermentation, the raw substrate undergoes many soaking and hydration stages and then cooking many potent toxins such as trypsin inhibitors and cyanogens in cassava declines. Aflatoxins present in peanut and grains is decreased (Steinkraus, 1983). Fermentation causes a decrease in the sugars and indigestible carbohydrates level in legumes, thus leading to reduce abdominal pain (Matarjemi and Nout, 1996).

Natural and induced fermentation assisted by *Lactobacillus plantarum* produces phenolic compounds in fermented cowpea flour followed by hydrolysis of complex polyphenols into simpler and more biologically active compounds (Duenas et al., 2005). *Lb. plantarum* also metabolizes *p*-coumaric, caffeic, ferulic, and *m*-coumaric acids (Rodriguez et al., 2009), and its growth in fresh olive mill wastewaters leads to hydrolysis of phenolic and depolymerization (Lamia and Moktar, 2003). The higher antioxidative activity has been reported in the samples fermented with *Lactobacillus rhamnosus* than samples fermented with *Saccharomyces cerevisiae*. For instance, the percentage of lipid peroxidation inhibition in liposomes was found to be 50.8% in unfermented barley, 52.4% in barley fermented with *S. cerevisiae*, and 60.9% in barley fermented with *Lb. rhamnosus* (Dordevic et al., 2010).

Several protein hydrolysates are produced from plant and animal sources using exogenous proteases, but few studies have been conducted for their production using microbial fermentation. The antioxidant and antibacterial activity of fermented protein hydrolysates was reported from delimed tannery fleshings (Balakrishnan et al., 2011). The proteolytic systems of LAB are involved in the hydrolysis of proteins during fermentation of foods like milk and meat products (Brink and HuisIntVeld, 1992). LAB (*Enterococcus faecium*) isolated from soy milk proteins have been found to produce fermented soy milk with multifunctional activities (Martinez-Villaluenga et al., 2012). An α-L-rhamnosidase producing *Lactobacillus* strains facilitates the hydrolysis of rhamnose-containing phenolic conjugates to increase in free polyphenols as well as flavonoid antioxidants in plant foods rich with rhamnose sugars (Beekwilder et al., 2009). The same has been done with *Lb. plantarum* as starter culture to the improvement of nutritional quality of fermented foods which is also relevant to other study of probiotic-riboflavin over producing *Lb. plantarum* (Arena et al., 2014). Juan and Chou (2009) reported the enhancement of phenolics, flavonoids, and antioxidant activity in black soybean extract due to the liberation of isoflavonoids from their corresponding glycosides by acids formed during the fermentation. The dominance of *Lactobacillus* species during cereal fermentation and the presence of other bacteria to enhance nutritional quality as well as safety challenges were identified (Oguntoyinbo et al., 2011; Oguntoyinbo and Narbad, 2012).

3.3.5 PRESERVATION

The preservative quality in food and beverages is due to antimicrobial metabolites formed during fermentation (Caplice and Fitzgerald, 1999). These metabolites include different organic acids such as acetic and propionic acids. LAB inhibits the growth of pathogenic organisms through acid production which decreases the pH (below 4) during fermentation, thus protecting the food from spoilage and poisoning (Ananou et al., 2007). Acids interfere with the maintenance of potential of cell membrane, inhibit active transport across membranes, and inhibit range of other metabolic functions (Doores, 1993). Antifungal properties of LAB have been observed (Schnürer and Magnusson, 2005). Thus fermentation increases the shelf life of fermented food. Microorganisms produce variety of antimicrobial proteins or peptides known as bacteriocins (Ross et al., 2002). Their mode of action involves cell wall synthesis inhibition or depolarizes the cell membrane (Abee et al., 1995). Besides the main function of fermentation being preservation, some still perceive it as "folk nutraceuticals" (Pieroni and Quave, 2006). The aromatic species like *Origanum vulgare* and *Ledum palustre* are added during fermentation of taar to enhance preservation as well as to develop specific taste (Kalle and Soukand, 2013). The purpose of preservation by fermentation is achieved by oxidation of carbohydrates into end-products like acids, alcohol, and CO_2. Most fermented foods rely on LAB for the fermentation and the end-products formed by carbohydrate catabolism give characteristic flavor, aroma, and texture besides preservation (Caplice and Fitzgerald, 1999).

3.4   TYPE OF FERMENTATION IN FOOD PRODUCTION AND PRESERVATION

Fermentation can occur spontaneously or can be induced. There are different types of fermentation used in food processing. Food fermentations can be classified either based upon categories, classes, or commodities. Some examples of food fermentations are described in the following sections.

3.4.1   ALCOHOLIC FERMENTATION

Yeast or yeast-like molds such as *Amylomyces rouxii* is involved in alcoholic fermentation (ALF) resulting in ethanol production. ALF is involved in the production of beverages such as wine, beer vodka, etc. and rising of bread dough. The substrates used for fermentation include honey, cereal grains, sap of palm, fruit juices, and grain malt, which contain sugars that can be fermented and are converted to ethanol by yeast. During the process, equal amount of CO_2 is also produced as a side product, and this process is carried out under anaerobic conditions (Zeuthen and Bogh-Sorensen, 2003).

3.4.2   LACTIC ACID FERMENTATION

Fermentation of lactic acid involves the transformation of sugars to lactic acid by organisms such as *Lactobacillus, Leuconostoc*, and *Streptococcus* bacteria. Lactic acid is the most important compound result from this reaction. Lactic acid-producing bacteria are most significant bacteria used in food fermentation and production. Sour milk is one of the most ancient lactic acid fermented food, in which the LAB will convert the milk sugar known as lactose to lactic acid resulting in sour or fermented milk. Dairy products for example yogurt, cheese, butter, and sour milk are also produced. LAF is used for the preservation of different vegetable foods (Steinkraus, 1983). The best example is of sauerkraut, produced by the action of LAB on cabbage (Pederson and Albury, 1969).

3.4.3   SOLID-STATE FERMENTATION

Solid-state fermentation (SSF) is a fermentation performed on a solid substrate acting both as support and nutrient source for the microorganism when there is no free flowing liquid (Pandey, 1992). SSF results in biomolecule manufacture utilized in food. These biomolecules are metabolites which are generated by microorganisms such as yeast or bacteria. This is an ancient process and different fungi are used in food production. Some examples are fermentation of rice and cheese by fungi. Industrial enzymes are produced commercially by SSF (Suryanarayan, 2003).

3.5 MICROBES IN FERMENTED FOOD

3.5.1 BEVERAGES

The process of fermentation for the production of beverages has been used since old times. Beverages are produced by long periods of storage which improves by natural way due to its physical and chemical actions. The natural aging of beverages is favorable and clearly studied by the researchers. The significant reactions are the slow oxidation of alcohol and aldehydes resulting in ester formation which increases the taste of the beverage (Bachmann et al., 1934). Fermented beverages are of two types, wine and beer, which are described in the following sections.

3.5.1.1 WINES

Wine is one of the most popular yeasts and LAB-fermented alcoholic beverages and is produced by the fermentation of fruit juices having fermentable sugar. The production of wine and beer involves the fermentation of sugar by yeasts like *S. cerevisiae* and *Saccharomyces ellipsoideus*, yielding ethyl alcohol and CO_2. Alcohol is an end product of fermentation, and depending on its concentration, it also acts as a preservative. The alcohol content of wines depends on the content of sugar, grapes, type of yeast, temperature, and O_2 level which occurs within the range of 12–15% alcohol by volume. Natural wines contain about 9–13% alcohol, and fortified wines have concentration of about 20% by volume (Cavalieri, 2003).

The alcoholic or primary fermentation in wine is a natural process performed by native yeasts, mainly belonging to *Saccharomyces* species. Here, yeast converts sugars to alcohol by ALF, whereas in most red and some white grape wines, there undergo a secondary fermentation known as malolactic fermentation (MLF) (Davis et al., 1985). In MLF, LAB such as *Oenococcus oeni* converts malic acid into lactic acid and confers the finest oenological malolactic characteristics (Alexandre et al., 2004; Fugelsang and Edwards, 2007; Massera et al., 2009). During these processes, various metabolites produced are responsible for the nutritional content, taste, and flavor of wine (Carrau et al., 2008; Son et al., 2009). The red wine samples were analyzed for *S. cerevisiae, O. oeni,* and DNA from the grape (*Vitis vinifera*) by Polymerase chain reaction denaturing gradient gel electrophoresis analysis with the set of universal primers. The *S. cerevisiae and V. vinifera* relatives were detected throughout the entire fermentation process, whereas *O. oeni* were detected only during the last stage of fermentation, and the *Propionibacterium acnes* were detected from the middle stage of ALF to the middle stage of MLF. Similar analysis was done for white wine, and it was observed that *S. cerevisiae* and *V. vinifera* relatives were found throughout the fermentation process. The *O. oeni* relatives were detected after the late stage of ALF; *P. acnes* after the middle stage of ALF and *Gluconacetobacter saccharivorans* was detected at the early and the middle stages of ALF (Kato et al., 2011).

The Chinese yellow rice wine is traditionally brewed from glutinous rice using starters like filamentous fungi, yeasts, and bacteria (Zhang et al., 2008; Rong et al., 2009). Studies conducted on Chinese rice wine fermentation has revealed huge microbial diversity (Mao, 2004; Xie et al., 2007; Rong et al., 2009; Shi et al., 2009). Fungal diversity in wheat Qu of Shaoxing rice wine included fungi like *Absidia corymbifera, Aspergillus fumigatus, Aspergillus oryzae, Rhizomucor pusillus, Rhizopus microsporus,* and *Rhizopus oryzae* (Xie et al., 2007). *Actinomycetales, Bacillales,* and *Lactobacillales* species were among the dominant bacterial species in wine starter of Fen Liquor (Shi et al., 2009). Similarly, *Monascus purpureus* predominanted in Hong Qu, and *Hansenula* spp., *Mucor* spp., *Pichia* spp., and *Saccharomyces* spp. were the major fungal species in Yao Qu (Chen et al., 2007; Ni et al., 2009).

3.5.1.2   BEERS

Beers produced from the products containing starch undergo enzymatic splitting, malting, and mashing so that the sugar becomes available to microorganism for fermentation. Yeasts are commonly used in combination with LAB as they have stabilizing and viscosity properties and are also used to produce CO_2 (Keukeleire, 2000). *S. cerevisiae* is the primary microorganism used in industry for fermentation of beverages and production of ethanol fuel. Lambic sour beer is one of the oldest brewed beers which are produced by a spontaneous fermentation process (De Keersmaecker, 1996). The sour taste of the beer is due to the metabolic activities of LAB, acetic acid bacteria, and various yeasts (Van Oevelen et al., 1977; Verachtert and Iserentant, 1995; Spitaels et al., 2014). A further study limbic beer showed three phases of fermentation. The first phase of initial one month showed microbial succession of *Enterobacteriaceae*. The second and main phase is dominated by *Pediococcus damnosus*, *S. cerevisiae*, and *Saccharomyces pastorianus* after two months followed by third that is, maturation phase after six months by the growth of *Dekkera bruxellensis* (Spitaels et al., 2014). Among the probiotics, *Lactobacillus fermentum KKL1* are responsible for preparation of rice beer and for its functional characteristics (Ghosh et al., 2015). The bulk residue of yeast biomass obtained from beer fermentation could be used for nutritional supplementation (rich in proteins, amino acids, and vitamin B) as well as for decontamination of mycotoxins (Olivon, 2013).

3.5.2 CEREAL-BASED FOODS PRODUCTS

Cereal grains have been the source of food of humans since ancient times. Grains can be processed into food by fermentation (Akinrele, 1970). "Ogi," cereal porridge, is made from maize, millet, and sorghum fermentation (Osungbaro, 2009). Most of cereal-based foods are obtained from maize, wheat, millet, sorghum, or rice. Fermented cereal products can either be liquid or solids (stiff gels). The use of single or mixed cultures of various microorganisms like LAB, yeast, and fungi are used for fermentation of cereal-based products such as rice vinegar, soy sauce, soybean-barley paste, natto, and tempeh (Murooka and Yamshita, 2008). *B. subtilis* is used for fermentation of natto. It is prepared from cooked soyabeans which contains health beneficiary compounds like aglycone, vitamin K_2, nattokinase, and superoxide dismutase, functioning to suppress low-density lipoprotein oxidation, reduction of DNA injury caused by cyclophosphamide. These even have antioxidant and antimutagenic effects (Wu and Chou, 2009). The fermentation of black soybean with *B. subtilis* showed higher antioxidant activity, total phenolic content, and total flavonoid content (TFC). Juan and Chou (2010) compared those of the nonfermented and fermented with *Lb. plantarum* (Wang et al., 2014). It has also been reported that wheat bran fermented with *Lactobacillus brevis* E95612 and *Kazachstania exigua* C81116 increased texture, antioxidant activities, peptides, total free amino acids content, and the in vitro digestibility of proteins (Coda et al., 2014).

Among different benefits of fermentation, one is to promote health for example by increasing antioxidant activities in chickpeas by fermentation with *Cordyceps militaris* (Xiao et al., 2014). The bacterial fermentation of cereals and legumes with LAB are common, but very few are fungal-fermented foods like *miso, tempe, shoyu,* etc. (Okada, 1988; Sasaki and Mori, 1991). Fungi promote the production of enzymes, degradation of antinutritive factors and improve the bioavailability of minerals during fermentation (Nout and Aidoo, 2002). In a study conducted by Meroth et al. (2003) to investigate the development of microorganisms during sourdough fermentations with the application of several different process parameters, it was observed that *Lactobacillus sanfranciscensis* was even dominant after 18 refreshments. Similarly, *Lactobacillus reuteri* stably persisted over 10 years or 50,000 generations (Ganzle and Vogel, 2002). This suggested variability regarding the availability of sugars and the amylolytic or peptidolytic enzymes in cereals, and also the selection of starter culture depends on the type of cereal (Vogelmann et al., 2009). Among all fermented foods (e.g., milk, meat, fish, soy, or wine), cereal-based products occupy the major space (Hammes et al., 2005). Fermented cereals have great role in human diet as to obtain nutraceuticals and hence offers a great potential for the food industry.

3.5.3 DAIRY-BASED FOOD PRODUCTS

Fermentation of milk and dairy products is very significant and is extremely important. Examples of dairy products that are fermented are yoghurt, kefir, cheese, butter, and sour milk. Fermented milk products are important source of probiotoics (Katz, 2001). A fermented milk product, yoghurt, has the ability to modulate the immune system.

FIGURE 3.1   General steps involved in fermentation of dairy-based food products.

Fermented dairy products are prepared from milk of almost all domesticated milch animals since past years. Several innovative categories of fermented milk products are becoming widespread all over the world (Table 3.2). The traditional fermentations are taking place as a result of the activities of natural flora present in the food or added from the surroundings. Over the period, researchers have tried to isolate and study the characters of such desirable organisms. Among the bacteria, the most important dominant group bringing fermentation is LAB (Table 3.3). The LAB are naturally accepted as generally regarded as safe for human consumption (Aguirre and Collins, 1993). The general procedure used for the preparation of fermented milk products is shown in Figure 3.1. During fermentation, certain physical and chemical changes occur in the milk due to the growth, and fermentative activities of LAB are used as starter cultures (Table 3.4).

TABLE 3.2   Microorganisms Involved in Fermentation of Dairy-based Food Products.

Microorganisms	Fermented Dairy-based Food Products
Lb. acidophilus	Acdiophilus milk
Lb. delbueckii sub sp. *bulgaricus*	Bulgarian butter milk
Lactococcus lactis sub sp. *lactis, Lc. lactis* sub sp. *cremoris, Lc. lactis* sub sp. *diacetylactis, Sc. thermophilus, Lb. delbueckii* sub sp. *bulgaricus, Priopionibacterium shermanii, Penicillium roqueforti*, etc.	Cheese
S. lactis sub sp. *diacetylactis, Sc. cremoris*	Cultured butter milk
Lb. lactis sub sp. *lactis, Lb. delbrueckii* subsp. *bulgaricus, Lb. plantarum, Streptococcus lactis, Sc. thermophiles, Sc. cremoris*	Curd
Sc. lactis, Leuconostoc spp., *Saccharomyces kefir, Torula kefir, Micrococci*	Kefir
Lb. acidophilus, Lb. bulgaricu, Saccharomyces, Micrococci	Kumiss
Lb. bulgaricus	Lassi
Sc. lactis, Sc. thermophiles, Lb. bulgaricus, Lactose fermenting yeast	Leben
Sc. thermophile, Lb. bulgaricus	Shrikhand
Lb. acidophilus, Sc. thermophiles, Lb. bulgaricus, Shrikhand	Yoghurt

TABLE 3.3 Some Important Strain of Lactic Acid Bacteria Used as Starter in Fermentation of Dairy-based Food Products.

Strain of Lactic Acid Bacteria	Temperature for Growth (°C)		
	Minimum	Optimal	Maximum
Lb. acidophilus	27	37	48
Lb. brevis	8	30	42
Lb. casei sub sp. casei	–	30	–
Lb. delbrueckii sub sp. Bulgaricus	22	45	52
Lb. delbrueckii sub sp. lactis	18	40	50
Lb. helveticus	22	42	54
Lb. kefir	8	32	43
Lc. lactis sub sp. cremoris	8	22	37
Lc. lactis sub sp. lactis	8	30	40
Lc. lactis sub sp. lactis biovar. diacetylactis	8	22–28	40
Leuconostoc mesenteroides sub sp. cremoris	4	22–28	37
Ln. mesenteroides sub sp. dextranicum	4	20–28	37
Sc. thermophilus	22	40	52

Source: Heller (2001).

TABLE 3.4 Physical and Chemical Changes in the Milk During Fermentation.

Increase	Decrease
Antibacterial substances	Lactose
Flavoring compounds	Protein
Free amino acids, lactic acid some organic acids	Fat
Glucose, galactose, and polysaccharides	Some vitamins
Some vitamins	

Source: Gandhi (2000).

Milk products also serve as the important delivery vehicles for probiotic bacteria. The probiotic bacteria have a long history of association with dairy products. This is because some of the same bacteria that are associated with fermented dairy products also make their homes in different sites on the human body, including mouth, gastrointestinal tract, etc. Some of these microbes, therefore, can play a dual role in transforming milk into a diverse array of fermented dairy products (yoghurt, cheese, kefir, etc.) and contributing to the important role of colonizing bacteria. Dairy products represent a major portion of the total European functional foods market and are at the forefront of probiotic developments. The growing interest of consumers towards therapeutic products has led to incorporation of probiotic cultures in different milk products. Within this sector, probiotic cultures have been incorporated in yoghurts and fermented milk products. The probiotic bacteria used in commercial products today are mainly members of the genera *Lactobacillus* and *Bifidobacterium* (Heller, 2001). In the dairy markets, yoghurt with its existing health image is well positioned to capitalize on the growth of healthy foods (Stanton et al., 2001).

Cultured dairy products have been providing vital importance in the human diet. However, further research into areas of the products as anticarcinogens, antitumor agents, and in the area of cholesterol will yield even more potential for cultured dairy products as an important component of human nutrition. However, careful selection of specific strains combined with proper production and handling procedures will be necessary to ensure that desired benefits are provided to consumers. At large, the opinion is in favor of probiotics and fermented foods. The challenge would be to find out the most suitable candidate organism for fermentation, select different protective and carrier media, evolve a suitable technology to design foods which contain and maintain large populations of viable bioactive organisms during processing and postharvest processing periods, and have longer shelf life. The products must also be readily acceptable as regular foods. Definitely, fermented dairy products containing probiotic combinations promise a healthy "functional food package" for improved long-term health benefits.

LAF of milk is used for the production of various products like yogurt, fresh cheeses, acidified dairy beverages, and desserts (Perina et al., 2015). A symbiotic culture of the homofermentative LAB, *Sc. thermophilus* and *Lactobacillus delbrueckii* subsp. *bulgaricus* in 1:1 ratio and incubating the milk at 37–451°C is used for yoghurt production (Morell et al., 2015). On the other hand, a coculture of yeasts and LAB are used to produce Kefir which is an alcohol-containing milk from Caucasian countries. Kombucha which is traditionally used for fermentation of sweetened black or green tea (Dufresne and Farnworth, 2000; Teoh et al., 2004) has been known to ferment milk to produce similar products to yoghurt or kefir (Milanovic et al., 2008, 2012; Ilicic et al., 2013). LAB for dairy fermentations based on their growth optimum are either mesophilic (with an optimum growth temperature between 201°C and 301°C) or thermophilic (with an optimum between 301°C and 451°C). Starter cultures of mesophilic lactococci are used in the manufacture of a broad array of cheese and butter, irrespective of milk type (Wouters et al., 2002).

3.5.4 MEAT-BASED FOODS PRODUCTS

Meat is consumed worldwide, however is greatly susceptible to contamination by pathogens. Therefore, it is necessary to preserve meat and maintain its stability. This is achieved by meat fermentation (Hugas et al., 2003). Using LAB for the fermentation of meat sausages preserve the meat and protect it from spoilage (Metaxopoulos et al., 2002). Fermented fish products can be stored for prolong periods by retaining their nutritional quality. More over different vegetables, legumes such as soy products and tea, coffee are also fermented.

The selected starter cultures (SSC) induces higher acidification as compared to natural starter cultures during fermentation as in case of Spanish chorizo added with *Micrococcaceae* (Banon et al., 2014), *Lactobacillus sakei* (Castellano et al., 2012), *Lb. plantarum*, and *Lb. rhamnosus* GG (Rubio et al., 2013), as well as in a traditional Turkish fermented sausage (Sucuk), with *Lb. plantarum* strains which were originally isolated from cheese (Kargozari et al., 2014). Similarly, use of *Lb. sakei* and *Staphylococcus equorum* as SSC for fermentation of Dacia sausage of Romania enhances its flavor, mastication attributes, and decreases biogenic amines (Ciuciu Simion et al., 2014). The microorganisms are added to meats to serve the following functions:

1. Safety assurance by the inactivation of pathogens,
2. Improvement in stability by inhibiting undesirable changes brought about by spoilage microorganisms or abiotic reactions,
3. Provision for diversity by modifying the raw material to obtain new sensory properties, and
4. Ensuring health benefits through positive effects on the intestinal flora.

In meat fermentations, LAB usually serves above functions except fourth, whereas other microorganisms, viz., catalase-positive cocci (*Staphylococcus, Kocuria*), yeasts (*Debaryomyces*), and moulds (*Penicillium*), are responsible for the desired sensory properties (Lucke, 2000). Although *Listeria monocytogenes* is usually found in raw meat, its potential for commercial sausage fermentation is low (Farber et al., 1993).

3.6 FACTORS FOR MICROBIAL GROWTH IN FERMENTED FOOD

Microbial growth is facilitated through the metabolism of food components that provide the energy required and cellular materials and substrate for many by-products. Microbial growth in laboratory media is also important for quantitative and qualitative detection of microbial quality of food. The ability of growth or cell multiplication of bacterial, yeasts, and molds is influenced by intrinsic and extrinsic environments of food (Tanaka et al., 1986).

3.6.1 INTRINSIC FACTORS

Intrinsic factors are those which are not dependent on external conditions, inherent, located within. Examples of intrinsic factors of food for microbial approaches in fermentations include available nutrients, biological structures like peel to inhibit microorganisms, growth factors and inhibitors, moisture content, oxidation-reduction potential, pH, physical structure of food, presence of antimicrobial agents and other microorganisms, and water activity (a_w).

3.6.1.1 NUTRIENT CONTENT

Microorganisms can use food as a source of nutrients and energy and also to derive chemical elements that constitute biomass. Microbial growth is accomplished through the synthesis of cellular components and energy. The inability of an organism to utilize a major component of a food material will limit its growth. All foods contain five major nutrient groups; they are carbohydrates, lipids, proteins, minerals, and vitamins. Fermentation is defined as an anaerobic process of degradation of organic nutrients through biological reactions to produce energy in the form of adenosine triphosphate (Jay, 2000). Fermentation plays an important role in producing food products with diversified taster, sensory qualities, and preserving the perishables.

3.6.1.1.1 Carbohydrates in Foods

All microorganisms are normally found in food metabolize glucose; however, their ability to utilize other carbohydrates differs considerably.

Food carbohydrates are metabolized by microorganisms principally to supply energy through several metabolic pathways. Microorganisms can also polymerize some monosaccharides to produce complex carbohydrates. Some of these carbohydrates from pathogens may cause health hazards may cause food spoilage. Carbohydrate metabolism profiles are extensively used in laboratories for biochemical identification of unknown microorganisms isolated from foods (Jay, 2000).

It has been reported that bacteria (*Acetovibrio, Bacillus, Bacteriodes, Cellulomonas, Clostridium, Erwinia, Microbispora, Ruminococcus, Streptomyces,* and *Thermomonospora*) and fungi (*Phanerochaete chrysosporium, Sclerotium rolfsii,* and species of *Aspergillus, Penicillium, Schizophyllum,* and *Trichoderma*) through enzymatic hydrolysis depolymerize polysaccharides to produce cellulases during fermentation (Sun and Cheng, 2002). *Eiseniabicyclis* and *Hijikiafusiforme* belonging to Phaeophyceae (brown algae) and employed for the production of carbohydrates can even be used as a feedstock for microbial fermentation to produce ethanol (Hirayama et al., 1998; Kuda et al., 1998; Kolb et al., 1999; USDOE, 2010).

3.6.1.1.2 *Proteins in Foods*

Proteins are of prime importance and are essential nutrient of human life and growth. In food preparation, proteins play important functional roles like forming gels, producing foams, binding water, and aiding browning. In addition, enzymes are protein molecules which catalyze a number of reactions that affect the characteristics of prepared foods. Protein exhibits buffering, denaturation, coagulation, and enzymatic functions. Microorganisms differ greatly in their ability to metabolize food proteins. Microorganisms produce extracellular proteinases and peptidases to hydrolyze large proteins and peptides to small peptides and amino acids before they can be transported inside the cells (Jay, 2000). Microorganisms can also metabolize different compounds found in foods. Production of specific metabolic products is used for laboratory identification of microbial isolates from food.

The precursors of sulfur compounds released during fermentation of sausages are sulfur-containing amino acids (cysteine and methionine), together with thiamine (Mottram, 1998). The thiamine content is the highest in pork, but the generation of free amino acids (major source of volatile compounds) during fermentation of dry fermented sausage is due to endogenous proteolytic activities (Mottram, 1991; Toldra et al., 2001). The methionine and cysteine are respectively degraded into methanethiol and hydrogen sulfide which generate other volatile sulfur compounds (Mottram, 1991). The volatile nitrogen compounds are also formed from amino acids through Strecker degradation (Meynier and Mottram, 1995).

3.6.1.1.3 Lipids in Foods

Lipids or fats act as primary tenderizing agents, found rich in foods of animal origin. Lipid content varies in fabricated or prepared foods. These are generally less-preferred substrates for microbial synthesis of energy and cellular materials. Oils are major components of salad dressings and mayonnaise. Fats may be heated to high temperatures and act as a medium of heat transfer in fried food. A number of flavor compounds are fat-soluble and are carried in the fat component of many food products. Fatty acids can be transported in cells and used for energy synthesis. Some microorganisms produce extracellular lipid oxidases, which oxidize unsaturated fatty acids to produce different aldehydes and ketones. In many foods, the action of these enzymes is associated with spoilage, and in others, it's used for desirable flavors. *Lb. acidophilus* strains are beneficial intestinal microorganisms and can metabolize cholesterol and are also believed to be able to reduce serum cholesterol levels in humans (Jay, 2000).

3.6.1.1.4 Minerals and Vitamins in Foods

Minerals and vitamins are required in small amounts for microorganisms for the supply of essential nutrients for growth. Microorganisms can synthesize vitamin-B, and many food products have these in essential amount (Jay, 2000). The flavor development in fermented sausages is due to activity of microbial and chemical reactions involving minerals (Toldra et al., 2001). The flavor is due to volatile sulfur and nitrogen compounds playing crucial role for aroma of sausage as their threshold values are low and having characteristic olfactive notes (Stahnke, 2002). It is reported that cooking meat at high temperature enhances formation of volatile sulfur and nitrogen compounds (Mottram, 1998). This volatile profiling due to application of heat is in contrast to that being reported for the dry-ripening process of meat where low temperatures are applied. A recent study has revealed the development of sulfur-containing odorants in cooked ham due to the thermal degradation of thiamine (Thomas et al., 2014). Copper plays an important role in plant metabolism, and especially in tea, it plays a functional role in polyphenol oxidase, an enzyme essential in fermentation during manufacturing of tea (Natesan and Ranganthan, 1990). Zinc (Zn) too has crucial role and its content in tea ranged from 23.7 to 122.4 mg/kg (Ashraf and Mian, 2008). The increase in Zn contents progresses with the withering time due to the gradual degradation of Zn-containing enzymes, proteins, and nucleic acid content of the leaves. Zn is responsible for the synthesis of nucleic acid and proteins and helps in utilization of phosphorus and nitrogen, thereby affecting the mineral constituents and chemistry of flavor as a whole (Wardlaw and Hampl, 2006).

3.6.1.2 MOISTURE CONTENT

Water is required in a possible form for the microorganisms. The moisture content of food should be controlled and is also one of the preservative strategies for food spoilage (Mossel et al., 1995). The effect of moisture in combination with heat on starch has been observed. Heat moisture treatment (HMT) affected significantly the total starch levels within the heat-moisture-treated fermented foxtail millet flour (FFMF) and heat-moisture-treated foxtail millet flour (FMF). The highest amount was observed in former with 51.10% followed in latter with 40.13%. These values were low in FFMF and FMF without HMT treatments which were 29.10% and 15.78%, respectively. Thus, HMT has greatly influenced the physicochemical properties of flour and has enhanced the starch yield (Chung et al., 2012; Wongsagonsup et al., 2008). The effect of moisture has also been studied by different researchers during microbial fermentation process of tea. When pretreated tea brick was placed for fungal fermentation in workshop for 2–3 weeks under controlled temperature and moisture, *Eurotium* spp. grew as the dominant one initially. But at the end of fermentation, it was predominant by yellow fungi (golden flora) denoting huge production of complex metabolites by fungi (Mo et al., 2008; Xu et al., 2011; Luo et al., 2013; Yue et al., 2014; Zhu et al., 2015).

3.6.1.3 OXIDATION-REDUCTION POTENTIAL

An oxidation-reduction reaction occurs as the result of a transfer of electrons between atoms or molecules. Redox potential exerts an important elective effect on the microflora of a food. In living cells, both the electron and hydrogen transfer reactions is an essential feature of the electron transport chain and energy generation by oxidative phosphorylation. The balance of various redox couples favors the oxidized state and thus has the tendency to accept electrons from the electrode creating positive potential. Although microbial growth can occur over a wide spectrum of redox potential, individual microorganisms are classified into several physiological groups on basis of the redox range over which they can grow and respond to O_2 (Leistner, 1995).

3.6.1.4 pH

The pH is the negative algorithm of the hydrogen ion concentration. The acidity and alkalinity of the environment has a profound effect on the activity and stability of macromolecules. Bacteria grow fastest in the pH range of 6.0–8.0, yeasts 4.5–6.0, and filamentous fungi 3.5–4.0. Many foods are at least slightly acidic. The acidity of product can have important implications for its microbial ecology and the rate and character of spoilage. The partial dissociation of weak acids plays an important part in their ability to inhibit microbial growth (Smelt et al., 1982).

pH is most important among all the internal parameters influencing food fermentation as it is closely related to microbial growth and the structural changes in phytochemicals. For instance, the breakdown of anthocyanin is dependent on the pH in the presence of O_2. It is also directly proportional to the level of pseudobase and inversely to the cation concentration (Su and Chien, 2007). Anthocyanins have been found to be stable at low pH (Cabrita et al., 2000; Nielsen et al., 2003) and showing highest stability along with the red flavylium cation stability around pH 1–2 (Nielsen et al., 2003). Structure of anthocyanins is another factor upon which its stability depends. This was proved in case of acylated anthocyanins being more stable than the nonacylated forms (Devi and Mohandas, 2012). pH also plays a major role in radical scavenging capacity of wine anthocyanins (Borkowski et al., 2005) as well as in their retention ability (Tagliazucchi et al., 2010). On the other hand, LAB isolated from Indonesian fermented foods shows probiotic behavior by tolerating the environment of bile salt and low pH and exerting antagonistic activity against foodborne pathogens similar to those of intestinal probiotics like *Lactobacillus acidophillus* and *Lactobacillus casei*. Other members of LAB isolated from fermented foods have also been shown to be adopted with the specific environments of high salt concentration and acidic condition as in case with *Ln. mesenteroides* in Kimchii and cane juice, *Leuoenos* in grape juice, and *Tetragenococcus halophillus* in soya sauce (Rhee et al., 2011).

3.6.1.5   PHYSICAL STRUCTURE OF FOOD

Food products are heterogeneous, and determination of the amount of each of the chemical components in food called the chemical composition is essential to determine the quality of food. Physical properties and stability are critical for safe and healthy food (Morris, 2000).

3.6.1.6   PRESENCE OF ANTIMICROBIAL AGENTS

Natural constituents like plant tissues such as pigments, alkaloids, and resins have antimicrobial properties. Herbs and spices have found to contribute to the microbiological stability of foods; however, their role in preservation is less, and in few cases, they can be source of microbial contamination, leading to spoilage. Antimicrobial components are present at varying concentrations in the natural product. Animal products also have found to have a range of nonspecific antimicrobial constituent. Both animal and plant products contain lysozyme which catalyze hydrolysis of glyosidic linkages in peptidoglycan. Lysozyme is most active against Gram-positive bacteria, where the peptidoglycan is more readily accessible, but can also kill Gram-negative with their protective outer membrane (Lund et al, 2000). LAB are known to produce different compounds antimicrobial in nature. These compounds restrict the growth of pathogenic microorganisms thus ensuring stability and safety of the fermented product (Holzapfel, 2002). The different sources of antimicrobials for kimchi fermentation are shown in Table 3.5.

TABLE 3.5   Natural Antimicrobial Agents for Controlled Fermentation of Kimchi.

Source	Effective against Microorganisms	Reference
Garlic chive extract	*Ln. mesenteroides, Pediococcus cerevisiae, Lb. plantarum,* and *S. cerevisiae*	Kim and Park (1995)
Mustard leaf, mustard oil, and mustard extract	*Lb. plantarum, Lb. brevis, P. cerevisiae,* and *Ln. mesenteroides*	Cole (1976), Park and Han (1994), and Hong and Yoon (1989)
Japanese apricot	LAB	Lee et al. (2002)
Magnolia vine seed oil	*B. subtilis, Escherichia coli,* and *Staphylococcus aureus*	Jung et al. (2000)
Citrus (orange extract, grapefruit extract, and tangerine extract)	*Lb. plantarum* and *Klebsiella pneumoniae*	Cho et al. (2005)

TABLE 3.5   *(Continued)*

Source	Effective against Microorganisms	Reference
Sappan lignum	LAB (*Lb. plantarum, Lb. sake, Ln. mesenteroides* and *Pediococcus pentosaceus*)	Lee et al. (1997)
Medicinal herbs such as amur corktree, baikal skullcap, cnidii rhizoma, and Chinese thoroughwax	*Lb. plantarum* and *Ln. mesenteroides*	Park et al. (2004)
Thyme, cardamom and cumin extract	*Lb. plantarum*	Kim et al. (1998)
Green tea leaf extract and green tea Catechins	*Ln. mesenteroides, Lb. plantarum,* and *P. cerevisiae*	Wee and Park (1997)
Essential oil of persimmon leaves	*Lb. plantarum* and *S. cerevisiae*	Park et al. (1994)

3.6.1.7   WATER ACTIVITY (A_w)

Water is a remarkable compound and is a very significant parameter, which is essential to understand the movement of water from the environment to cytoplasm and vice versa. Water exhibits cogitative property and depends on the number of molecules or ions present in the solution. The capability of the growth of microorganisms depends on the water activity in the environment. The range of water activities (a_w) allowing growth is influenced by other physicochemical and nutritional conditions. It is essential to transport nutrients, remove water materials, carry out enzymatic reactions, synthesize cellular materials, and biochemical reactions (Jay, 2000; Morris, 2000).

The measurement of water activity (a_w) is achieved by measuring the water content if the shape of the isotherm has been established. Water activity can be measured by measuring the equilibrium relative humidity of the atmosphere in contact with the sample. Microorganisms have an optimum, maximum, and minimum a_w level for their growth (Table 3.6). Microorganisms actually grow better at reduced water activity and are described as halophilic, osmophilic, and xerophilic (Mossel and Thomas, 1988). The analysis of a_w is very useful to control spoilage and also to enhance the growth of desirable ones in food bioprocessing.

TABLE 3.6   Levels of Water Activity (a_w) in Microbial Groups.

Microbial groups	Water activity (a_w) values (minimum levels)
Bacterial spoilage (generally)	0.90
Gram negative bacteria	0.93
Gram positive bacteria	0.90
Halophilic bacteria	0.75
Moulds	0.80
Xerophilic moulds	0.6
Yeast	0.85
Osmophilic yeast	0.6–0.7

3.6.2   EXTRINSIC FACTORS

Extrinsic factors are those which do not inherent and act from the outside. Examples of extrinsic factors of food for microbial approaches in fermentations include addition of nitrites, CO_2 or O_2, casing and packaging to inhibit microbial growth, gaseous atmosphere, low pH, relative humidity, and temperature.

3.6.2.1   TEMPERATURE

The growth of microorganisms occurs over a wide range of temperature (i.e., from 80 to 100°C). No microorganism is capable to grow in this range; bacteria and molds are limited to a temperature span of 35°C and 30°C, respectively. In food microbiology, there are three classes of microbes according to temperature range, which are: (1) psychrotrophes (7–30°C), (2) mesophiles (20–45°C), and (3) thermophiles (55–65°C) for growth and development. Mesophilic and psychrotrophic organisms are of significant importance. Mesophilic organisms with optimum temperature of 37°C grow more quickly than psychotropic as they are capable of growing at temperatures at or less than 7°C (44.6°F). Thus, the spoilage of food stored in mesophilic range is rapid than under chilled conditions (Loss and Hotchkiss, 2002), whereas psychrotrophic provide an estimation of the product's shelf life as well as are responsible for spoiling refrigerated foods.

One of the important features of slow and eventual cessation of microbial growth at low temperatures is mainly considered to be due to changes in the membrane structure that affect the uptake and supply of nutrients to enzyme system within the cell. Microorganisms usually respond to growth at lower temperatures by increasing the amount of unsaturated fatty acids in their membrane lipids, and that psychrotrophs have higher levels of unsaturated fatty acids than mesophiles. Increasing the degree of unsaturation in a fatty acid decreases its melting point so that membranes containing higher levels of unsaturated fatty acid will remain fluid and hence functional lower temperatures. As the temperature increases above the optimum range, the growth rate declines much more sharply as a result of irreversible denaturation of proteins and the thermal breakdown of cell's plasma membrane. At temperatures above the maximum the growth, these changes are sufficient to kill the organism.

The effect of temperature has significant role in fermentation. It enhances fermentation process by increasing microbial growth and enzymatic activity. The phenolic compounds responsible for the antioxidant activity of foods derived from plants are greatly affected by change in temperature during fermentation of these foods. By increasing temperature, the viscosity of solvent decreases, solute molecules move faster; as a result, more polyphenols are dissolved leading to increment in polyphenol content which ultimately increases diffusion coefficient and stability of the system (Ajila et al., 2011). For example, increasing the fermentation temperature from 30°C to 34°C and 38°C subsequently increased polyphenols, and reverse was observed in case of bush tea (Hlahla et al., 2010). But in case of red grape pomace peels, polyphenolic content, color, and antioxidative activity are fairly heat-stable (Su and Silva, 2006). However, a temperature above 100°C is not recommended due to loss of antioxidative activity (Larrauri et al., 1997). The β-glucosidase activity of *Lb. casei* is maximal at 35°C (Coulon et al., 1998), whereas that of *Lb. plantarum* is reduced by 50% when raised to 50°C for 5 min (Sestelo et al., 2004). Hence, it can be concluded that optimal temperature is required for the optimum antioxidative activity by fermentation.

3.6.2.2 RELATIVE HUMIDITY

Relative humidity is a measure of water activity of gas phase. When food commodities with low water activity are stored in an atmosphere of high relative humidity, water will transfer from gas phase to the food. The time required may be quite long for bulk commodity to increase in water activity, but condensation may occur on surfaces giving rise to localized regions of high water activity.

Once microorganisms have started to grow and become physiologically active, they usually produce water as an end product of respiration (Farber, 1991). Thus, they increase the water activity of their own immediate environment so that microorganisms are able to grow and spoil a food, which was initially considered to be microbiologically stable. The storage of fresh fruit and vegetables requires very careful control of relative humidity.

3.6.2.3   CARBON DIOXIDE (CO_2) AND OXYGEN (O_2)

O_2 is the most vital gas in contact with food and comprises 21% of earth's atmosphere. Its occurrence and influence on redox potential are significant determinants of the microbial associations that develop their rate of growth. The inhibitory effect of CO_2 on microbial growth is applied in modified atmosphere packing food and is beneficial consequence of its use at elevated pressures in carbonated mineral water and soft drinks. The molds and oxidative Gram negative bacteria are most sensitive; Gram negative bacteria are most resistant, and yeasts show considerable tolerance of high CO_2 levels and dominate the spoilage of microflora. Other factors includes changes in physical properties of plasma membrane adversely affecting solute transport, inhibition of key enzymes involving carboxylation/decarboxylation reactions in which CO_2 is a reactant, and reaction with protein amino groups causing changes in their properties and activities (Loss and Hotchkiss, 2002).

3.7   FUTURE PERCEPTIVE AND CHALLENGES

Over the decades, science and its advancement have demonstrated and proved the fundamental methods that can protect human health and wellbeing. The current research is advanced with the designing and development of new products with greater awareness for environmental impact. The technological advancement of probiotics is based on its viability and stability which is of great demand. Probiotic foods with specific strains with enhanced shelf life should be produced. The demand for probiotics is increasing, and hence, new technologies and formulation procedures are required for effective and functional healthy properties. Future technology should have novel developments which broaden the challenges in the different range of food products thus maintaining health and prevention of disease.

3.8 CONCLUSION

Food preservation methods have been used for safety and to maintain edible products for consumption. New infectious diseases are emerging and adapted to infect humans in its own unique way. Besides these emerging and reemerging pathogens adaptation to the human environment, the changes in human lifestyle, population, and globalization of food supply have created new outbreaks for pathogens to infect humans. Preservation procedures are intended to decrease the growth of microorganisms which are affected by several extrinsic and intrinsic foods. A combination of several procedures and processes can be used to make the food safer and to extend the shelf life of food products. Fermented foods denotes approximately one-third of food consumption. Fermentation plays a role from basic yoghurt and bread to beer and wine fermentation and is involved not only in the production of these compounds but also provides functional properties to food products and constituents. Different microbial strains, optimum fermentation conditions, and several advanced techniques are used to produce fermented products such as dairy (cheese, butter), vegetables (pickles, olives, and sauerkraut), meat, beverages, bread, vinegar, and other organic acids. Besides these advantages, fermentation technologies are sensitive. There is a risk of adulteration and intoxication; therefore, it should be carefully controlled. A widespread suppression or eradication of the foodborne agents is not feasible. Continuous research efforts and newer strategies are essential for better understanding to control foodborne pathogens and consumer education should be consistent which will create a quick response to the new and reemerging foodborne pathogens.

KEYWORDS

- **Fermentation**
- **microbial growth**
- **preservation**

REFERENCES

Abee, T.; Krockel, L.; Hill, C. Bacteriocins: Modes of Action and Potentials in Food Preservation and Control of Food Poisoning. *Int. J. Food Microbiol.* **1995**, *28*, 169–185.

Adams, M. R. Topical Aspects of Fermented Foods. *Trends Food Sci. Technol.* **1990**, *1*, 141–144.

Aguirre, M.; Collins, M. D. Lactic Acid Bacteria and Human Clinical Infection. *J. Appl. Bacteriol.* **1993**, *75*(2), 95–107.

Ajila, C. M.; Brar, S. K.; Verma, M.; Tyagi, R. D.; Valero, J. R. Solid-State Fermentation of Apple Pomace Using *Phanerocheate chrysosporium*–Liberation and Extraction of Phenolic Antioxidants. *Food Chemi.* **2011**, *126*(3), 1071–1080.

Akinrele, I. A. Fermented Studies on Maize During Preparation of Traditional African Starch Cake Food. *J. Sci. Food Agric.* **1970**, *21*, 619–625.

Alexandre, H.; Costello, P. J.; Remize, F.; Guzzo, J.; Guilloux-Benatier, M. *Saccharomyces cerevisiae-oenococcus* oeni Interactions in Wine, Current Knowledge and Perspectives. *Int. J. Food Microbiol.* **2004**, *93*, 141–154.

Ananou, S.; Maqueda, M.; Martínez-Bueno, M.; Valdivia, E. Biopreservation, an Ecological Approach to Improve the Safety and Shelf-Life of Foods, *Commun. Curr. Res. Educ. Top. Trends Appl. Microbiol.* **2007**, *1/2*, 475–485.

Andrade, M. J.; Cordoba, J. J.; Sanchez, B.; Casado, E. M.; Rodriguez, M. Evaluation and Selection of Yeasts Isolated from Dry-Cured Iberian Ham by Their Volatile Compound Production. *Food Chem.* **2009a**, *113*, 457–463.

Andrade, M. J.; Rodriguez, M.; Casado, E. M.; Bermudez, E.; Cordoba, J. J. Differentiation of Yeasts Growing on Dry-Cured Iberian Ham by Mitochondrial DNA Restriction Analysis, RAPD-PCR and their Volatile Compounds Production. *Food Microbiol.* **2009b**, *26*, 578–586.

Aquilanti, L.; Santarelli, S.; Silvestri, G.; Osimani, A.; Petruzzelli, A.; Clementi, F. The Microbial Ecology of a Typical Italian Salami During its Natural Fermentation. *Int. J. Food Microbiol.* **2007**, *120*, 136–145.

Arena, M. P.; Pasquale Russo, P.; Capozzi, V.; Lopez, P.; Fiocco, D.; Spano, G. Probiotic Abilities of Riboflavin-Overproducing *Lactobacillus* strains: A Novel Promising Application of Probiotics. *Appl. Microbiol. Biotechnol.* **2014**, *98*(17), 7569–7581.

Asefa, D. T.; Moretro, T.; Gjerde, R. O.; Langsrud, S.; Kure, C. F.; Sidhu, M. S.; Nesbakken, T.; Skaar, I. Yeast Diversity and Dynamics in the Production Processes of Norwegian Dry-Cured Meat Products. *Int. J. Food Microbiol.* **2009**, *133*, 135–140.

Ashraf, W.; Mian, A. A. Levels of Selected Heavy Metals in Black Tea Varieties Consumed in Saudi Arabia. *Bull. Environ. Contam. Toxicol.* **2008**, *81*, 101–104.

Bachmann, J. A.; Alameda, Wilkins, R.; Calif, H.. Method of Treatment for Fermented and Distilled Beverages and the Like. US 2086891 A, Patent Application, October 23, 1934, Serial No. 749, 574.

Balakrishnan, B.; Prasad, B.; Rai, A. K.; Velappan, S. P.; Subbanna, M. N.; Narayan, B. In Vitroantioxidant and Antibacterial Properties of Hydrolysed Proteins of Delimed Tannery Fleshings: Comparison of Acid Hydrolysis and Fermentation Methods. *Biodegradation.* **2011**, *22*, 287–95.

Banon, S.; Serrano, M.; Bedia, M. Use of Micrococcaceae Combined with a Low Proportion of Lactic Acid Bacteria As a Starter Culture for Salami Stuffed in Natural Casing. *CyTA-J. Food.* **2014**, *12*, 160–165.

Barbosa, T. M.; Levy, S. B. The Impact of Antibiotic Use on Resistance Development and Persistence. *Drug Resist. Updates.* **2000**, *3*, 303–311.

Beekwilder, J.; Marcozzi, D.; Vecchi, S.; de Vos, R.; Janssen, P.; Francke, C.; van Hylckama Vlieg, J.; Hall, R. D. Characterization of Rhamnosidases from *Lactobacillus plantarum* and *Lactobacillus acidophilus*. *Appl. Environ. Microbiol.* **2009**, *75*, 3447–3454.

Borkowski, T.; Szymusiak, H.; Gliszczynska-Swiglo, A.; Rietjens, I. M. C. M.; Tyrakowska, B. Radical Scavenging Capacity of Wine Anthocyanins is Strongly pH-dependent. *J. Agric. Food Chem.* **2005**, *53*(14), 5526–5534.

Brink, T. B.; HuisIntVeld, J. H. J. Application of Metabolic Properties of Lactic Acidbacteria. In *Les Bacteries Lactiques;* Novel, G.; Le Querler, J. F., Eds.; Ardie Normandie: Caen, France, 1992 pp. 67–76.

Cabrita, L.; Fossen, T.; Andersen, O. M. Colour and Stability of the Six Common Anthocyanidin 3-Glucosides in Aqueous Solutions. *Food Chem.* **2000**, *68*(1), 101–107.

Campbell-Platt, G. Fermented Foods of the World–A Dictionary and Guide', Butterworths: London, 1987, ISBN: 0-407-00313-4.

Caplice, E.; Fitzgerald, G. F. Food Fermentations: Role of Microorganisms in Food Production and Preservation. *Int. J. Food Microbiol.* **1999**, *50*, 131–149.

Carrau, F. M.; Medina, K.; Farina, L.; Boido, E.; Henschke, P. A.; Dellacassa, E. Production of Fermentation Aroma Compounds by *Saccharomyces cerevisiae* Wine Yeasts: Effects of Yeast Assimilable Nitrogen on Two Model Strains, *FEMS Yeast Res.* **2008**, *8*, 1196–1207.

Castellano, P.; Aristoy, M.; Sentandreu, M.; Vignolo, G.; Toldra, F. *Lactobacillus sakei* CRL1862 Improves Safety and Protein Hydrolysis in Meat Systems. *J. Appl. Microbiol.* **2012**, *113*, 1407–1416.

Cavalieri D.; McGovern P.E.; Hartl D. L.; Mortimer R. and Polsinelli M. Evidence of *S. cerevisiae* Fermentation in Ancient Wine. *J. Mol. Evol.* **2003**, *57*(1), 226–32.

Chen, X. H.; Zhang, W.; Fang, L.; Mao, A. N.; Ni, L. The Isolation of Predominant Fungi in Fujian White Koji and Their Characteristics. *J. Fuzhou Univ. (Nat. Sci.)* **2007**, *35*, 635–640.

Cho, S. H.; Lee, S. C.; Park, W. S. Effect of Botanical Antimicrobial Agent-Citrus Products on the Quality Characteristics During Kimchi Fermentation. *Korean J. Food Preserv.* **2005**, *12*, 8–16.

Chung, H. J.; Cho, A.; Lim, S. T. Effect of Heat-Moisturetreatment for Utilization of Germinated Brown Rice in Wheatnoodle. *LWT Food Sci. Technol.* **2012**, *47*(2), 342–347.

Ciuciu Simion, A. M.; Vizireanu, C.; Alexe, P.; Franco, I.; Carballo, J. Effect of the Use of Selected Starter Cultures on Some Quality, Safety and Sensorial Properties of *Dacia sausage*, a Traditional Romanian Dry-sausage Variety. *Food Control.* **2014**, *35*, 123–131.

Coast, J.; Smith, R. D. Antimicrobial Resistance: Cost and Containment. *Expert Rev. Anti-Infect. Ther.* **2003**, *1*, 241–251.

Coda, R.; Rizzello, C. G.; Curiel, J. A.; Poutanen, K.; Katina, K. Effect of Bioprocessing and Particle Size on the Nutritional Properties of Wheat Bran Fractions. *Innov. Food Sci. Emerg. Technol.* **2014**, *25*, 19–27.

Cole, R. A. Isothiocyanates, Nitriles and Thiocyanates as Products of Autolysis of Glucosinolates in Cruciferae. *Phytochemistry.* **1976**, *15*, 759–762.

Condon, S. Responses of Lactic Acid Bacteria to Oxygen. *FEMS Microbiol. Rev.* **1987**, *46*, 269–280.

Coulon, S.; Chemardin, P.; Gueguen, Y.; Arnaud, A.; Galzy, P. Purification and Characterization of an Intracellular b-glucosidase from *Lactobacillus casei* ATCC 393. *Appl. Biochem. Biotechnol.* **1998**, *74*, 105–114.

Davis, C. R.; Wibowo, D.; Eschenbruch, R.; Lee, T. H.; Fleet, G. H. Practical Implications of Malolactic Fermentation, A Review. *Am. J. Enol. Viticult.* **1985**, *36*, 290–301.

De Keersmaecker, J. The Mystery of Lambic Beer. *Sci. Am.* **1996**, *275*, 74–81.

De Vuyst, L.; Vandamme, E. J. *Bacteriocins of Lactic Acid Bacteria.* Blackie Academic and Professional: London, 1994a.

De Vuyst, L.; Vandamme, E. J. Antimicrobial Potential of Lactic Acid Bacteria. In *Bacteriocins of Lactic Acid Bacteria*; De Vuyst, L.; Vandamme, E. J., Eds.; Blackie Academic and Professional: London, 1994b, pp. 91–149.

Devi, P. S. S. M.; Mohandas, S. The Effects of Temperature and pH on Stability of Anthocyanins from Red Sorghum (Sorghum bicolor) Bran. *Afr. J. Food Sci.* **2012**, *6*(24), 567–573.

Doores, S. *Organic Acids.* In Davidson, P. M, Branen, A.L., Eds.; Marcel Dekker: New York, 1993, pp. 95–136.

Dordevic, T. M.; Siler-Marinkovic, S. S.; Dimitrijevic-Brankovic, S. I. Effect of Fermentation on Antioxidant Properties of Some Cereals and Pseudo Cereals. *Food Chem.* **2010**, *119*(3), 957–963.

Duenas, M.; Fernandez, D.; Hernandez, T.; Estrella, I.; Munoz, R. Bioactive Phenolic Compounds of Cowpeas (*Vigna sinensis* L.). Modifications by Fermentation with Natural Microflora and with *Lactobacillus plantarum* ATCC 14917. *J. Sci. Food Agric.* **2005**, *85*(2), 297–304.

Dufresne, C.; Farnworth, E. Tea, Kombucha and Health: A review. *Food Res. Int.* **2000**, *33*, 409–421.

Egervarn, M.; Roos, S.; Lindmark, H. Identification and Characterization of Antibiotic Resistance Genes in *Lactobacillus reuteri* and *Lactobacillus plantarum*. *J. Appl. Microbiol.* **2009**, *107*, 1658–1668.

Farber, J. M. Microbiological Aspects of Modified Atmosphere Packaging Technology–A Review. *J. Food Prot.* **1991**, *54*, 58–70.

Farber, J. M.; Daley, E.; Holley, R.; Usborne, W. R., Survival of *Listeria monocytogenes* during the Production of Uncooked German, American and Italian-style Fermented Sausages. *Food Microbiol.* **1993**, *10*, 123–132.

Fugelsang, K. C.; Edwards, C. G. Fermentation and Post-fermentation Processing. In *Wine Microbiology Practical Applications and Procedures*. Springer: New York, USA, 2007, pp. 115–138.

Gandhi, D. N. Fermented Dairy Products and Their Role in Controlling Food Borne Diseases" In *Food Processing: Biotechnological Applications*; Marwaha, S. S.; Arora, J. K., Eds.; Asiatech Publishers Inc.: New Delhi, 2000, pp. 209–220.

Ganzle, M. G.; Vogel, R. F. Contribution of Reutericyclin Production to the Stable Persistence of *Lactobacillus reuteri* in an Industrial Sourdough Fermentation. *Int. J. Food Microbiol.* **2002**, *80*, 31–45.

Ge, B.; Jiang, P.; Han, F. Identification and Antimicrobial Susceptibility of Lactic Acid Bacteria from Retail Fermented Foods. *J. Food Prot.* **2007**, *70*, 2606–2612.

Ghosh, K.; Ray, M.; Adak, A.; Halder, S. K.; Das, A.; Jana, A.; Parua (Mondal), S.; Vagvolgyi, C.; Mohapatra, P. K. D.; Pati, B. R.; Mondal, K. C. Role of Probiotic *Lactobacillus fermentum* KKL1 in the Preparation of a Rice Based Fermented Beverage. *Bioresour. Technol.* **2015**, *188*, 161–168.

Hammes, W. P.; Brandt, M. J.; Francis, K. L.; Rosenheim, J.; Seitter, M. F. H.; Vogelmann, S. A. Microbial Ecology of Cereal Fermentations. *Trends Food Sci. Technol.* **2005**, *16*, 4–11.

Hansen, E. B. Commercial Bacterial Starter Cultures for Fermented Foods of the Future. *Int. J. Food Microbiol.* **2002**, *78*, 119–131.

Heller, K. J. Probiotic Bacteria in Fermented Foods: Product Characteristics and Starter Organisms. *Am. J. Clin. Nutr.* **2001**, *73*(2), 374S–379S.

Hirayama, K.; Rafter, J. The Role of Lactic Acid Bacteria in Colon Cancer Prevention: Mechanistic Considerations. *Antonie van Leeuwenhoek.* **1999**, *76*, 391–394.

Hirayama, S.; Ueda, R.; Ogushi, Y.; Hirano, A.; Samejima, Y.; Hon-Nami, K.; Kunito, S. Ethanol Production from Carbon Dioxide by Fermentative Microalgae Source: Advances in Chemical Conversions for Mitigating Carbon Dioxide. *Book Ser. Stud. Surf. Sci. Catal.* **1998**, *114*, 657–660.

Hlahla, L. N.; Mudau, F. N.; Mariga, I. K. Effect of Fermentation Temperature and Time on the Chemical Composition of Bush Tea (*Athrixia phylicoides* DC.). *J. Med. Plants Res.* **2010**, *4*(9), 824–829.

Holzapfel, W. H. Appropriate Starter Culture Technologies for Small-Scale Fermentation in Developing Countries. *Int. J. Food Microbiol.* **2002**, *75*, 197–212.

Hong, W. S.; Yoon, S. The Effects of Low Temperature Heating and Mustard Oil on the Kimchi Fermentation. *Korean J. Food Sci. Technol.* **1989**, *21*, 331–337.

Hosono, A.; Kashina, T.; Kada, T. Antimutagenic Properties of Lactic Acid Cultured Milk on Chemical and Faecal Mutagens. *J. Dairy Sci.* **1986**, *69*(9), 2237–2242.

Hugas, M.; Garriga, M.; Aymerich, M. T. Functionality of Enterococci in Meat Products. *Int. J. Food Microbiol.* **2003**, *88*(2–3), 223–233.

Ilicic, M.; Milanovic, S.; Caric, M.; Vukic, V.; Kanuric, K.; Ranogajec, M.; Hrnjez, D. The Effect of Transglutaminase on Rheology and Texture of Fermented Milk Products. *J. Texture Stud.* **2013**, *44*, 160–168.

Jay, J. M. *Modern Food Microbiology.* 6th ed. Aspen: Gaithersburg (MD), 2000, pp. 679.

Juan, M. Y.; Chou, C. C. Enhancement of Antioxidant Activity, Total Phenolic and Flavonoid Content of Black Soybeans By Solid State Fermentation with *Bacillus subtilis* BCRC 14715. *Food Microbiol.* **2009**, *27*, 586–591.

Juan, M. Y.; Chou, C. C. Enhancement of Antioxidant Activity, Total Phenolic and Flavonoid Content of Black Soybeans By Solid State Fermentation with *Bacillus subtilis* BCRC 14715. *Food Microbiol.* **2010**, *27*, 586–591.

Jung, G. T.; Ju, I. O.; Choi, J. S.; Hong, J. S. The Antioxidative, Antimicrobial and Nitrite Scavenging Effects of Schizandra Chinensis RUPRECHT (Omija) seed. *Korean J. Food Sci. Technol.* **2000**, *32*, 928–935.

Kalle, R.; Soukand, R. *Eesti Looduslikud Toidutaimed. Kasutamine18. Sajandist Tanapaevani.* Varrak: Tallin, 2013.

Kargozari, M.; Moini, S.; Akhondzadeh Basti, A.; Emam-Djomeh, Z.; Gandomi, H.; Revilla Martin, I.; Ghasemlou, M.; Carbonell-Barrachina, A. A. Effect of Autochthonous Starter Cultures Isolated from Siahmazgi Cheese on Physicochemical, Microbiological and Volatile Compound Profiles and Sensorial Attributes of Sucuk, a Turkish Dry-fermented Sausage. *Meat Sci.* **2014**, *97*, 104–114.

Kato, S.; Ishihara, T.; Hemmi, H.; Kobayashi, H.; Yoshimura, T. Alterations in D-amino Acid Concentrations and Microbial Community Structures During The Fermentation of Red and White Wines. *J. Biosci. Bioeng.* **2011**, *111*(1), 104–108.

Katz, F. Active Cultures Add Function to Yoghurt and Other Foods. *Food Technol.* **2001**, *55*, 46–49.

Keukeleire, D. D. Fundamentals of Beer and Hop Chemistry, *Quimica Nova.* **2000**, *23*(1), 108–112.

Khedid, K.; Faid, M.; Mokhtari, A.; Soulaymani, A.; Zinedine, A. Characterization of Lactic Acid Bacteria Isolated from the One Humped Camel Milk Produced in Morocco. *Microbiol. Res.* **2009**, *164*(1), 81–91.

Kim, O. M.; Kim, M. K.; Lee, S. O.; Lee, K. R.; Kim, S. D. Antimicrobial Effect of Ethanol Extracts from Spices Against *Lactobacillus plantarum* and *Leuconostoc mesenteroides* Isolated from Kimchi. *J. Korean Soc. Food Sci. Nutr.* **1998**, *27*, 455–460.

Kim, S. J.; Park, K. H. Antimicrobial Activities of the Extracts of Vegetable Kimchi Stuff. *Korea J. Food Sci. Technol.* **1995**, *27*, 216–220.

Kolb, N.; Vallorani, L.; Stocchi, V. Chemical Composition and Evaluation of Protein Quality by Amino Acid Score Method of Edible Brown Marine Algae Arame (eiseniabicyclis) and hijiki (hijikiafusiforme). *Acta Alimentaria.* **1999**, *28*, 213–222.

Kuda, T.; Goto, H.; Yokoyama, M.; Fujii, T. Fermentable Dietary Fiber in Dried Products of Frown Algae and Their Effects on Cecal Microflora and Levels of Plasma Lipid in Rats. *Fish. Sci.* **1998**, *64*, 582–588.

Lamia, A.; Moktar, H. Fermentative Decolorization of Olive Mill Wastewater by *Lactobacillus plantarum*. *Process Biochem.* **2003**, *39*(1), 59–65.

Larrauri, J. A.; Ruperez, P.; Saura-Calixto, F. Effect of Drying Temperature on the Stability of Polyphenols and Antioxidant Activity of Red Grape Pomace Peels. *J. Agric. Food Chem.* **1997**, *45*(4), 1390–1393.

Lee, S. H.; Choi, J. S.; Park, K. N.; Im, Y. S.; Choi, W. J. Effects of Prunus mume Sie. Extract on Growth of Lactic Acid Bacteria Isolated from Kimchi and Preservation of Kimchi. *Korean J. Food Preserv.* **2002**, *9*, 292–297.

Lee, S. H.; Choi, W. J.; Jo, O. K.; Son, S. J. Antimicrobial Activity of Ethanol Extract of *Caesalpina sappan* L. and Effect of the Extract on the Fermentation of Kimchi. *J. Food Sci. Technol.* **1997**, *9*, 167–171.

Leistner, L. Principles and Applications of Hurdle Technology. In *New Methods of Food Preservation*; Gould, GW, Ed.; Blackie Academic; Professional: London, 1995, pp. 1–21.

Loss, C. R.; Hotchkiss, J. H. Inhibition of Microbial Growth by Low-Pressure and Ambient Pressure Gasses. In *Control of Foodborne Microorganisms*; Juneja, V. K.; Sofos, J. N., Eds.; Marcel Dekker: New York, 2002, pp. 245–279.

Lucke, F. K. Utilization of Microbes to Process and Preserve Meat. *Meat Sci.* **2000**, *56*, 105–115.

Lund, B. M.; Baird-Parker, T. C.; Gould, G. W. In *The Microbiological Safety and Quality of Foods*; Lund, B. M.; Baird-Parker, T. C.; Gould, G. W. Eds.; Aspen: Gaithersburg (MD), 2000, Vol. 1; p. 2.

Luo, Z.; Du, H.; Li, L.; An, M.; Zhang, Z.; Wan, X. Fuzhuanins A and B: The B-ring Fission Lactones of Flavan-3-ols from Fuzhuan Brick-Tea. *J. Agric. Food Chem.* **2013**, *61*, 6982–6990.

Mao, Q. Z. Microbial Changes and Functions in the Production of Wine Starter. *Liquor-Making Sci. Technol.* **2004**, *5*, 44–46 (In Chinese).

Marshall, B. M.; Ochieng, D. J.; Levy, S. B. Commensals: Underappreciated Reservoir of Antibiotic Resistance. *Microbe.* **2009**, *4*, 231–238.

Martinez-Villaluenga, C.; Torino, M. I.; Martin, V.; Arroyo, R.; Garcia-Mora, P.; Pedrola I. E. Multifunctional Properties of Soy Milk Fermented by *Enterococcus faecium* Strains Isolated from Raw Soy Milk. *J. Agric. Food Chem.* **2012**, *60*, 10235–10244.

Massera, A.; Soria, A.; Catania, C.; Krieger, S.; Combina, M. Simultaneous Inoculation of Malbec (*Vitis vinifera*) Musts with Yeast and Bacteria: Effects on Fermentation

Performance, Sensory and Sanitary Attributes of Wines. *Food Technol. Biotechnol.* **2009**, *47*, 192–201.

Matarjemi, Y.; Nout, M. J. R. Food Fermentation: A Safety and Nutritional Assessment'. *Bull. World Health Org.* **1996**, *74*(6), 553–559.

McFeeters, R. F. Fermentation Microorganisms and Flavor Changes in Fermented Food. *J. Food Sci.* **2004**, *69*, 35–37.

Meroth, C. B.; Walter, J.; Hertel, C.; Brandt, M. J.; Hammes, W. P. Monitoring the Bacterial Population Dynamics in Sourdough Fermentation Processes by Using PCR-denaturing Gradient Gel Electrophoresis. *Appl. Environ. Microbiol.* **2003**, *69*, 475–482.

Metaxopoulos, J.; Mataragas, M.; Drosinos, E. H. Microbial Interaction in Cooked Cured Meat Products Under Vacuum or Modified Atmosphere at 4; deg; C. *J. Appl. Microbiol.* **2009**, *93*(3), 363–373.

Meynier, A.; Mottram, D. S. The Effect of pH on the Formation of Volatile Compounds in Meat-Related Model Systems. *Food Chem.* **1995**, *52*, 361–366.

Milanovic, S.; Kanuric, K.; Vukic, V.; Hrnjez, D.; Ilicic, M.; Ranogajec, M.; Milanovic, M. Physicochemical and Textural Properties of Kombucha Fermented Dairy Products. *Afr. J. Biotechnol.* **2012**, *11*(9), 2320–2327.

Milanovic, S.; Loncar, E.; Djuric, M.; Malbasa, R.; Tekic, M.; Ilicic, M.; Durakovic, K. Low Energy Kombucha Based Beverages. *Acta Period. Technol.* **2008**, *39*, 37–46.

Mo, H.; Zhu, Y.; Chen, Z. Microbial Fermented Tea–A Potential Source of Natural Food Preservatives. *Trends Food Sci. Technol.* **2008**, *19*(3), 124–130.

Morell, P.; Hernando, I.; Llorca, E.; Fiszman, S. Yogurts with an Increased Protein Content and Physically Modified Starch: Rheological, Structural, Oral Digestion and Sensory Properties Related to Enhanced Satiating Capacity. *Food Res. Int.* **2015**, *70*, 64–73.

Morris, J. G. The Effect of Redox Potential. In *The Microbiological Safety and Quality of Food*; Lund, B. L.; Baird-Parker, T. C.; Gould G. W., Eds.; Aspen: Gaithersburg (MD), 2000, Vol. 1, pp 235–250.

Mossel, D. A. A.; Thomas, G. Securite microbioligique des plats prepares refrigeres: recommendations en matiere d'analyse des risques, conception et surveillance du processus de fabrication. *Microbiol. Aliements Nutr.* **1988**, *6*, 289–309.

Mossel, D. A. A.; Corry, J. E. L.; Struijk, C. B.; Baird, R. M. Essentials of the Microbiology of Foods: A Textbook for Advanced Studies. John Wiley and Sons: Chichester (England), 1995, p 699.

Mottram, D. S. Meat. In *Volatile Compounds in Foods and Beverages*; Maarse, H., Ed.; Marcel Dekker Inc.: New York, 1991, pp. 107–177.

Mottram, D. S. Flavor Formation in Meat and Meat Products: A Review. *Food Chem.* **1998**, *62*, 415–424.

Murooka, Y.; Yamashita, M. Traditional Healthful Fermented Products of Japan. *J. Ind. Microbiol. Biotechnol.* **2008**, *35*, 791–798.

Natesan, S.; Ranganthan, Y. Contents of Various Elements in Different Parts of the Tea Plant and in Infusion of Black Tea From Southern India. *J. Sci. Food Agric.* **1990**, *51*, 125–139.

Ni, L.; Wu, Z. G.; Zhang, W. Physiological and Biochemical Properties of Monascus from Qu of Fujian Rice Wine. *J. Fuzhou Univ. (Nat. Sci. Ed.)* **2009**, *37*, 929–934 (In Chinese).

Nielsen, I. L. F.; Haren, G. R.; Magnussen, E. L.; Dragsted, L. O.; Rasmussen, S. E. Quantification of Anthocyanins in Commercial Black Currant Juices by Simple High-Performance Liquid Chromatography. Investigation of Their pH Stability and Antioxidative Potency. *J. Agric. Food Chem.* **2003**, *51*(20), 5861–5866.

Nout, M. J. R.; Aidoo, K. E. Asian Fungal Fermented Food. In *The Mycota;* Osiewacz, H. D., Ed.; Springer-Verlag: NewYork, 2002, pp 23–47.

Nout, M. J. R. Rich Nutrition from the Poorest–Cereal Fermentations in Africa and Asia. *Food Microbiol.* **2009**, *26*(7), 685–692.

Nuraida, L.; Owens, J. D. Sweet, Sour, Alcoholic Solid Substrate Fungal Fermentations. In *Indigenous Fermented Foods of South-east Asia*, Owens, J. D. Ed.; CRC Press: USA, 2014, pp. 137–155.

Nuraida, L.; Owens, J. D.; Bakar, J. A. Lactic Vegetable and Fruit Fermentations. In *Indigenous Fermented Foods of Southeast Asia*; Owens, J. D. Ed.; CRC Press: USA, 2014, pp. 185–209.

O'Brien, T. F. Emergence, Spread, and Environmental Effect of Antimicrobial Resistance: How Use of an Antimicrobial Anywhere Can Increase Resistance to Any Antimicrobial Anywhere Else. *Clin. Infect. Dis.* **2002**, *34*, S78–S84.

Oguntoyinbo, F. A.; Narbad, A. Molecular Characterization of Lactic Acid Bacteria and In Situ Amylase Expression During Traditional Fermentation of Cereal Foods. *Food Microbiol.* **2012**, *31*, 254–262.

Oguntoyinbo, F. A.; Tourlomousis, P.; Gasson, M. J.; Narbad, A. Analysis of Bacterial Communities of Traditional Fermented West African Cereal Foods Using Culture Independent Methods. *Int. J. Food Microbiol.* **2011**, *145*, 205–210.

Okada, N. Temph–Indonesian Fermented Soya Bean Food. *Shokuryo.* **1988**, *27*, 65–93.

Olivon, B. Os paises que mais produziram cerveja em 2012, 2013. Available at http:// exame.abril.com.br/economia/noticias/os-paises-que-mais-produziram-cerveja-em 2012 (accessed in Dec 2013).

Osungbaro, T. O. Physical and Nutritive properties of Fermented Cereal Foods. *Afr. J. Food Sci.* **2009**, *3*(2), 23–27.

Oyewole, O. B. Lactic Fermented Foods in Africa and Their Benefits. *Food Control.* **1997**, *8*(5–6), 289–297.

Ozturk, I. Presence, Changes and Technological Properties of Yeast Species During Processing of Pastrima, a Turkish Dry Cured Meat Products. *Food Control.* **2015**, *50*, 76–84.

Pandey, A. Recent Process Developments in Solid-State Fermentation. *Process Biochem.* **1992**, *27*, 109–117.

Panesar, P. S.; Panesar, R. S.; Singh, S, Kennedy, J. F.; Kumar, H. Microbial Production, Immobilization and Applications of β-D-Galactosidase. *J. Chem. Technol. Biotechnol.* **2006**, *81*(4), 530–543.

Paredes-López O.; Harry G. I. Food Biotechnology Review: Traditional Solid-State Fermentations of Plant Raw Materials–Application, Nutritional Significance and Future Prospects. *CRC Crit. Rev. Food Sci. Nutr.* **1988**, *27*, 159–187.

Park, H. J.; Han, Y. S. Effect of Mustard Leaf on Quality and Sensory Characteristics of Kimchi. *J. Korean Soc. Food Nutr.* **1994**, *23*, 618–624.

Park, M. K.; Jung, K. S.; In, M. J. Effects of *Scutellaria baicalensis* and *Phellodendron amurense* Extracts on Growth of Lactic Acid Bacteria and Kimchi Fermentation. *J. Korean Soc. Food Sci. Nutr.* **2004**, *33*, 420–426.

Park, S. K.; Kang, S. G.; Chung, H. J. Effects of Essential oil in Astringent Persimmon Leaves on Kimchi Fermentation. *Korean J. Appl. Microbiol. Biotechnol.* **1994**, *22*, 217–221.

Pederson, C. S.; Albury, M. N. The Sauerkraut Fermentation. New York State Agriculture. *Exp. Station Bull.* **1969**, *824*, 87.

Perina, N. P.; Granato, D.; Hirota, C.; Cruz, A. G.; Bogsan, C. S. B.; Oliveira, M. N. Effect of Vegetal-Oil Emulsion and Passion Fruit Peel-Powder on Sensory Acceptance of Functional Yogurt. *Food Res. Int.* **2015**, *70*, 134–141.

Pieroni, A.; Quave, C. L. Functional Foods or Food-medicines? On the Consumption of Wild Plants among Albanians and Southern Italians in Lucania. In *Eating and Healing: Traditional Food as Medicine*; Pieroni, A.; Price, L. L., Eds.; Haworth Press: Binghamton, 2006, pp 101–129.

Pruitt, K. M.; Tenovuo, J.; Mansson-Rahemtulla, B.; Harrington, P.; Baldone, D. C. Is Thiocyanate Peroxidation at Equilib-Rium In Vivo? *Biochim. Biophys. Acta.* **1986**, *870*, 385–391.

Purrinos, L.; Garcia Fontan, M. C.; Carballo, J.; Lorenzo, J. M. Study of the Counts, Species and Characteristics of the Yeast Population During The Manufacture of Dry Cured Lacon. Effect of Salt Level. *Food Microbiol.* **2013**, *34*, 12–18.

Rantsiou, K.; Urso, R.; Iacumin, L.; Cantoni, C.; Cattaneo, P.; Comi, G.; Cocolin, L. Culture-Dependent and Independent Methods to Investigate the Microbial Ecology of Italian Fermented Sausages. *Appl. Environ. Microbiol.* **2005**, *71*, 1977–1986.

Ray, B. Bacteriocins of Starter Culture Bacteria as Food Biopreservative. In *Food Biopreservatives of Microbial Origin*; Ray, B.; Daeschel, M., Eds.; CRC Press: Florida, 1992, Vol. 8., pp. 177–205.

Reiter, B.; Harnulv, B. G. Lactoperoxidase Antibacterial System: Natural Occurrence, Biological Functions and Practical Application. *J. Food Prot.* **1984**, *47*, 724–732.

Rhee, S. J.; Lee, J. E.; Lee, C. H. Importance of Lactic Acid Bacteria in Asian Fermented Foods. *Microb. Cell Fact.* **2011**, *10*(1), S5.

Rodriguez, H.; Curiel, J. A.; Landete, J. M.; delas Rivas, B.; de Felipe, F. L.; Gomez-Cordoves, C. Food Phenolics and Lactic Acid Bacteria. *Int. J. Food Microbiol.* **2009**, *132*(2–3), 79–90.

Rong, R. J.; Li, Z. M.; Wang, D. L.; Bai, Z. H.; Li, H. Y.; Rong, R. F. Research Progress on Microorganisms in Chinese Liquor Qu. *China Brewing.* **2009**, *6*, 5–8 (In Chinese).

Ross, R. P.; Morgan, S.; Hill, C. Preservation and Fermentation: Past, Present and Future. *Int. J. Food Microbiol.* **2002**, *79*, 3–16.

Rubio, R.; Aymerich, T.; Bover-Cid, S.; Guardia, M. D.; Arnau, J.; Garriga, M. Probiotic Strains *Lactobacillus plantarum* 299V and *Lactobacillus rhamnosus* GG as Starter Cultures for Fermented Sausages. *Lwt Food Sci. Technol.* **2013**, *54*, 51–56.

Sasaki, M.; Mori, S. The Flavors of Soy Sauce. *Nippon Jyozo Kyokaishi.* **1991**, *86*, 913–922.

Schieberle, P. Quantitation of Important Roast-Smelling Odorants in Popcorn by Stable Isotope Dilution Assays and Model Studies on Flavor Formation During Popping. *J. Agric. Food Chem.* **1995**, *43*, 2442–2448.

Schnürer, J.; Magnusson, J. Antifungal Lactic Acid Bacteria as Biopreservatives. *Trends Food Sci. Technol.* **2005**, *16*(1–3), 70–78.

Sestelo, A. B. F.; Poza, M.; Villa, T. G. β-Glucosidase Activity in a *Lactobacillus plantarum* Wine Strain. *World J. Microbiol. Biotechnol.* **2004**, *20*(6), 633–637.

Shi, J. H.; Xiao, Y. P.; Li, X. R.; Ma, E. B.; Du, X. W.; Quan, Z. X. Analyses of Microbial Consortia in the Starter of Fen liquor. *Lett. Appl. Microbiol.* **2009**, *48*, 478–485.

Smelt, J. P. P. M.; Raatjes, J. G. M.; Crowther, J. C.; Verrips, C. T. Growth and Toxin Formation by Clostridium Botulinum at Low pH values. *J. Appl. Bacteriol.* **1982**, *52*, 75–82.

Son, H. S.; Hwang, G. S.; Park, W. M.; Hong, Y. S.; Lee, C. H. Metabolomic Characterization of Malolactic Fermentation and Fermentative Behaviors of Wine Yeasts in Grape Wine. *J. Agric. Food Chem.* **2009**, *57*, 4801–4809.

Spitaels, F.; Wieme, A. D.; Janssens, M.; Aerts, M.; Daniel, H. M.; Van Landschoot, A.; De Vuyst, L.; Vandamme, P. The Microbial Diversity of Traditional Spontaneously Fermented Lambic Beer. *PLoS One.* **2014**, *9*(4), e95384.

Stahnke, L. H. Flavor Formation in Fermented Sausage. In *Research Advances in the Quality of Meat and Meat Products*; Toldra, F., Ed.; Research Signpost: India, 2002.

Stanton, C.; Gardiner, G.; Meehan, H.; Collins, K.; Fitzgerald, G.; Lynch, P. B.; Ross R. P. Market Potential for Probiotics. *Am. J. Clin. Nutr.* **2001**, *73*(2) 476S–483S.

Steinkraus, K. H. Lactic Acid Fermentation in the Production of Foods From Vegetables, Cereals and Legumes. *Antonie van Leeuwenhoek.* **1983**, *49*, 337–348.

Steinkraus, K. H. Nutritional Significance of Fermented Foods. *Food Res. Intern.* **1994**, *27*(3), 259–267.

Steinkraus, K. H. Fermentations in World Food Processing. *Comprehen. Rev. Food Sci. Food Saf.* **2002**, *1*, 24–32.

Su, M. S.; Silva, J. L. Antioxidant Activity, Anthocyanins, and Phenolics of Rabbiteye Blueberry (*Vaccinium ashei*) by-Products As Affected by Fermentation. *Food Chem.* **2006**, *97*(3), 447–451.

Su, M. S.; Chien, P. J. Antioxidant Activity, Anthocyanins, and Phenolics of Rabbiteye Blueberry (*Vaccinium ashei*) Fluid Products as Affected by Fermentation. *Food Chem.* **2007**, *104*(1), 182–187.

Sun, Y.; Cheng, J. Hydrolysis of Lignocellulosic Materials for Ethanol Production: A Review. *Bioresour. Technol.* **2002**, *82*, 1–11.

Suryanarayan, S. Current Industrial Practice in Solid State Fermentations for Secondary Metabolite Production: The Biocon India Experience. *Biochem. Eng. J.* **2003**, *13*, 189–195.

Swain, M. R.; Anandharaj, M.; Ray, R. C. Fermented Fruits and Vegetables of Asia: A Potential Source of Probiotics. *Biotechnol. Res. Int.* **2014**, *19*, 1–19, Article ID 250424.

Tagliazucchi, D.; Verzelloni, E.; Bertolini, D.; Conte, A. In vitro Bioaccessibility and Antioxidant Activity of Grape Polyphenols. *Food Chem.* **2010**, *120*(2), 599–606.

Tamang, J. P. Naturally Fermented Ethnic Soybean Foods of India. *J. Ethnic Foods.* **2015**, *2*, 8–17.

Tanaka N, Traisman E, Plantong P, Finn L, Flom W, Meskey L, Guggisberg J. Evaluation of Factors Involved in Antibotulinal Properties of Pasteurized Process Cheese Spreads. *J. Food Prot.* **1986**, *49*(7), 526–531.

Teoh, A. L.; Heard, G.; Cox, J. Yeast Ecology of Kombucha Fermentation. *Int. J. Food Microiol.* **2004**, *95*, 119–126.

Teuber, M.; Meile, L.; Schwarz, F. Acquired Antibiotic Resistance in Lactic Acid Bacteria from Food. *Antonie Van Leeuwenhoek.* **1999**, *76*, 115–137.

Thomas, C.; Mercier, F.; Tournayre, P.; Martin, J. L.; Berdague, J. L. Identification and Origin of Odorous Sulphur Compounds in Cooked Ham. *Food Chem.* **2014**, *155*, 207–213.

Toldra, F.; Sanz, Y.; Flores, M. Meat Fermentation Technology. In *Meat Science and Applications*; Hui, Y. H.; Nip, W. K.; Rogers, R. W.; Young, W. A., Eds.; Marcel Dekker, Inc.: New York, Basel, 2001.

USDOE. National Algal Biofuels Technology Roadmap. US Department of Energy, Office of Energy Efficiency and Renewable Energy, Biomass Program, 2010. Visit (http://biomass. energy.gov) for more information.

Van Oevelen, D.; Spaepen, M.; Timmermans, P.; Verachtert, H. Microbiological Aspects of Spontaneous Wort Fermentation in the Production of Lambic and Gueuze. *J. Inst. Brewing.* **1977**, *83*, 356–360.

Verachtert, H.; Iserentant, D. Properties of Belgian Acid Beers and Their Microflora. Part I. The Production of Gueuze and Related Refreshing Acid Beers. *Cerevisia, Belgian J. Brewing Biotechnol.* **1995**, *20*, 37–41.

Villani, F.; Casaburi, A.; Pennacchia, C.; Filosa, L.; Russo, F.; Ercolini, D. Microbial Ecology of the Soppressata of Vallo di Diano, A Traditional Dry Fermented Sausage from Southern Italy, and In Vitro and In Situ Selection of Autochthonous Starter Cultures. *Appl. Environ. Microbiol.* **2007**, *73*, 5453–5463.

Vogelmann, S. A.; Seitter, M. F. H.; Singer, U.; Brandt, M. J.; Hertel, C. Adaptability of Lactic Acid Bacteria and Yeasts to Sourdoughs Prepared from Cereals, Pseudocereals and Cassava and Use of Competitive Strains as Starters. *Int. J. Food Microbiol.* **2009**, *130*, 205–212.

Wang, C. Y.; Wu, S. J.; Shyu, Y. T. Antioxidant Properties of Certain Cereals as Affected by Food-Grade Bacteria Fermentation. *J. Biosci. Bioeng.* **2014**, *117*(4), 449–456.

Wardlaw, G. M.; Hampl, S. J. *Trace Minerals. Perspectives in Nutrition*, seventh ed. McGraw-Hill Ryerson: Whitby, Canada, 2006.

Wee, J. H.; Park, K. H. Retardation of Kimchi Fermentation and Growth Inhibition of Related Microorganisms by Tea Catechins. *Korean J. Food Sci. Technol.* **1997**, *29*, 1275–1280.

Wongsagonsup, R.; Varavinit, S.; BeMiller, J. N. Increasing slowly Digestible Starch Content of Normal and Waxy Maizestarches and Properties of Starch Products. *Cereal Chem.* **2008**, *85*(6), 738–745.

Wouters, J. T. M.; Ayad, E. H. E.; Hugenholtz, J.; Smit, G. Microbes from Raw Milk for Fermented Dairy Products. *Int. Dairy J.* **2002**, *12*(2–3), 91–109.

Wu, C. H.; Chou, C. C. Enhancement of Aglycone, Vitamin K2 and Superoxide Dismutase Activity of Black Soybean Through Fermentation with *Bacillus subtilis* BCRC 14715 at Different Temperatures. *J. Agric. Food Chem.* **2009**, *57*, 10695–10700.

Xiao, Y.; Xing, G.; Rui, X.; Li, W.; Chen, X.; Jiang, M.; Dong, M. Enhancement of the Antioxidant Capacity of Chickpeas by Solid State Fermentation with Cordyceps Militaris SN-18. *J. Funct. Foods.* **2014**, *10*, 210–222.

Xie, G. F.; Li, W. J.; Lu, J.; Cao, Y.; Fang, H.; Zou, H. J. Isolation and Identification of Representative Fungi from Shaoxing Rice Wine Wheat Qu Using a Polyphasic Approach of Culture-Based and Molecular-Based Methods. *J. Inst. Brewing.* **2007**, *113*, 272–279.

Xu, A.; Wang, Y.; Wen, J.; Liu, P.; Liu, Z.; Li, Z. Fungal Community Associated with Fermentation and Storage of Fuzhuan Brick-Tea. *Int. J. Food Microbiol.* **2011**, *146*(1), 14–22.

Yue, Y.; Chu, G.; Liu, X.; Tang, X.; Wang, W.; Liu, G. TMDB: A Literature Curated Database for Small Molecular Compounds Found from Tea. *BMC Plant Biol.* **2014**, *14*, 143.

Zeuthen, P.; Bogh-Sorensen, L. B. *Food Preservation Techniques*. Woodhead Publishing: London, 2003.

Zhang, Z. Y.; Chang, X. X.; Zhong, Q. D. Liquor Qu Fungus System and Enzymatic System Character and Microbial Dynamic Variety During Vintage. *Liquor Making.* **2008**, *5*, 24–29 (In Chinese).

Zhu, Y. F.; Chen, J. J.; Ji, X. M.; Hu, X.; Ling, T. J.; Zhang, Z. Z. Changes of Major Tea Polyphenols and Production of Four New B-Ring Fission Metabolites of Catechins from Post-Fermented Jing-Wei Fu Brick Tea. *Food Chem.* **2015**, *170*, 110–117.

CHAPTER 4

MICROBIAL INTOXICATION IN DAIRY FOOD PRODUCTS

ALAA KAREEM NIAMAH* and DEEPAK KUMAR VERMA

Department of Food Science, College of Agriculture, University of Basrah, Basra City, Iraq

Corresponding author. E-mail: alaakareem2002@hotmail.com

CONTENTS

ABSTRACT

The safety and quality of milk and dairy products depend on the health and condition of the system used in the production of milk. To reduce the health risks, milk and dairy products require a healthy animal feed system, which begins and ends with cooling milk. Milk must transfer under hygienic conditions to begin manufacturing processes. Thermal processes help to produce healthy milk because it inhibits the growth of pathogenic bacteria. The pasteurization is used to eliminate common pathogenic microorganisms (including *Mycobacterium bovis*, commonly responsible for tuberculosis and *Coxiella burnetii*, causing fever). It can increase the shelf life of milk by breaking down almost all molds, yeasts, and common spoilage bacteria, but spore cells are resistant to heat pasteurization. Microbial intoxication form the most serious risks that cause economic losses in milk products and health damage affecting human life, so a strict health system in the production of milk and dairy products must be followed.

4.1 INTRODUCTION

Milk and dairy products form a significant part of the human diet. They are very important for the growth and development of the body; it is all due to richness of nutrients such as proteins, fats, carbohydrate, vitamins, and minerals especially for iron element. They are also important because milk- and dairy-based products are susceptible to rapid growth of microbes. In some instances, this rapid growth of microbes may be beneficial, whereas in others, it is undesirable and not acceptable. Milk and dairy-based products are vulnerable to spoilage or contamination with pathogens or microbial toxins. Therefore, the microbial study of milk and dairy-based products is today's key interest and industrial need. This chapter will not address all the pathogenic microbes that are of concern in all dairy foods. However, selected pathogens will be described which illustrate typical organism types (e.g., *Salmonella*, *Listeria* and pathogenic *Escherichia coli*) of common concern in dairy food manufacturing. Food-borne illness caused by microorganisms is generally classified into two categories: food infection (food-borne infection) and food intoxication (food-borne intoxication). The term food poisoning is often used very loosely and includes food-borne diseases resulting from both food infection and food intoxication (Frazier and Westhoff, 2006).

Part of the bacteria in udder form, the skin of the cow, and the vessels used for collection get transmitted to milk product. Milking fully healthy animals under aseptic conditions results in a bacteria count as low as 2×10^2–4×10^2 CFU/ml. Collected with somewhat less care, it may contain 2×10^3–6×10^3; with non-ideal conditions (e.g., vessel milking is not clean or unhealthy animals), the counts in fresh milk may range from 3×10^4 CFU/ml to 1×10^5 CFU/ml or more, indicating a high degree of contamination (Marth and Steele, 2001).

Even though, the dairy products are consumed daily by most individuals in the United States. Milk, cheese, and ice cream are still among the safest foods and have most recently accounted for last years than 1.5% of all foodborne illness cases reported annually (Bean et al., 1996). Milk-borne outbreaks of the disease were since the start of the history of the dairy industry. Bacterial infections, including diphtheria, scarlet fever, tuberculosis, and typhoid fever, prevailed before World War II, and we are almost always associated with to the consumption of raw milk. However, two outbreaks in 1985—the first involving up to 85 deaths in southern California from *Listeria*-contaminated cheese and the second in the Chicago area in which more than 16,000 cases of salmonellosis were traced to one particular brand of pasteurized milk— reaffirm the need for continued vigilance by the dairy industry to safeguard public health (Zheng and Kaiser, 2009). This study used data from 1990 to 2006 for the Centers for Disease Control and Prevention Annual Listings of Disease Outbreaks and the Foodborne Outbreak Database to create the base pandemic characteristics of line for the spread of diseases associated with liquid milk were reported and found 83 milk-borne outbreaks between 1990 and 2006, resulting in 3621 from diseases. The average number of diseases in the outbreak of 43.6 (range of disease: 2–1644) was associated with the consumption of unpasteurized milk with 55.4% of reported outbreaks. *Campylobacter* spp., *E. coli*, *Campylobacter* spp., and *Salmonella* spp. caused 10.8%, 51.2%, and 9.6%, respectively, of reported outbreaks. Private houses accounted 41.0% of the outbreak sites. The number of patients, the site of the outbreak, and the causes of the underlying properties can indicate to the potential case of deliberate contamination. In 2007, one pasteurized milk-borne outbreak caused by *Listeria* has been flagged as the arrival can be compared with the projected outbreak profile (Newkirk et al., 2011). The increased prevalence of milk products is directly proportional to the diseases transmitted by milk and dairy products (Fig. 4.1).

Additional new foodborne pathogens include *E. coli* O157:H7, *Aeromonas hydrophila*, *Aeromonas caviae*, *Aeromonas sobria*, *Mycobacterium* spp., vancomycin-resistant enterococci, nongastric *Helicobacterpylori*, *Enterobacter sakazakii*, nonjejuni/coli species of *Campylobacter*, and non O157 Shiga toxin-producing *E. coli* (Jay et al., 2005).

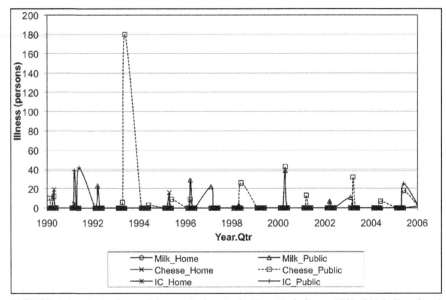

FIGURE 4.1   Dairy-borne Disease Outbreaks in New York State, 1990–2006. (Reprinted from Zheng, Y., and Kaiser H. M, Dairy-borne Disease Outbreak and Milk Demand: A Study Using Outbreak Surveillance Data, Volume 38, Issue 3 December 2009, pp. 330-337. With permission from Cambridge University Press.)

4.2   MICROBIAL INTOXICATION

Illness in this case is by swallowing preforming bacteria or mold toxins because of the growth in foods. The poison has to be present in contaminated foods. Once the microorganisms grew and produced toxin, viable cells are not needed during food consumption for the disease to occur, for example, Staphylococcal food poisoning, botulism (Table 4.1).

Food intoxication results from consumption of toxins (or poisons) produced in food by bacterial growth. Toxins, not bacteria, cause illness. Toxins may not alter the appearance, odor, or flavor of food and food production. Common kinds of bacteria that produce toxins include *Staphylococcus aureus* (Enterotoxin) and *Clostridium botulinum*, botulism (neurotoxin-paralysis). See Table 4.1 for more information on its prevention. In the case of *Clostridium perfringens*, illness is caused by toxins released in the gut when large numbers of vegetative cells are eaten (Galanos and Freudenberg, 1993).

A toxic infection is caused when microbial grows on food and food products and is consumed by humans (like infection); the microbial produces a toxin in the body. It is important to understand that, under the right set of circumstances, anyone can become ill due to eating contaminated foods. A healthy adult may be without symptoms or may have gastro-intestinal symptoms. In most cases, the healthy adult host will recover in a few days. However, the risks and dangers associated with foodborne illness are much greater for the elderly, infants, pregnant women, and people who have a weakened immune system.

TABLE 4.1 The Difference between Food Borne Infections and Intoxication.

Subject	Infections	Intoxication
Toxin type	Bacteria, parasite, viral	Natural, chemical, microbial
In intestine	Invade and or multiply in lining of intestine	No invasion or multiplication
Incubation period	Longer, hours to days	Shorter, minutes to hours
Symptoms	Diarrhea, nausea, vomiting, abdominal cramps, fever	Diarrhea, nausea, vomiting, diplopia, weakness, rasp failure, motor dysfunction
Spread	Spreads from person to person	Nonspreads from person to person
Factors	Inadequate cooking, cross contamination, poor personal hygiene, bare hand contact	Improper handling temperatures, inadequate cooking

4.3 MICROBES CAUSES INTOXICATION IN DAIRY FOOD

4.3.1 ASPERGILLUS SPP. ASPERGILLUS FLAVUS OR ASPERGILLUS PARASITICUS

Aspergillus is a large genus composed of more than 180 accepted anamorphic species, with teleomorphs described in nine different genera. The genus is subdivided in seven subgenera, which in turn are further divided into sections (Pitt et al., 2000). Some species of *Aspergillus* genes were Aflatoxins production. Aflatoxins are a group of closely related heterocyclic compounds produced predominantly by two filamentous fungi, *Aspergillus flavus*, and *Aspergillus parasiticus*.

A. flavus and *A. parasiticus* were belonging to *Trichocomaceae* family. Discrimination of *A. flavus* and *A. parasiticus* from nearly all other species is easy. Both grow rapidly on standard identification media such as Potato dextrose agar or Malt extract agar and yellow green conidia production from mold colonies which are in some time uncolored.

A. flavus and *A. parasiticus* produced Aflatoxin under different culture conditions like temperature (20–40°C), incubation time (7–21 days), and the pH (2.0–6.0) in two different synthetic media (Ritter et al., 2011). Aflatoxin M_1 is the principal hydroxylated B_1 metabolite present in milk of cows fed with a diet contaminated with B_1 depicted in Figure 4.2.

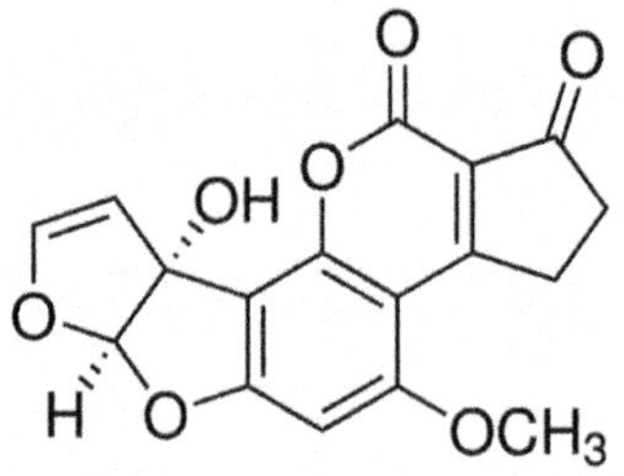

FIGURE 4.2   The chemical structure of Aflatoxin M_1.

4.3.2 *BACILLUS SPP. BACILLUS CEREUS*

The *B. cereus* is one of the members of *Bacillaceae* family, be Gram-positive, rod-shaped, 1.0–1.2×3.0–7.0 μm, beta hemolytic bacterium, endospores formation, aerobic grow, acid production nongas from glucose, motility, heat resistance, and growth on cheap sources (Logan and De Vos, 2009). Some strains are harmful to humans and cause foodborne illness, *B. cereus* can cause vomiting or diarrhea and, in some cases, both (Table 4.2). When *B. cereus* grows and produces "emetic toxin" in food, it can cause vomiting and produce another toxin "diarrheal toxin." It is arising human pathogen that causes infection the stomach and intestines and is now the third leading cause of food poisoning mass accidents in European Union, after *Salmonella* spp. and *S. aureus* (Hernaiz et al., 2003; Ramarao and Sanchis, 2013). The conditions for toxins production were maximum at 20–25°C, in a range of

TABLE 4.2   Characters of Emetic and Diarrheal.

Character	Emetic	Diarrheal
Syndrome	Nausea, vomiting, malaise, occasionally followed by diarrhea	Abdominal pain, watery diarrhea, occasional nausea
Incubation	0.5–6 h	8–16 h
Dose	Large numbers in the range of 10^5–10^8 (CFU/g)	Total cells 10^5–10^7 (CFU/g)

10–40°C, with an optimum pH 6–7, emetic toxin stable between pH 2 and pH 9, and oxygen required for production of emetic toxin (Agata et al., 2002).

4.3.3 CAMPYLOBACTER SPP.

Campylobacter is a genus of Gram-negative, 0.2–0.8×0.5–5.0 μm, micro-aerophilic, oxidase-positive, nonfermentative, nonspore forming, bacteria spiral, and motile through bipolar or unipolar flagella in bacteria cell. Most species of *Campylobacterare* cause disease and infect humans and animals. *Campylobacter jejuni* and *Campylobacter coli* have been implicated in human disease, and they are most common (Park, 2002).

   Campylobacter is transmitted to milk by clinical or subclinical mastitis or fecal contamination during the milking process. Cattle frequently harbor *campylobacter* as commensal in their gastrointestinal tract and *campylobacter* spp. in raw milk, most commonly derived from secondary faucal contamination during the milking process (Oliver et al., 2005). The average number of *Campylobacter* in milk ranged from 1 to 100 CFU/100 ml. According to various surveys in different countries, the *Campylobacter* spp. has found no survival at commercial pasteurization temperatures. Survival is about fivefold higher in refrigerated tissues than in frozen tissues. All *Campylobacter* spp. caused an infectious disease called Campylobacteriosis. Most people who become ill with campylobacteriosis get diarrhea, cramping, abdominal pain, and fever within 2 to 5 days after exposure to the organism, and diarrhea may be bloody and can be accompanied by nausea and vomiting (CDC, 2005).

4.3.4 CLOSTRIDIUM SPP. CLOSTRIDIUM PERFRINGENS AND CLOSTRIDIUM BOTULINUM

Clostridium is a genus of Gram-positive, rod-shaped, 0.3–1.9×2.0–10.0 μm, obligate anaerobes, motile, or nonmotile. When motile, cells usually are peritrichous. The majority of species form oval or spherical endospores that usually distend the cell. The genes belonging to the *Clostridiaceae* family contain around 100 species (Rainey et al., 2009).

   C. perfringens is a Gram-positive, rod-shaped, anaerobic, spore-forming bacterium of the genus *Clostridium. C. perfringens* is ever-present in nature and can be found as a normal component of decaying vegetation, marine sediment, the intestinal tract of humans and other vertebrates, insects, and soil. It prefers to grow in conditions with very little or no oxygen, and under ideal conditions can multiply very rapidly. Some strains of *C. perfringens* produce a toxin in the intestine that causes illness, and it is one of the most common causes of foodborne illness. The persons infected with *C. perfringens* develop diarrhea and abdominal cramps within 6–24 h. The illness usually begins suddenly and lasts for less than 24 h. Persons infected with *C. perfringens* usually do not have fever or vomiting (Grass et al., 2013).

C. botulinum is a Gram-positive, rod-shaped, 0.5–2.4 1.7–22.0 μm, anaerobic, spore-forming, motile bacterium with the ability to produce the neurotoxin botulinum (Rainey et al., 2009). The botulism as a result of the growth of *Cl. botulinum* bacteria in packaged foods is very severe illness; botulism is potent neurotoxins, which can be formed in food during growth bacteria. *Cl. botulinum* can produce seven different types of botulism, and very low concentrations from the botulism can cause symptoms. The appearance of symptoms usually takes in 12–36 h after eating, but depends on the amount of poison eating, and could take longer than that when the concentration of the toxin is less. Symptoms include initial diarrhea and vomiting followed by neurological effects including lack of visibility, weakness, difficulty talking, breathing, and swallowing. Dairy foods involved in outbreaks are usually in formula milk powder provisionally implicated in a case of infant botulism in the United Kingdom (Johnson et al., 2005).

4.3.5 CRONOBACTER SPP. CRONOBACTER SAKAZAKII

Cronobacter sakazakii is a Gram-negative, rod-shaped, nonspores forming, motile peritrichous, yellow-pigment product, pathogenic bacterium and which can survive in very dry conditions. Before 2007, it was called *E. sakazakii*. *Cronobacter* genes are one of the members of *Enterobacteriaceae* family. Seven species are found within the *Cronobacter* genes (Iversen et al., 2008). Majority of cases by *Cronobacter* are in adults, and in addition to that which is associated with a rare cause of invasive infections in children with rates of death historically high (40–80%) (Bowen and Braden, 2006). It has been found in a variety of dry foods, for example, infant formula, skimmed milk powder, and milk powder, and it usually causes sepsis or severe meningitis for infants (<12 months old).

C. sakazakii causes foodborne illnesses by the consumption of a variety of foods such as milk powder, cheese, and particularly infant foods. In this study, the presence of *C. sakazakii* was investigated in 60 milk powder, 50 whey powder, and 50 white cheeses. *C. sakazakii* was identified from milk powder 5% (3/60) and white cheese 4% (2/50) (Gokmen et al., 2010). In August 2004, all neonatal units were advised to cease using powdered infant formula, tube replaced by the prepared "ready-to-feed" infant formula that is recommended by the Ministry of Health and the New Zealand Food Safety Authority. Three strains from *Cronobacter* spp., isolated from 77 powder infant formula samples and study of isolates survival in dairy products after 24 month, found that no *Cronobacter* was detected in liquid milk, but it was prevalent in the prefinal product, and packaged final product was 3.70% and 4.35%, respectively (Songzhe et al., 2011).

4.3.6 *ESCHERICHIA COLI*

E. coli is a Gram-negative, facultative anaerobic, bacilli, 1.1–1.5×2.0–6.0 µm, motile, catalase, indole and methyl red positive, oxidase negative, nonspore form of the genus *Escherichia* that is commonly found in the lower intestine of warm-blooded organisms, and the *Escherichia* is one of the members of *Enterobacteriaceae* family (Singleton, 2004). More than 700 serotypes of *E. coli* have been identified. *E. coli* O157:H7 is an enterohemorrhagic serotype of the bacterium *E. coli* and a cause of illness, typically through consumption of contaminated food. But some types of *E. coli* were K vitamin production in lower intestine of human and animals.

Forty-eight samples from local dairy products (white soft cheese, curls cheese, yoghurt, and local kaimer) in Basrah city in Iraq were collected, high number of *E. coli* O157:H7 bacteria was found in white soft cheese and curls cheese samples than yoghurt and Local kaimer samples, the numbers of bacteria were low (Niamah and Al-Sahlany, 2011). Strains of *E. coli* are classified as different which cause human diseases by type of symptoms that cause diseases in humans. Moreover, it can be divided into six groups, although the properties are not private and may be shared by more than one group. Shiga-toxigenic *E. coli* (STEC) also called verocytotoxic-producing *E. coli* is one of the types of diseases. It causes symptoms ranging from mild to severe and bloody diarrhea. Approximately, 10% of patients (especially children and the elderly) with the infection can lead to life-threatening disease, such as the hemolytic uremic syndrome (HUS). Enterohemorrhagic *E. coli* (EHEC) are a subset of STEC commonly associated with a bloody diarrhea and HUS, which produce cytotoxins known as Shiga-like toxins or verotoxins. With respect to public health, *E. coli* (O157:H7) strain is very important EHEC serotype related to foodborne disease, leading to a high incidence of EHEC infections and deaths every year (Hussein and Sakuma, 2005).

4.3.7 LISTERIA MONOCYTOGENES

Listeria monocytogenes is a Gram-positive, bacilli, short rod, 0.4–0.5×0.5–2.0 μm, motile by peritrichous flagella, aerobic and facultative anaerobic, nonspore forming, catalase positive, oxidase negative, not acid-fast, capsules not formed, and a beta hemolysis product when grown on blood agar. Temperature limit of growth is 0–45°C, with optimal growth at 30–37°C. These do not survive when heated at 60°C for 30 min (Mclauchlin and Rees, 2009). It is one of the members of *Listeria* genus, and the genes belong to the *Listeriaceae* family which includes seven species—*L. monocytogenes*, *L. innocua*, *L. seeligeri*, *L. welshimeri*, *L. ivanovii*, *L. grayi*, and *L. ivanovii*. *L. monocytogenes* is the bacteria that cause the infection listeriosis. The bacteria are frequently found in the environment in milk, dairy production, and the dairy-processing factories. During the period of 2002–2004, the study revealed the presence of *L. monocytogenes* in nine dairy factories in five counties of Moldova territory. The purpose of this study was to identify the main cause, that is, the way of pollution of dairy products. A total of 196 samples were analyzed from dairy products, 20.4% were identified as contaminated with *Listeria* spp., out of which 3.57% with *L. monocytogenes* (Minea et al., 2005). Listeriosis caused mainly inflammations in the central nervous system (meningitis, meningitis brain, inflammation of cerebral, and brain abscess) and the bacteria those which are immunocompromised. The incubation period in humans is 21 days, with a permanent diarrhea ranging from 1 to 3 days. Patients present with a high fever, diarrhea, muscle aches, headaches and digestive stiff neck, nausea, confusion and loss of balance, or convulsions (Hof, 1996).

4.3.8 SALMONELLA SPP.

Salmonella is a genus of Gram negative, bacilli, 0.5–0.7×1.0–3.0 μm, motile with peritrichous flagella, facultative anaerobes, catalase, methyl red and Simmons citrate positive, oxidase, indole, H_2S producing, urea, and Voges Proskauer negative. D-Glucose is fermented with the production of acid and usually gas (Müller, 2012). It is one of the members of *Enterobacteriaceae* family and comprises two species only, *S. bongori* and *S. enterica*. They cause diseases such as typhoid fever and paratyphoid fever, and presence of these in food causes food poisoning (Ryan and Ray, 2004). McManus and Lanier (1987) have showed the prevalence of *Salmonella* species; in bulk tank, milk has been reported ranging from <1% to almost 9%. This study was conducted to determine the prevalence of *Salmonella* species in 600 samples from milk and locally processed milk products traded for human consumption and assess the risk factors associated with *Salmonella* milk contamination in Kanam, Nigeria. The study utilized microbiological culture and isolation as well as questionnaire analysis. The study revealed an overall prevalence of 8.7% (52/600), 0.2% (1/600), and 3.0% (18/600) of fresh milk (madara), full creamed milk (kindirmo), and skimmed milk (nono), respectively (Karshima et al., 2013). In 1985, the salmonellosis outbreak in the United States was *Salmonella typhimurium* in milk from the Hill farm Dairy in Melrose Park.

Salmonellosis is a type of food poisoning caused by *Salmonella* spp. The salmonellosis is more common in the summer than in the winter. Symptoms of salmonellosis include diarrhea, fever, and abdominal cramps. They develop 12–72 h after infection, and the illness usually lasts for 4–7 days. Most people recover without treatment. But diarrhea and dehydration may be so severe that it is necessary to go to the hospital. Older adults, infants, and those who have impaired immune systems are at highest risk (Santos et al., 2001).

4.3.9 STAPHYLOCOCCUS SPP. STAPHYLOCOCCUS AUREUS

Staphylococcus is a genus Gram-positive bacteria, which appears singly, in pairs, in tetrads, in short chains (3–4 cells), and forms in grape-like clusters, 0.5–1.5 μm in diameter; growth can also occur in a 6.5% NaCl solution, nonmotile, nonspore-forming, usually catalase-positive and oxidase-negative, and unencapsulated or limited capsule formation. *Staphylococcus* spp. can be differentiated from other aerobic and facultative anaerobic. The *Staphylococcus* genus includes at least 40 species, and it is one member from *Staphylococcaceae* family (Schleifer and Beel, 2009).

S. aureus group includes two species (*S. aureus*, *S. simiae*) is catalase and coagulase enzymes production, nonmotile, nonspore forming, facultative anaerobe, fermentation of glucose, produces mainly lactic acid, ferments mannitol and golden yellow colony on agar. de Oliveira et al. (2011) in their study verified the presence of *S. aureus* in milk consumed in Reconcavo Baiano, Brazil, in which 34 out of 50 samples of raw milk showed contamination by *S. aureus* corresponding to 68% from total isolates. *S. aureus* causes most staph infections (pronounced "staph infections") including (1) skin infections, (2) pneumonia, (3) food poisoning, (4) toxic shock syndrome, and (5) blood poisoning (bacteremia). According to European Food Safety Authority in Europe, a total of 5550 outbreaks of food-borne illnesses in 2009 was reported, which approximately affected 49,000 people and causing

46 deaths. From these, 293 were caused by *Staphylococcus* spp. outbreaks. The bacterial toxins, fourth most common causative agent of the disease in outbreaks, came from foodborne diseases (Schelin et al., 2011).

S. aureus food poisoning (SFP) symptoms include diarrhea, vomiting, and abdominal pain. They are normal, but life-threatening. Most cases of SFP infection do not need treatment because the disease gets cured by own. Most patients recover from food poisoning within 2 days. According to Food and Drug Administration (FDA), SFP associated deaths are very rare. But there is an increased risk for these complications among older persons and children and persons with weak immune system (Food and Drug Administration (FDA), 2012).

SFP like symptoms results to severe state of gastroenteritis. Symptoms quickly in some cases may appear in as little as 30 min. However, it usually takes up to 6 h for the symptoms to show (Dinges et al., 2000). Symptoms of SFP include diarrhea, vomiting, and nausea for patients with mild spasm in abdomen, and in general, most men are recovering within 1–3 days.

4.3.10 YERSINIA SPP. YERSINIA ENTEROCOLITICA

Yersinia spp. are Gram-negative, bacilli, 0.5–1.0×1.0–2.0 μm, facultative anaerobes and belong to *Enterobacteriaceae* family. The *Yersinia* genus consists of 11 species: *Y. enterocolitica*, *Y. frederiksenii*, *Y. intermedia*, *Y. pseudotuberculosis*, *Y. kristensenii*, *Y. pestis*, *Y. bercovieri*, *Y. rohdei*, *Y. ruckeri*, and *Y. aldovae*. Among them, only *Y. pestis*, *Y. pseudotuberculosis*, and some of *Y. enterocolitica* strains cause diseases to humans and animals, whereas other types of wild origin can serve as opportunists. However, *Yersinia* strains can be isolated from clinical sources, so be diagnosed to species level.

Y. enterocolitica is one of the members of *Yersinia* genus, Gram-negative, facultatively anaerobic, bacilli to coccobacilli sharp, nonmotile, nonspore-forming, facultative intracellular bacterium, and *Y. enterocolitica*. It can be divided into four major groups, two of which are sucrose negative and the rest are rhamnose positive (Swaminathan et al., 1982). Rahimi et al. (2014) showed that 52 (9.42%) and 28 (5.07%) of the total 552 milk and dairy samples collected from Isfahan Province, Iran were positive for the existence of *Yersinia* species and *Y. enterocolitica*, respectively. A total of 24 of 28 *Y. enterocolitica* isolates by culture were positive in PCR test (4.59%). During a year period, 150 in each food group, a total of 750 samples including ice cream, raw milk, feta cheese, and chicken were collected from markets located in the northeast region of Turkey (Kars, Ardahan, and Iğdır). These were analyzed for determining of *Yersinia* spp. incidence. A total of 57 samples (7.6%) were evaluated as positive for *Yersinia* spp., and 18 (2.4% in total) of them, isolated from six feta cheese, four ice cream, two chicken, and two raw milk samples, were identified as pathogenic *Y. enterocolitica* (Güven et al., 2010).

Yersiniosis is an infectious disease caused by a bacterium of the genus *Yersinia*. In the United States, most yersiniosis infections among humans are caused by *Y. enterocolitica*. Incubation period is 4–6 days (range, 1–14 days). The symptoms include fever, abdominal pain (may mimic appendicitis), and diarrhea (may be bloody and can persist for several weeks). Necrotizing enterocolitis has been described in young infants. Common symptoms in children are fever, abdominal pain, diarrhea (which is often bloody), and in older children and adults, the most common symptoms are right-sided abdominal pain and fever. In Ontario during 2003–2009, there was an average of two to three confirmed cases of yersiniosis reported per 100,000 persons each year (Heymann, 2008).

4.4 MICROBIAL TOXINS PRODUCTIONS IN DAIRY FOOD

4.4.1 ENDOTOXINS

Endotoxins, the biological activity that is potential within the lipopolysaccharide (LPS), are the major glycolipids of most Gram-negative bacterial outer membranes; LPS comprises three parts depicted in Table 4.3. Endotoxins' productions from Gram-negative bacteria are the pathogenic *Vibrio*, *Salmonella* spp., *Shigella* spp. *Yersinia enterocolitica*, *Enterobacter agglomerans*, and *E. coli*, member of the normal human intestinal flora, etc. (Arduino et al., 1989; Bui, 2012).

TABLE 4.3 Parts of Lipopolysaccharide (LPS).

S. no.	Lipopolysaccharide (LPS)	Remark
1.	Polysaccharide or O antigen	Repetitive glycan polymer contained within a LPS is referred to as O side-chain of the bacteria
2.	Core oligosaccharide	Contains an oligosaccharide component that attaches directly to lipid A and commonly contains sugars such as heptose and 3-deoxy-D-mannooctulosonic acid (also known as KDO, keto-deoxyoctulosonate) (Hershberger and Binkley, 1968)
3.	Lipid A	Phosphorylated glucosamine disaccharide decorated with multiple fatty acids (Tzeng et al., 2002)

4.4.1.1 *PROPERTIES OF ENDOTOXINS*

They are weakly toxic, rarely fatal (fatal dose on the order of hundreds of micrograms); symptoms of endotoxins are fever, diarrhea, vomiting, stable at 100°C for 1 h, and the M wt. between 10 and 1000 kDa. The bacterial chromosome targets endotoxin production, which leads to weak immune reactions; and most enzymes effect on endotoxins. LPSs are made up of a hydrophobic lipid (lipid A, which is responsible for the toxic properties of the molecule), a hydrophilic core polysaccharide chain, and a hydrophilic *O*-antigenic polysaccharide side chain. The polysaccharide chain is longer and more difficult than hydrolysis. A variety of basic units of saccharide, in LPS produced from *E. coli* O111:B4, was in found glycerol, D-mannos, galactose and glucose as depicted in Figure 4.3.

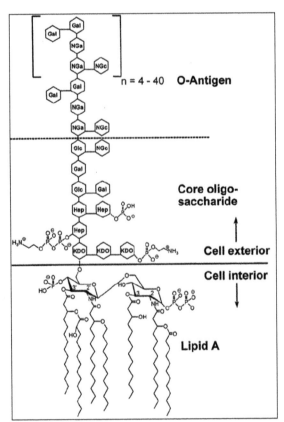

FIGURE 4.3 Chemical structure of endotoxin from *E. coli* O111:B4 according to Ohno and Morrison (1989). (Hep) L-glycerol-D-manno-heptose; (Gal) galactose; (Glc) glucose; (KDO) 2-keto-3-deoxyoctonic acid; (NGa) *N*-acetyl-galactosamine; (NGc) *N*-acetyl-glucosamine.

4.4.2 EXOTOXINS

Exotoxin is a poisonous substance secreted by certain bacteria. In their purest form, they are the most potent poisons known and are the active agents in diphtheria, tetanus, and botulism. Exotoxin can cause damage to the host by destroying cells or disrupting normal cellular metabolism (Patočka and Středa, 2006). There are three types of exotoxins: enterotoxins, neurotoxins, and cytotoxins. These give an indication of the site of action. Enterotoxins act on lining of gastrointestinal. Neurotoxins act on the function of neurons. Cytotoxins damage the functioning of host cells. Cholera, diphtheria, and tetanus are the sum of diseases that are caused by exotoxins (Popoff and Poulain, 2010). Exotoxins are highly antigenic. These can stimulate the immune system. By stimulating the immune system, they produce antitoxins to neutralize the toxin.

4.4.2.1 PROPERTIES OF EXOTOXINS

It is produced from Gram-positive and Gram-negative bacteria, synthesized in cytoplasm; may or may not secreted proteins, highly antigenic, inactivated by heat; except, *S. aureus* secretes an exotoxin, which causes gastroenteritis, and has the ability to form toxoid and plasmid charge of exotoxin production.

4.4.3 AFLATOXINS

Aflatoxins are metabolic secondary products, toxins fungi, produced by *Asperigullus* spp. (*A. bombycis*, *A. flavus*, *A. nomius*, *A. ochraceoroseus*, *A. parasiticus*, and *A. pseudotamarii*). The name was given around 1960 after the discovery that the death source of more than 100,000 turkey poultry's in the South of England was *A. flavus* toxins (Wannop, 1961). A total of 14 different types of Aflatoxin are found in nature. The major types are Aflatoxin B1 and B2, produced by *A. flavus* and *A. parasiticus*; Aflatoxin G1 and G2, produced by *A. parasiticus*; Aflatoxin M1 and M2; metabolite of Aflatoxin B1 and B2; aflatoxicol; Aflatoxin P1; and Aflatoxin Q1 (Fig. 4.4) (Smith and Rachel, 1991).

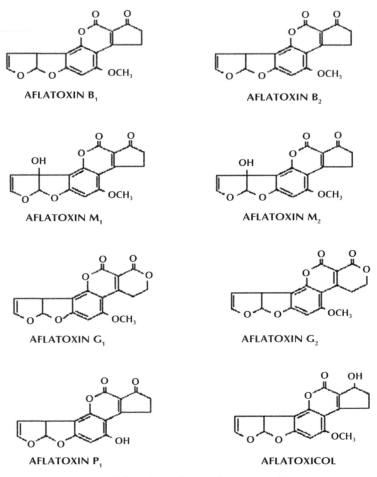

FIGURE 4.4   Some types of Aflatoxin production from *Aspergillus* spp.

Aflatoxins are heat-stable compounds in a dry environment, very good heat stability during thermal processing of food and animal feeds, and partial and variable concentration decrease are observed during autoclaving or roasting, boiling or cooking, and extrusion or fermentation processes. Aflatoxin B1 is metabolized by the microsomal mixed function oxidase system in the liver of human or animal, leading to the formation of highly reactive intermediates, one of which is 2,3-epoxy Aflatoxin B1. Binding of these reactive intermediates to DNA results in disruption of transcription and abnormal cell proliferation, leading to mutagenesis or carcinogenesis. Aflatoxin types also inhibit oxygen uptake in the tissues by acting on the electron transport chain and inhibiting various enzymes resulting in decreased ATP product.

After entering in the human and animal body, Aflatoxins may be metabolized by the liver to a reactive epoxide intermediate or hydroxylase to become the less harmful Aflatoxin M1. Aflatoxin M1 present in the milk of dairy cows fed a diet contaminated with Aflatoxin B1 (Neal et al., 1998).

The studies of Aflatoxins contamination in milk and dairy products indicated the Aflatoxin M1. It is considered to pose certain hygienic risks for human health which is formed when animals ingest feed contaminated with Aflatoxin B1. The amount of Aflatoxin M1 is found in milk depending on several factors, such as animal breed, lactation period, mammary infections, milking types, etc. The M1 could be detected in milk in 12–24 h; after 72 h, the M1 concentration in milk decreases to an undetectable level. The heating treatment of dairy productions, for example, pasteurization did not effect on M1 level (Aliabadi et al., 2013).

The Aflatoxins react with nucleic acids after first being converted to an epoxide by a cytochrome P450. The epoxide reacts with guanine in DNA and RNA leading to deprivation. The primary effect is to inhibit protein and DNA synthesis in the most active tissues, including the liver, the intestines, and the bone marrow (Fig. 4.5). The damage they do to DNA can be mutagenic, a GC to AT mutation, and also carcinogenic with liver cancer a common long-term effect of exposure.

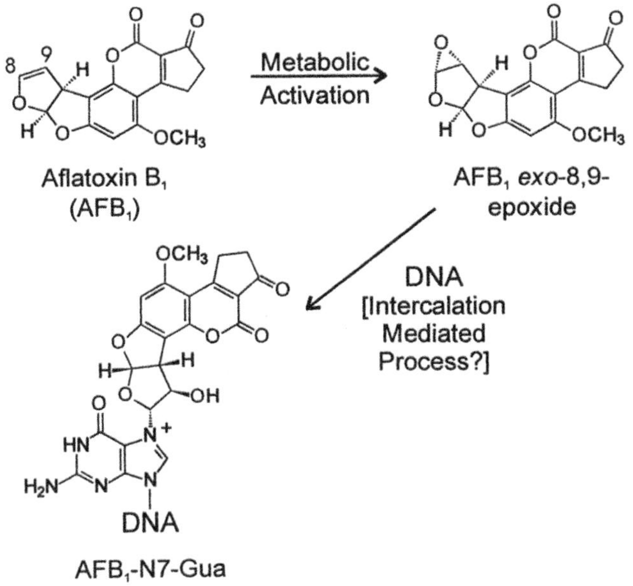

FIGURE 4.5 Reaction Aflatoxin B1 with DNA cells. (Adopted from Kobertz et al., 1997).

4.5 FACTORS RESPONSIBLE FOR GROWTH OF MICROBIAL INTOXICATION IN DAIRY FOOD

4.5.1 TEMPERATURE

Temperature is probably the most critical factor affecting the growth of bacteria in foods. Most disease-causing bacteria grow within a temperature range between 4°C and 60°C. This is commonly referred to as the "Temperature Danger Zone"; it is the temperature range in which food-borne bacteria below 4°C will not necessarily inhibit pathogenic bacteria but will slow down their growth. At temperatures above 60°C, pathogenic bacteria will not grow.

The pasteurization and sterilization temperatures will ensure that pathogenic bacteria are destroyed in milk and dairy products. Spores cell were stable after pasteurization processes, but sterilization temperature kill all life types. Some types of toxin-producing bacteria can grow at 45–50°C (Table 4.4), but it inhibits growth in pasteurization temperature.

TABLE 4.4   Approximate Temperature Requirements for Microorganism (°C).

Microorganism	Minimum	Optimum	Maximum
Campylobacter spp.	32	42–45	45
Clostridium botulinum types A & B	10–12	30–40	50
Clostridium botulinum type E	3–3.3	25–37	45
Clostridium perfringens	12	43–47	50
Enterohemorrhagic Escherichia coli	7	35–40	46
Listeria monocytogenes	0	30–37	45
Salmonella spp.	5	35–37	45–47
Staphylococcus aureus growth	7	35–40	48
Staphylococcus aureus toxin	10	40–45	46

4.5.2 pH

pH is a measure of hydrogen ion [H$^+$] concentration in environments. The higher the pH reading, the more alkaline and oxygen rich the fluid is. The lower the pH reading, the more acidic and oxygen deprived the fluid is. The pH range is from 0 to 14, with 7.0 being neutral. Anything above 7.0 is alkaline, and anything below 7.0 is considered acidic.

Microbes are sensitive to the hydrogen ion concentration they find in their environment. Large proteins, such as enzymes, are affected by pH. Their shape changes (they denature) and the very often brings about an alteration of the ionic charges on the molecule. In general, most bacteria grow best around neutral pH values (6.5–7.0), but lactic acid bacteria and acetic acid bacteria can growth in pH 5.0 and 6.0, growth of yeasts are in pH (4.5–6.5) and molds (3.5–6.8). The milk pH is 6.4–6.8 that optimum for most pathogenic microbial; dairy fermentation products are low and inappropriate in terms of pH for the growth of most of the bacteria (Table 4.5) (Kosikowski and Mistry, 1997).

TABLE 4.5   pH Values in Milk and Dairy Products.

Names of product	pH values
Milk	6.4–6.8
Cheese	5.1–5.9
Yoghurt	4.25–4.6
Sweet butter	6.7–6.85
Sour butter	4.6–5.0
Sour cream	4.6–4.8
Sweet cream	6.5–6.7
Kefir	4.2–4.6
Ice cream	6.2–6.7

4.5.3   MOISTURE OR WATER ACTIVITY (A_w)

Nutrients must enter the bacterial cell through its cell wall; therefore, nutrients must be soluble so that they can be carried into the cell by any free (or unbound) water that is available in the environment where the bacteria are living (such as in food). The measure of this available free water is known as the water activity (a_w). Its ratio of partial pressure of water vapour in the substance is divided by the state of the standard partial vapour pressure of water. In food science field, water activity, in practice, as is measured in equilibrium relative humidity (ERH) can be obtained by the following formula (Young and Cauvain, 2000):

$$\text{ERH } (\%) = a_w \times 100$$

Water activity in foods is an important property which can be used for forecasting food safety, stability, and quality. The water activity effects on chemical stability of foods such as decreasing Maillard reactions, caramelization reaction, and oxidation reactions of lipid. It increases desirable activity of some enzymes and vitamins in food interactions and improves the physical properties of food such as texture (Rockland and Beuchat, 1987).

The Hazard Analysis and Critical Control Point programs are used the water activity as a critical control point of food samples. Samples of food products are taken periodically from the production area, and testing to ensure water activity values is within the specified range for food safety and quality. The water activity (a_w) of milk is 0.993 (Singh et al., 1997) and dairy products are 0.2–0.94. Water activity in milk is adequate for most microbes, whereas dairy products are sometimes not suitable for this reason that reduces the growth of microorganisms (Table 4.6).

TABLE 4.6   Minimum of a_w Values for Microbial Growth.

Microbial names	Minimum a_w for growth
Clostridium botulinum A	0.95
Clostridium botulinum B	0.94
Clostridium botulinum E	0.97
Pseudomonas fluorescens	0.97
Clostridium perfringens	0.95
Escherichia coli	0.95
Salmonella spp.	0.95
Vibrio cholerae	0.95
Bacillus cereus	0.95
Listeria monocytogenes	0.92
Bacillus subtilis	0.91
Staphylococcus aureus	0.86
Campylobacter coli	0.97
Campylobacter jejuni	0.98
Vibrio parahaemolyticus	0.94
Yersinia enterocolitica	0.96
Aspergillus flavus	0.78–0.84
Most molds	0.80

Source: Gustavo et al. (2007).

Pathogenic bacteria need a water supply to survive, the amount of water in milk and dairy products can be reduced by processes such as drying or adding salt or sugar. Water activity, less than 0.85, decreases the potential of bacteria to grow (Fox and McSweeney, 1998). In general, the requirement of the water activity of microorganisms decreases in the following order: bacteria > yeast > mold. Below 0.60, any microbiological growth is possible. So the dried dairy product such as milk powder is more stable and safe shelf compared with wet or semiwet foods.

4.5.4 REDOX POTENTIAL

Redox potential (Eh) is a measure to reduce tendency of chemical species to acquire electrons. Reduction potential is measured in volts (V) or millivolts (mV). The element or compound loses electrons, and is said to be oxidized, while becoming a pillar to gain electrons reduced. Aerobic microorganisms need positive values Eh for growth, but the anaerobic microorganisms needs negative values Eh for growth.

In dairy food, several factors effect on Eh values as: redox couples present ratio of oxidant to reluctant, pH, poising capacity, availability of oxygen, and microbial activity. The Eh value of milk was −0.3 V in pH 6.6; when *E. coli* and *Streptococcus lactis* grow in milk, the Eh value became +0.3 V and −0.22 V, respectively, after 24 h (Frazier and Whittier, 1931). Tachon et al. (2010) demonstrated that the electron transport chain is responsible for the decrease in the milk Eh from 300 mV to 220 mV when the oxygen concentration reaches zero or under anaerobic conditions. The Eh values in milk and dairy products were various because of the different concentrations of mineral elements depicted in Table 4.7. The values of Eh were low effect on microbial growth in dairy production.

TABLE 4.7 Redox Potential (Eh) Values in Milk and Dairy Products.

Names of product	Redox potential (V)
Raw milk	+0.3
Pasteurized milk	+0.1
Processed cheese	+0.05
Yoghurt	−0.1
Emmental cheese	−0.3

Source: Belitz et al. (2009).

4.5.5 NUTRIENTS

Like all other living things, microorganisms too need water, a source of carbon, as energy source, and a source of nitrogen, minerals, vitamins, and physical factors for the growth and function normally. Since then, the foods are a rich source of these compounds, and thus can be used by microorganisms as well. Because of these reasons, various food products such as malt extract, yeast extract, beef extract, peptone, tryptone, tomato juice, sugar, starch, and minerals and vitamins are included in the means of microbial media. The milk and dairy products' content availability essential for the growth of microorganisms to the elements are considered media appropriate for the growth of different microorganisms.

All dairy products contain elements such as proteins, fats, carbohydrates, lactose, etc. (Table 4.8). For growth, these elements are ideal environments for the production of toxins.

TABLE 4.8 Nutrition Values of Milk and Some Dairy Products.

Name product	Protein %	Fat %	Carbohydrate %	Moisture %	Ash %
Raw milk	3.2	3.9	4.8	87.8	0.3
Cottage cheese	11.12	4.30	3.38	80.5	0.7
Yoghurt	16	4	4	75.4	0.6
Butter	1	81	0	17.7	0.3
Whipped cream	6	33	4	53	4

4.6 DIRECTIONS FOR FUTURE RESEARCH PROSPECTIVE

1. The use of modern techniques in the detection of bacteria and bacterial toxins in dairy products such as RT-PCR, TOF-TOF, and GC–MS.
2. Antimicrobial activity of natural extraction using dairy products model media: efficacy, synergistic potential, and interactions with dairy products components.
3. Study of new mycotoxins in milk and dairy products and effect on human and cattle animals.
4. New type milk such as donkey milk is used in dairy products, for example, cheese manufacture and its affect on microbial growth and toxins production.

5. Probiotic bacteria are used in bio-yogurt and fermented dairy products to reduce the concentrations of Aflatoxin.
6. Study of industry and storage conditions creates the best conditions limiting contamination.

4.7 CONCLUSIONS

Milk and dairy products from the food-rich nutrients encourage the growth of microorganisms and toxins production. Thermal processes such as pasteurization and sterilization limit microbial growth and inhibit the production of toxins; storage conditions and transport better reduce damage of dairy products and microbiological contamination of toxin-producing. Feeding animals producing milk free of mycotoxins further help in reducing concentrations of these toxins in milk and milk products.

KEYWORDS

- dairy products
- microbial intoxication
- endotoxins
- exotoxins
- aflatoxins

REFERENCES

Agata, N.; Ohta, M.; Yokoyama, K. Production of *Bacillus cereus* Emetic Toxin (Cereulide) in Various Foods. *Int. J. Food Microbiol.* **2002,** *73*, 23–27.

Aliabadi, M. A.; Alikhani, F.; Mohammadi, M.; Darsanaki, R. K. Biological Control of Aflatoxins. *Eur. J. Exp. Biol.* **2013,** *3*, 162–166.

Arduino, M. J.; Bland, L. A.; Tipple, M. A.; Aguero, S. M.; Favero, M. S.; Jarvis, W. R. Growth and Endotoxin Production of *Yersinia enterocolitica* and *Enterobacter agglomerans* in Packed Erythrocytes. *J. Clin. Microbiol.* **1989,** *27*, 1483–1485.

Bean, N. H.; Goulding, J. S.; Lao, C.; Angulo, F. J. Surveillance of Foodborne Disease Outbreaks—United States, 1988–1992. *Morbid Mortal Weekly Rep.* 1996, *45*(SS-5), 1.

Belitz, H. D.; Grosch, W.; Schieberle, P. *Food Chemistry.* Springer Science & Business Media: London, 2009.

Bowen, A. B.; Braden, C. R. Invasive *Enterobacter sakazakii* Disease in Infants. *Emerg. Infect. Dis.* **2006,** *12,* 1185–1189.

Bui, A. *Structural Characteristics of Bacterial Endotoxins.* Ph.D. Thesis, University of Pecs, 2012.

CDC, *Campylobacter* infections. Adtlanta, GA: Separtment of Health and Human Services, Centers for Disease Control, Division of Bacterial and Mycotic Diseases, 2005. http://www.cds.gov/ncidod /dbmd/dese asein fo/campylobacter_g.htm.

Center for Science in the Public Interest (CSPI). A review of Foodborne illness in America from 2002–2011, Washington, DC. 2014. cspinet.org/reports/outbreakalert2014.pdf.

de Oliveira, L. P.; Barros, L. S. S.; Silva, V. C.; Cirqueira, M. G. Study of *Staphylococcus aureus* in Raw and Pasteurized Milk Consumed in the Reconcavo Area of the State of Bahia, Brazil. *J. Food Proc. Technol.* **2011,** *2,* 128–133.

Dinges, M.; Orwin, P.; Schlievert, P. Exotoxins of *Staphylococcus aureus. Clin. Microbiol. Rev.* **2000,** *13,* 16-34.

Food and Drug Administration (FDA). *Staphylococcus aureus,* 2012. http://www.fda.gov/food/foodsafety/foodborneillness/foodborneillnessfoodbornepathogensnaturaltoxins/badbugbook/ucm070015.htm.

Fox, P. F.; McSweeney, P. L. H. *Dairy Chemistry and Biochemistry.* Blackie Academic & Professional: London, 1998.

Frazier, W. C.; Westhoff, D. C. *Food Microbiology.* Fifth Ed.; Tata McGraw Hill Publishing Co. Ltd, New Delhi, 2006.

Frazier, W. C.; Whittier, E. O. Studies on the Influence of Bacteria on the Oxidation-Reduction Potential of Milk. *J. Bacteriol.* **1931,** *21,* 239–251.

Galanos, C.; Freudenberg, M. A. Mechanisms of Endotoxin Shock and Endotoxin Hypersensitivity. *Immunobiology.* **1993,** *187,* 346–356.

Gokmen, M.; Tekinsen, K. K.; Gurbuz, Ü. Presence of *Enterobacter sakazakii* in Milk Powder, Whey Powder and White Cheese Produced in Konya. *Kafkas Üniversitesi Veteriner Fakültesi Dergisi,* **2010,** *16,* S163–S166.

Grass, J. E.; Gould, L. H.; Mahon, B. E. Epidemiology of Foodborne Disease Outbreaks Caused by *Clostridium perfringens,* United States 1998–2010. *Foodborne Pathogens Dis.* **2013,** *10,* 131–136.

Gustavo V. B. C.; Anthony J. F.; Shelly J. S.; Theodore P. L. *Water Activity in Foods Fundamentals and Applications.* Blackwell Publishing Professional: Iowa, USA, 2007.

Güven, A.; Sezer, Ç.; Aydin, B. D.; Oral, N. B.; Vatansever, L. Incidence and Pathogenicity of *Yersinia enterocolitica* Isolates from Foods in Turkey. *Kafkas Univ. Vet Fak. Derg.* **2010,** *16,* S107–S112.

Hernaiz, C.; Picardo, A.; Alos, J. I.; Gomez-Garces, J. L. Nosocomial Bacteremia and Catheter Infection by Bacillus cereus in an Immunocompetent Patient. *Clin. Microbiol. Infect.* **2003,** *9,* 973–975.

Hershberger, C.; Binkley, S. B.. Chemistry and Metabolism of 3-Deoxy-d-Mannooctulosonic Acid. I. Stereo Chemical Determination. *J. Biol. Chem.* **1968,** *243,* 1578–1584.

Heymann, D. L. *Control of Communicable Diseases Manual;* 19th ed.. American Public Health Association: Washington D.C., 2008.

Hof, H. *Listeria monocytogenes. Baron's Medical Microbiology;* fourth ed.; University of Texas Medical Branch: USA, 1996.

Hussein, H. S.; Sakuma, T. Invited Review: Prevalence of Shiga Toxin-Producing *Escherichia coli* in Dairy Cattle and Their Products. *J. Dairy Sci.* **2005,** *88,* 450–465.

Iversen, C.; Mullane, N.; McCardell, B.; Tall, B. D.; Lehner, A.; Fanning, S.; Stephan, R.; Joosten H. *Cronobacter* gen. nov., A New Genus to Accommodate The Biogroups of *Enterobacter sakazakii*, and Proposal of *Cronobacter sakazakii* gen. nov., *comb.* nov., *Cronobacter malonaticus* sp. nov., *Cronobacter turicensis* sp. nov., *Cronobacter muytjensii* sp. nov., *Cronobacter dublinensis* sp. nov., *Cronobactergenomospecies*, and of three subspecies, *Cronobacter dublinensis* subsp. *dublinensis* subsp. nov., *Cronobacter dublinensis* subsp. *lausannensis* subsp. nov. and *Cronobacter dublinensis* subsp. *Lactaridi* subsp. nov. *Int. J. Syst. Evol. Microbiol.* **2008**, *58*, 1442–1447.

Jay, J.; Loessner, M.; Golden, D. A. *Modern Food Microbiology;* seventh ed.; Springer: New York, 2005.

Johnson, E. A.; Tepp, W. H.; Bradshaw, M.; Gilbert, R. J.; Cook, P. E.; McIntosh, E. D. G. Characterization of *Clostridium botulinum* Strains Associated with an Infant Botulism Case in the United Kingdom. *J. Clin. Microbiol.* **2005**, *43*, 2602–2607.

Karshima, N. S.; Pam, V. A.; Bata, S. I.; Dung P. A.; Paman N. D. Isolation of *Salmonella* species from Milk and Locally Processed Milk Products Traded for Human Consumption and Associated Risk Factors in Kanam, Plateau State, Nigeria. *J. Anim. Prod. Adv.* **2013**, *3*, 69–74.

Kobertz, W. R.; Wang, D.; Wogan, G. N.; Essigmannn, J. M. An Intercalation Inhibitor Altering the Target Selectivity of DNA Damaging Agents: Synthesis of Site-Specific Aflatoxin B1 adducts in a p53 Mutational Hotspot. *Proc. Natl. Acad. Sci.* **1997**, *94*, 9579–9584.

Kosikowski, F. V.; Mistry, V. V. *Cheese and Fermented Milk Products;* third ed.; Westport, F V Kosikowski Llc: USA, 1997.

Logan, N. A.; De Vos, P. *Genus Bacillus. In: Bergey's Manual of Systematic Bacteriology,* Vol. 3; second ed.; Springer Dordrecht: New York, 2009.

Marth, E. H.; Steele, J. L. *Applied Dairy Microbiology;* second ed.; Marcel Dekker: New York, 2001.

Mclauchlin, J.; Rees, C. E. D. *Genus Listeria. In: Bergey's Manual of Systematic Bacteriology, Vol. 3;* second ed.; Springer Dordrecht: New York, 2009.

McManus, C.; Lanier, J. M. *Salmonella, Campylobacter jejuni* and *Yersinia enterocolitica* in Raw Milk. *J. Food Protect.* **1987**, *50*, 51–54.

Minea, L.; Drug, O.; Cîmpeanu, C.; Vasilov, M.; Bucşă, D.; Mircea, D.; Gafencu, C.; Puiu-Berizinţu, L.; Cibi, M.; Iaşi, I. S. P.; Botoşani, D. S. P.; Galaţi, D. S. P.; Neamţ, D. S. P.; Bacău, D. S. P.; Iaşi, D. S. P. The Main Sources of *Listeria monocytogenes* Contamination in Milk Processing Plants. *J. Prevent. Med.* **2005**, *13*, 43–51.

Müller, K. *Genetic and Phenotypic Characteristics of Importance for Clonal Success and Diversity in Salmonella.* Ph.D. Thesis, Technical University of Denmark, Denmark, 2012.

Neal, G. E.; Eaton, D. L.; Judah, D. J.; Verma, A. Metabolism and Toxicity of Aflatoxins M1 and B1 in Human-Derived In Vitro Systems. *Toxicol. Appl. Pharmacol.* **1998**, *151,* 152–158.

Newkirk, R.; Hedberg, C.; Bender, J. Establishing a Milk Borne Disease Outbreak Profile: Potential Food Defense Implications. *Food Borne Pathog. Dis.* **2011**, *8*, 433–437.

Niamah, A. K.; Al-Sahlany, S. T. Detecting for *E. coli* O157:H7 in Dairy Products which Were Locally Processed and Found in Basra City Markets, *Basrah. J. Agric. Sci.* **2011**, *24*, 290–299.

Ohno, N.; Morrison, D. C. Lipopolysaccharide Interaction with Lysozyme. Binding of Lipopolysaccharide to Lysozyme and Inhibition of Lysozyme Enzymatic Activity. *J. Biol. Chem.* **1989**, *264*, 4434–4441.

Oliver, S. P.; Jayarao, B. M.; Almeida, R. A. Foodborne Pathogens in Milk and the Dairy Farm Environment: Food Safety and Public Health Implications. *Foodborne Pathog. Dis.* **2005**, *2*, 115–129.

Park, S. The Physiology of *Campylobacter* Species and its Relevance to Their Role as Foodborne Pathogens. *Int. J. Food Microbiol.* **2002**, *74*, 177–188.

Patočka, J.; Středa, L. Protein Bio Toxins of Military Significance. *Acta Med.* **2006**, *49*, 3–11.

Pitt, J. I.; Samson, R. A.; Frisvad, J. C. List of Accepted Species and Their Synonyms in the Family *Trichocomaceae*. In *Integration of Modern Taxonomic Methods for Penicillium and Aspergillus Classification;* Samson, R. A., Pitt, J. I., Eds.; Harwood Academic Publishers: Amsterdam, 2000; pp 9–79.

Popoff, M. R.; Poulain, B. Bacterial Toxins and the Nervous System Neurotoxins And Multi Potential Toxins Interacting with Neuronal Cells. *Toxins.* **2010**, *2*, 683–737.

Rahimi, E.; Sepehri, S.; Dehkordi, F. S.; Shaygan, S.; Momtaz, H. Prevalence of *Yersinia* Species in Traditional and Commercial Dairy Products in Isfahan Province. *Jundishapur J. Microbiol.* **2014**, *7*, 1–6.

Rainey, F. A.; Hollen, B. J.; Small, A. *Genus Clostridium. Bergey's Manual of Systematic Bacteriology, Vol. 3*; second Ed.; Springer Dordrecht: New York, 2009.

Ramarao, N.; Sanchis, V. The Pore-Forming Haemolysins of *Bacillus cereus*: A Review. *Toxins.* **2013**, *5*, 1119–1139.

Ritter, A. C.; Hoeltz, M.; Noll, I. B. Toxigenic Potential of *Aspergillus flavus* Tested in Different Culture Conditions. *Ciênciae Tecnologia de Alimentos.* **2011**, *31*, 623–628.

Rockland, L. B.; Beuchat, L. R. *Water Activity: Theory and Applications to Food;* second ed.; Marcell Dekker: New York, 1987.

Ryan, K. J.; Ray, C. G. *Sherri's Medical Microbiology*; fourth ed.; McGraw Hill: New York, USA, 2004.

Santos, R. L.; Zhang, S.; Tsolis, R. M.; Kingsley, R. A.; Adams, L. G.; Baumler, A. J. Animal Models of *Salmonella infections*: Enteritis versus Typhoid Fever. *Microbes Infect.* **2001**, *3*, 1335–1344.

Schelin, J.; Wallin-Carlquist, N.; Cohn, M. T.; Lindqvist, R.; Barker, G. C.; Rådström, P. The Formation of *Staphylococcus aureus* Enterotoxin in Food Environments and Advances in Risk Assessment. *Virulence.* **2011**, *2*, 580–592.

Schleifer, K. H.; Beel, J. A. *Genus Staphylococcus. Bergey's Manual of Systematic Bacteriology, Vol. 3*; second ed.; Springer Dordrecht: New York, 2009.

Singh, H.; McCarthy, O. J.; Lucey, J. A. Physicochemical Properties of Milk. In *Advanced Dairy Chemistry;* Fox, P. F., Ed.; Chapman & Hall: London, 1997; pp 469–518.

Singleton, P. *Bacteria in Biology, Biotechnology and Medicine;* fifth ed.; Wiley: USA, 1999; pp 444–454.

Smith, J. E.; Rachel, S. H. *Mycotoxins and Animal Foods.* CRC Press: UK, 1991.

Songzhe, F. U.; Jianxin, G.; Ying, L.; Haiying, C. Isolation of *Cronobacter* spp. Isolates from Infant Formulas and Their Survival in the Production Process of Infant Formula. *Czech J. Food Sci.* **2011**, *79*, 391–399.

Swaminathan, B.; Harmon, M. C.; Mehlman, I. J. *Yersinia enterocolitica. J. Appl. Bacteriol.* **1982**, *52*, 151–183.

Tachon, S.; Brandsma, J. B.; Yvon, M. NoxE NADH Oxidase and the Electron Transport Chain are Responsible for the Ability of *Lactococcus lactis* to Decrease the Redox Potential of Milk. *Appl. Environ. Microbiol.* **2010**, *76*, 1311–1319.

Tzeng, Y. L.; Datta, A.; Kolli, V. K.; Carlson, R. W.; Stephens, D. S. Endotoxin of *Neisseria meningitidis* Composed Only of Intact Lipid A: Inactivation of the Meningococcal 3-Deoxy-D-Manno-Octulosonic Acid Transferase. *J. Bacteriol.* **2002,** *184*, 2379–2388.

Wannop, C. C. The Histopathology of Turkey "X" Disease in Great Britain. *Avian Dis.* **1961,** *5,* 371–381.

Young, L.; Cauvain, S. P. *Bakery Food Manufacture and Quality: Water Control and Effects.* Blackwell Science: Oxford, 2000.

Zheng, Y.; Kaiser, H. M. Dairy-Borne Disease Outbreak and Milk Demand: A Study Using Outbreak Surveillance Data. *Agric. Resour. Econ. Rev.* **2009,** *38,* 330–337.

CHAPTER 5

MICROBIAL SPOILAGE IN MILK AND MILK PRODUCTS: POTENTIAL SOLUTION, FOOD SAFETY, AND HEALTH ISSUES

DEEPAK KUMAR VERMA*, DIPENDRA KUMAR MAHATO, SUDHANSHI BILLORIA, MANDIRA KAPRI, P. K. PRABHAKAR, AJESH KUMAR V., and PREM PRAKASH SRIVASTAV

Department of Agricultural and Food Engineering, Indian Institute of Technology, Kharagpur 721302, West Bengal, India

*Corresponding author. E-mail: deepak.verma@agfe.iitkgp.ernet.in; rajadkv@rediffmail.com

CONTENTS

ABSTRACT

The chapter outlines are the basic cause and sources of different microbiological contamination in the milk and milk products. The pre- and post-processing causes of contamination and their possible solutions have been proposed to mitigate the possible foodborne-related contaminations and outbreaks to guarantee the food safety and health issues of the consumers.

5.1 INTRODUCTION

Milk is the nutritious fluid secreted by all the mammalian species to fulfill the nutritional requirements of their neonates, and today it has become a major part of human diet. Milk being diverse in its composition, high water activity, and neutral pH has made it highly prone to different microbiological contaminations which ultimately spoil the raw milk as well as the milk products (Hassan and Frank, 2011). Therefore, proper care has to be taken starting from the feed material of the animals to milking equipments, health conditions, and other environments (Quigley et al., 2013). Raw milk is prone to both Gram-positive and Gram-negative bacteria (Champagne et al., 1994; Munsch-Alatossava and Alatossava, 2006; Hantsis-Zacharov and Halpern, 2007) as well as yeasts (Cocolin et al., 2002), so it is subjected to sterilization techniques to reduce the potential risk and guarantee the food safety and proper health of the consumers.

5.2 MICROBIAL SPOILAGE

Microorganisms as well as the released toxins are major threat for food industries as they significantly affect the productivity and present risk regarding the food safety and well-being of humans and the animals (Driehuis et al., 2008; Fallah, 2010; Brayden, 2012).

5.2.1 SOURCES OF MICROBIAL SPOILAGE

5.2.1.1 RAW MILK CONTAMINATION

The raw milk is contaminated either from the feed material or improper handling of equipment and the animals during and after milking. The concentrated feed of corn silage and grass are given to the animals during winter when grazing is less. These grain-base feed provide a risk for aflatoxin contamination (Fink-Gremmels, 2008; Duarte et al., 2013). These are toxins produced by many strains of *Aspergillus flavus*, *Aspergillus parasiticus*, and *Aspergillus nomius* that grow in vegetal products and can be present in milk and its derivates due to animals feeding on contaminated plants. Aflatoxin B1 (AFB1) is a carcinogenic and mutagenic difuranocoumarin derivative and is considered the most potent natural toxin, whereas aflatoxin M1 (AFM1) is the hydroxylated metabolite of AFB1 which is found in milk and in other dairy products, when lactating animals are fed with contaminated feedstuffs (Prandini et al., 2009). Even AFM1 toxin has been detected in the milk of contaminated lactants (Pietri et al., 1997; Sibanda et al., 1999; Turconi et al., 2004; Prandini et al., 2009). The transfer of AFB1 from feed to milk depends on various factors such as nutritional and physiological. For example, they are biotransformation capacity of liver feeding regimens, health of the animal, milk production, rate of digestion, and rate of ingestion (Duarte et al., 2013). Aflatoxins are subjected to the thermo stable but are never destroyed even by thermal processing like pasteurization and sterilization techniques. Children are at higher risk than adults because they consume more milk and dairy products than adults do. The contamination of milk and milk products are a serious issue of health hazard for consumers (Ruangwises and Ruangwises, 2010). The contamination of milk and dairy products with aflatoxins and their concentration depends and varies on the geographic location, climatic conditions, and economic status of the country.

5.2.1.2 CONTAMINATION IN MILK PRODUCTS

The main factors responsible for the contamination in milk and dairy products are climatic variations of the region (Rahimi et al., 2010). The high temperature and humidity in tropical and subtropical regions is responsible for the growth of *Aspergillus* species in the feed materials of the animals (Decastelli et al., 2007; Bilandzic et al., 2014). The drying and storage of feed should be taken care to avoid the moisture levels in them as this favors fungal development and the aflatoxin production as well (Prandini et al., 2009; Duartea et al., 2013).

5.2.2 MICROBIAL SPOILAGE IN MILK PRODUCTS

5.2.2.1 CHEESE

Cheese is made from milk by coagulating the casein present in milk by using the enzyme rennet. The important phase during cheese production is ripening process in which bacteria and fungi develop on the cheese surface and are responsible for the organoleptic and textural characteristics of cheese. The ripening is achieved due to the proteolytic and lipolytic activities of these microorganisms leading to the production of ammonia and sulfur compounds (Schornsteiner et al., 2014). During ripening, starter cultures are added, whereas some others grow naturally (Irlinger and Mounier, 2009; Montel et al., 2014). In the meanwhile, cheese surface remains aerobic, whereas the core part becomes anaerobic. The surface ripened cheeses are further of two types: bacteria-ripened (e.g., Limburger, Munster, Tilsit, and Appenzeller) and mold-ripened (e.g., Camembert, Brie) (Brennan et al., 2002; Mounier et al., 2005; Dolci et al., 2009).

Late blowing defect (LBD) is one the major cause of spoilage in semi-hard and hard cheeses characterized by unpleasant aroma and cracks in the cheese paste. It is mainly caused due to the production of butyric acid, CO_2, and H_2 by lactate metabolism by *Clostridium* species like *butyricum* (major cause), *sporogenes*, and *beijerinckii* (Klijn et al., 1995; Cocolin et al., 2004; Le Bourhis et al., 2005, 2007; Garde et al., 2011). The LBD cases with high level of butyric acid have been reported in cheese like in Saint-Nectaire (Mayenobe et al., 1983), Gouda (Mayenobe et al., 1983; Klijn et al., 1995), Grana Padano (Cocolin et al., 2004), Emmental, Gruyere, Comte, Beaufort, and Ossau-Iraty (Le Bourhis et al., 2005), and Manchego (Garde, et al., 2012).

Mozzarella cheese is used as a table cheese. Traditional Mozzarella cheese has high moisture content (usually ranged from 50% to 60%) and is usually dipped into a tap water, brine, and whey. This liquid medium supports the growth of psychrotrophic bacteria (Cabrini and Neviani, 1983; Mauro et al., 2005). *Pseudomonas* spp. and coliforms are responsible for proteolysis, discolorations, pigmentation, and off-flavors in Mozzarella cheese (Cabrini and Neviani, 1983; Massa et al., 1992; Cantoni et al., 2003a,b; Cantoni et al., 2006a,b; Cantoni and Bersani, 2010). High-moisture Mozzarella cheese is a soft (50–60% moisture) and unripened cheese manufactured from cow's milk. Around 66 potential spoilage strains have been identified from high moisture Mozzarella cheese which belongs to Pseudomonas, Acinetobacter, and Rahnella species (Baruzzi et al., 2012). Discoloration in High-moisture

Mozzarella cheese is observed due to microbial infestation like reddish discoloration by *Pseudomonas putida* (Soncini et al., 1998), bluish discoloration by *Pseudomonas fluorescens biovar IV* and *Pseudomonas libanensis* (Cantoni et al., 2003a,b), yellow-purple spots by *Pseudomonas gessardii* (Cantoni et al., 2006a,b), and greenish and fluorescent discoloration by *P. fluorescens* (Franzetti and Scarpellini, 2007). Cheese is very prone to mold growth (Pitt and Hocking, 2009), and some of them produce mycotoxins like Sterigmatocystin produced by *Antilles pinktoe Versicolor*, which is harmful for consumers' health (Northolt et al., 1980). Therefore, proper monitoring should be done during the critical phases to maintain proper quality in the final product (Coppola et al., 2008).

5.2.2.2   CULTURED BUTTERMILK

Buttermilk is the liquid released during churning of cream in the butter-making process. Though it is considered a low value by product and usually dried for animal feed material less than for human consumption (Fauquant et al., 2014), it has gained attention because chemically, it is skim milk rich in fat globule membrane (SMFGM) (Corredig et al., 2003; Rombaut et al., 2006; Sodini et al., 2006). Buttermilk provides a nutritious and favorable medium for the growth of spoilage and pathogenic microorganisms like *Listeria*, *Brucella*, *Mycobacterium*, and *Salmonella*. SMFGM has found to contain polar lipids which act as membrane receptors for pathogens by facilitating in pathogen-induced signaling pathways. Besides this, polar lipids are amphiphilic in nature, having hydrophobic tail and hydrophilic head and overall negative charge facilitates bacterial attachment to its surface, and hence, spoilage occurs very rapidly (Busscher et al., 1998).

5.2.2.3   DAHI

Dahi is among one of the traditional fermented milk products very common in Indian subcontinent countries like India, Nepal, Pakistan, Bangladesh, and Bhutan. It is a similar product to yoghurt, but unlike yoghurt where a specific mixed culture of *Lactobacillus delbrueckii* subsp. *bulgaricus* and *Streptococcus thermophilus* is used as a starter culture, there is no such specific starter culture for dahi. Instead, it contains various strains of lactic acid bacteria (LAB), and a small amount of curd is added to the milk for the purpose of fermentation. Dahi has been produced by the use of specific culture of *Lactococcus lactis* (Yadav et al., 2006) and a mixed culture of *lactobacilli* and *lactococci* (Yadav et al., 2007). The spoilage of dahi occurs due to microbial contamination which adversely affects consumer's health (Aziz, 1985; Kober et al., 2007). The common food borne pathogens reported are *Escherichia coli* 0157:H7, *Listeria monocytogenes*, and *Yersinia enterocolitica* (Morgan et al., 1993; Mead et al., 1999). *L. monocytogenes* have been reported in other fermented milk products as well like yogurt, kefir, and labneh (Garrote et al., 2000; AI-Kadamany et al., 2002; Gulmez and Guven, 2003). A food-poisoning case by consuming dahi due to staphylococcal contamination has been reported (Ghose and Chattoraj, 1963). Hence, there is a need for a specific starter culture for dahi to avoid contamination and health issues.

The *Staphylococcus aureus* has even been detected in the cheese made from raw or pasteurized milk due to slow acidification of curd (Coveney et al., 1994; Le Loir et al., 2003). Enterotoxin A produced by *S. aureus* can be detected in cheese sample if the initial population of the *S. aureus* is 103 CFU/ml (Meyrand et al., 1998). *S. thermophilus* and *Enterococci* are the dominant LAB, and *E. faecalis* and *E. faecium* have been reported in all types of curd and the cheese made from them as well (Bottazzi et al., 1976; Senini et al., 1997).

5.2.2.4   SOUR/CULTURED CREAM

Cream is a dairy-based product made from butterfat layer deposited on the top of milk before homogenization. Custard cream which is made from milk and eggs (Arakawa et al., 2008) may contain sugar, food grade color, flavors, and modified starch in some proportions (de Wijk et al., 2003). This cream has been widely used for many other preparations like cream puffs, pastries, and cakes. The cream has shorter shelf life compared to milk (Dieu and CuQ, 1989) and is highly susceptible to pathogenic microorganisms like *E. coli* 0157:H7, *Salmonella typhmurium*, *L. monocytogenes*, and *Bacillus cereus* (Oliver et al., 2005; Bennett et al., 2013). Talking individually, *E. coli* 0157:H7 has diarrheal and hemolytic uremic syndrome in human (Centers for Disease Control and Prevention, 2011); *S. typhmurium* had infected some people in Australia after consuming the infected custard cream (SADHA, 2012); *L. monocytogenes* causes listeriosis (Centers for Disease Control and Prevention, 2008), and *B. cereus* produces emetic toxins and causes diarrhea, and their spores are not even destroyed by pasteurization of milk (Wong et al., 1988; Kim et al., 2011).

Staphylococcal food poisoning (SFP) is caused by intake of *S. aureus* contaminated food products, and the bacteria can easily be transmitted to the food chain by handling with contaminated hands (Argudin et al., 2010). SFP outbreaks are very common in foods like salads, bakery products, meat and meat products, and milk and milk products (Hennekinne et al., 2012). The pastry confectionary products are very susceptible to molds, yeasts, and bacterial growth, and their ordinary quality often leads to food poisoning (Minguella, 1981; Michard et al., 1986). Staphylococcal intoxication and salmonellosis have been occurred in the United States, Canada, and Europe by the intake of contaminated pies and cream-filled pastries (Buening-Pfaue, 1978; Shankaran and Leela, 1983; Todd et al., 1983; Ferron and Michard, 1993). As we know that salmonellosis is very common with milk (Sharp, 1987; De Buyser et al., 2001; Newkirk et al., 2011) and ice cream (Cowden et al., 1989a,b; Hennessy et al., 1996; Daniels et al., 2002; Di Pietro et al., 2004), so the contamination of *Salmonella* should be prevented at any cost during the processing of milk and ice cream (Blaser, 1996; Wang et al., 2004).

5.2.2.5   YOGHURT

Yoghurt is lactic acid fermented milk product similar to curd but have symbiotic cultures of two LABs, viz., *S. thermophilus* and *Lb. delbrueckii* subsp. *bulgaricus* (Codex, 2003). It has been found that there occurs a mutual interaction between these two cultures in a 1:1 ratio which is referred to as protocooperation (Tamine and Robinson, 1985). It is protocooperation that is responsible for the production of lactic acid, flavor as well as textural properties of the final product (Courtin and Rul, 2004). For production of yoghurt, the culture used is either a small amount from previous batch or commercially prepared one.

The acidic condition of yoghurt is supposed to inhibit the microbial growth (Jay, 1992), but even many outbreaks of infection have been reported (Morgan et al., 1993). The *E. coli* O157:H7 in raw milk and those that escapes pasteurization process finally contaminate the milk products. *E. coli* O157:H7 have escaped the acidic condition of apple cider (Besser et al., 1993), mayonnaise (Weagan et al., 1994), and yoghurt (Morgan et al., 1993) which suggests that the acidic condition no longer can resist their growth. It has been observed that *E. coli* O157:H7 inoculated into yoghurt survived few hours at 30–40°C, and up to 1–8 days in refrigerated condition (Massa et al., 1997). The cases of AFM1 contamination in milk and milk products are due to the feeding of lactating mammals with the AFM1-contaminated feed materials which is later converted into cytotoxic and genotoxic AFM1 form (Cole and Cox, 1981).

5.3 POTENTIAL SOLUTION FOR MICROBIAL SPOILAGE IN MILK PRODUCTS

Growth of spore-forming bacteria in milk and milk products adversely affects their quality and even pose a greater toward food poisoning as they produce several heat-labile enterotoxins (Ehling-Schulz et al., 2004; Stenfors Arnesen et al., 2008; Ehling-Schulz et al., 2011). To mitigate/eradicate these problems, several techniques have been followed as well as proposed to reduce food loss due to microbial spoilage.

5.3.1 PASTEURIZATION

Pasteurization is among one of the techniques used to kill the spores of the mold and inactivates the pathogenic and spoilage bacteria present in the milk (Doyle and Marth, 1975). During pasteurization, milk is heated up to 72°C and held for 15 s at this temperature and then cooled to a temperature around 35°C. In modern heat exchanger, the incoming cold milk is passed countercurrent to the hot pasteurized milk to conserve the energy (Knight et al., 2004). *S. thermophilus* can escape pasteurization process and attach to the cooler side of heat exchange plates leading to the formation of biofilm (Flint et al., 1999; de Jong et al., 2002). These have detrimental effect if their number exceeds 105 ml^{-1} in the pasteurized milk (Hup et al., 1980; Bouman et al., 1982). These biofilms are major threat to food industries as a source of contamination, so either the surface properties of the exchanger are altered or the surface area is reduced to prevent contamination (Notermans et al., 1991; Carpentier and Cerf, 1993; Zottola and Sasahara, 1994; Wong and Cerf, 1995; Kumar and Anand, 1998).

5.3.2 STERILIZATION

Sterilization is a process which involves the heating around 120°C for 10–20 min to remove the microorganisms completely. It is widely used for concentrated milk and ready-to-drink beverages with high protein contents.

Gram-positive bacteria of Firmicutes are sporeformers which are resistant to chemicals (Russell, 1990), pH changes (Blocher and Busta, 1983), heat, osmotic shock, and ultraviolet light penetration (Roberts and Hitchins, 1969) and germinate into vegetative cells upon arrival of favorable conditions (Russell, 1990). These are usually found in silage (Vissers et al., 2007a,b,c), soil (Barash et al., 2010), forage, animal feces (Princewell and Agba, 1982) and can be removed only by ultra-high temperature (UHT) and commercial sterilization (Cox, 1975). On the other hand, mild heat treatments like thermization activate spore germination (Griffiths et al., 1988; Hanson et al., 2005) and make the bacteria resistant to even severe heat exposure which is known as "Thermoduric" bacteria (Gleeson et al., 2013). *S.* expresses a number of virulence factors like staphylococcal enterotoxins (SEs), SEs-like, exfoliative toxin A and B, and toxic shock syndrome toxin-1 genes which are responsible for causing food poisoning (Argudin et al., 2010). *S. aureus* crosscontamination could be avoided by sterilization. The botulism outbreak in the United Kingdom in 1989 was due to insufficient heat sterilization of yoghurt (Critchley et al., 1989).

5.3.3 DEHYDRATION

Dehydration refers to removal of water content by the application of heat. It lowers the water activity (a_w) levels which inhibits the growth of microorganisms. Drying is a widely used unit operation in the dairy industries. The survival of bacteria in spray dried product is lower than in freeze-dried product (Chavez and Ledeboer, 2007; Santivarangkna et al., 2008). The major factors responsible are the inlet and outlet air temperatures during spray drying. When microorganisms are exposed to sublethal temperature, they develop resistance, and the phenomena is known as heat shock response which has been observed in *Salmonella typhimurium*, *Vibrio parahaemolyticus*, *L. monocytogenes*, and *E. coli* O157:H7 (Bunning et al., 1990; Farber and Brown, 1990; Mackey and Derrick, 1990; Linton et al., 1992; Murano and Pierson, 1992; Jorgensen et al., 1999; Chang et al., 2004; Lin and Chou, 2004). *Cronobacter sakazakii* which is a major cause for foodborne illness has been obtained in milk, powdered milk, and cheese products (Leclercq et al., 2002; Forsythe, 2005; Restaino et al., 2006; Shaker et al., 2008). Heat shock enhances the viability of *C. sakazakii* during spray–drying and freeze–drying; hence, hygienic standards should be implemented to prevent contamination.

5.3.4 USE OF ANTIMICROBIAL AGENTS

In recent days, the consumers are aware of the food safety and quality issues because many incidences of foodborne illnesses have been reported across the globe. Hence, food industries make use of many different antimicrobial agents to inhibit the growth of fungus and bacteria. In this regard, the extract of mango kernel has been found to be effective against a wide range of foodborne pathogenic bacteria. When 3000 ppm of methanolic extract of mango seed kernel was added to raw cow milk, it inhibited the total bacterial count and coliform growth for 6 h of incubation at room temperature (Abdalla et al., 2007). It also enhances the oxidative stability of fresh cheese and ghee (Dinesh et al., 2000; Parmar and Sharma, 1990; Puravankara et al., 2000). The lactoferrin present in milk is effective against a wide range of bacteria and viruses (Leonnerdal, 2011). The lactoferrin hydrolysate inhibited the growth of *E. coli* O157:H7 and *L. monocytogenes* in UHT milk (Murdock and Matthews, 2002) and pathogenic bacteria in a dairy product such as rehydrated powder infant milk formula (Shimazaki, 2000). Chitosan has been reported to inhibit the growth of *Salmonella* and *Staphylococcus* spp. in raw milk (Tsai et al., 2000). Lysozyme naturally present in avian eggs and mammalian milk is generally recognized as safe (GRAS) for direct addition to foods (FDA, 1998). The white lysozyme of hen eggs is a bacteriolytic enzyme used as a preservative for milk and dairy products as well as for fruits and vegetables, fish and fish products, and meat and meat products (Cegielska-Radziejewska et al., 2009). Similarly, milk-derived bioactive compounds like casein and whey proteins have antimicrobial properties (Phelan et al., 2009; Schanbacher et al., 1998). The antimicrobial activity of reuterin along with bacteriocins is effective against foodborne pathogens in milk stored at low temperature (Arques et al., 2011). Reuterin has been used to control Gram-positive and Gram-negative pathogens in milk and milk products (El-Ziney et al., 1999; Arques et al., 2011). The use of edible oil from cinnamon bark, cinnamon leaf, and clove could be used as natural antimicrobials in milk-based drinks (Cava et al., 2007). About 1% clove, cinnamon, thyme, and bay oil inhibits the growth of *L. monocytogenes* and *S. enteritidis* in both low fat and full fat cheese (Smith et al., 2010). These edible oils can be nanoencapsulated into different food matrices as antimicrobial delivery systems for protection against foodborne pathogens (Zou et al., 2012). The reduction of *L. monocytogenes* has been reported by nanoencapsulation of thymol in milk (Xue et al., 2013).

5.3.5 USE OF GRAS SUBSTANCES

GRAS substances are those chemical substances that are supposed to be safe when added to food products by an American Food and Drug Administration.

5.3.5.1 ACT AS A FOOD ADDITIVE

According to Codex Alimentarius, a food additive is "any substance not normally consumed as a food itself and not normally used as a typical ingredient of the food, whether or not it has nutritive value, the intentional addition of which to food for a technological (including organoleptic) purpose in the manufacture, processing, preparation treatment, packing, packaging, transport, or holding of such food results, or may be reasonably expected to result (directly or indirectly) in it or its by-products becoming a component of or otherwise affecting the characteristics of such foods." The additives are added to different food stuffs to maintain quality by reducing spoilage (Table 5.1). The food additives are essential for food industry to meet the increasingly challenging market and legal demands (Saltmarsh, 2013). Nowadays, consumers are very much aware of food safety, and food additives are among the most controversial one (Aoki et al., 2010).

TABLE 5.1 Food Additives in Milk and Milk Products.

Additive	Product	GRAS/E no. status	Reference
Antioxidants			
β-Carotene	Dairy products	GRAS/E160a	Smith and Hong-Shum (2011)
Tocopherols	Dairy products	GRAS/E306	Smith and Hong-Shum (2011)
Antimicrobials			
Essential oils	Dairy products	Not GRAS/No E number	Burt (2004)
Lactoferrin	Baby formulas	Not GRAS/No E number	Giteru et al. (2015)
Lactoperoxidase	Milk	Not GRAS/No E number	Batt and Tortorello (2014)
Natamycin	Gorgonzola cheese	Not GRAS/E235	Oliveira et al. (2007)
Nisin	Dairy products	Not GRAS/E234	Lacroix (2011)
Pediocin	Dairy products	Not GRAS/No E number	Lacroix (2011)
Poly-L-lysine	Custard creams	Not GRAS/No E number	Baines and Seal (2012)

TABLE 5.1 *(Continued)*

Additive	Product	GRAS/E no. status	Reference
Reuterin	Cheese, cottage cheese	Not GRAS/No E number	Langa et al. (2013); Gomez-Torres et al. (2014)
Colorants			
Annatto	Dairy products	Approved/E160b	Scotter (2009)
Carminic acid	Dairy products	Approved/E120	MacDougall (2002)
Carotenoids	Milk	Not GRAS/No E number	Baines and Seal (2012)
Chlorophylls	Dairy products	Not GRAS/E140	MacDougall (2002)
Curcumin	Yoghurt, Dairy products, Ice creams	Approved/E100	Hendry and Houghton (1996); MacDougall (2002)
Sweeteners			
Erythritol	Fermented milk	Not GRAS/E968	Baines and Seal (2012)
Steviol glycosides	Dairy products, ice cream, Frozen desserts	Not GRAS/E960	O'Brien-Nabors (2001); Baines and Seal (2012)
Tagatose	Yoghurts, frostings, ice cream	Not GRAS/No E number	O'Brien-Nabors (2001); Dobbs and Bell (2010); Baines and Seal (2012)

5.3.5.2 ACT AS A FOOD STABILIZERS AND EMULSIFIERS

Food is a complex and heterogeneous system containing many different chemical types and species. There occurs a complex interaction among proteins, lipids, carbohydrates, and electrolytes in food matrices like milk, cream, yoghurt, cheese, and beverages. So stabilizers play a major role to make interaction well balanced and overall stable system (Samant et al., 2007). Different compounds as shown in Table 5.2 are used as stabilizers in specific food matrix. Even edible hydrocolloids are used as thickeners, stabilizer, gelling agents, syneresis control, emulsifiers, or suspension stability in the food industries (Lucey, 2002; Nikoofar et al., 2013), and they also determine flavor and aromas release (Zhao et al., 2009). The most widely used polysaccharide emulsifiers in food applications are gum arabic, modified starches, modified celluloses, pectin, and some galactomannans (Dickinson, 2003; Garti and Reichman, 1993).

TABLE 5.2   Use of Stabilizer in Different Food Matrix.

Stabilizer	Food matrix	Purpose	Reference
Exopolysaccharides	Yoghurt, cheese, fermented cream milk based desserts	Firmness and creaminess	Duboc and Mollet (2001)
Gelatin	Corn–milk yoghurt	Firmness	Supavititpatana et al. (2008)
Guar gum	Yoghurt	Firmness	Hassan et al. (2015)
Locust bean gum	Ice cream	Viscousness	Barak and Mudgil (2014)
Pectin	Yoghurt	Compactness	Everett and McLeod (2005)
Xanthan	Yoghurt	Firmness	Everett and McLeod (2005)
β-Lactoglobulin	Dressings	Creaminess	Christiansen et al. (2004)
κ-Carrageenan	Vanilla ice cream	Reduction of hardness and iciness	Parvar et al. (2013)

5.4   FOOD SAFETY AND HEALTH ISSUES OF MICROBIAL SPOILAGE

To improve the quality and safety of food products, the concept of predictive microbial modeling among food manufacturers (Buchanan, 1993; McMeekin et al., 2002) and regulatory agencies (Manfreda and De Cesare, 2014) has arisen through which one can easily predict the behavior of microorganisms during processing, storage, and distribution (McMeekin et al., 2002; Nauta, 2002). It can further be helpful during Hazard Analysis Critical Control Point planning, process design, and product reformulation (Nauta, 2002; McMeekin et al., 2006; Halder et al., 2010). *L. monocytogenes* is the causal agent of human and animal listeriosis after ingestion of contaminated food. Several outbreaks of listeriosis and sporadic cases have been reported in North America and Europe due to intake of dairy products. In European countries, the annual incidence of reported that listeriosis cases varies between 0.3 and 7.5 cases/million inhabitants (Swaminathan and Gerner-Smidt, 2007). Hence, it is challenging for food safety issue in the dairy industry. Although pasteurization is supposed to reduce the foodborne pathogens in milk and dairy foods cheeses being an exception where there is no heat treatment and the psycrothrophic bacteria like *L. monocytogenes* easily contaminate it (Gould et al., 2014; Oliver et al., 2005). The control measures for *L. monocytogenes* in the dairy industry are a continuous and never-ending task, suggesting for permanent and diligent vigilance, monitoring, and corrective action. Also a deeper knowledge about the potential hazards and its control measures is required in the dairy industry to ensure food safety.

5.5   CONCLUSION

Milk and milk products have become integral parts in human diet as a source of essential nutrients. There are different hurdles in the way of processing of these products in the industry to the consumption by consumers. All the risk factors need to be taken care of to guarantee the food safety and health hazards and possible foodborne outbreaks so that pure and healthy products are accessible to every individual. Besides this, an intense research is required in the field of dairying and animal husbandry for proper health of animals and human beings.

KEYWORDS

- Aflatoxin
- antimicrobial
- bacteria
- contamination
- dairy
- dairy product
- diarrhea
- feed material
- fermentation
- flavor
- food additive
- food borne
- food poisoning
- food safety
- fungi
- hazard

- health
- health issues
- lactic acid
- microbial spoilage
- microbiological contamination
- microorganism
- milk
- milk product
- organoleptic
- pasteurization
- pathogen
- pathogenic
- poisoning
- potential solution
- processing
- quality
- spoilage
- stabilizers
- sterilization
- toxins

REFERENCES

Abdalla, A. E.; Darwish, S. M.; Ayad, E. H.; El-Hamahmy, R. M. Egyptian Mango By-product 2: Antioxidant and Antimicrobial Activities of Extract and Oil from Mango Seed Kernel. *Food Chem.* **2007**, *103*(4), 1141–1152.

Al-Kadamany, E.; Toufeili, I.; Khattar, M.; Abou-Jawdeh, Y.; Harakesh, S.; Haddad, T. Determination of Shelf Life of Concentrated Yogurt (labneh) Produced by In-bag Straining of Set Yogurt Using Hazard Analysis. *J. Dairy Sci.* **2002**, *85*, 1023–1030.

Aoki, K.; Shen, J.; Saijo, T. Consumer Reaction to Information on Food Additives: Evidence from an Eating Experiment and a field Survey. *J. Econ. Behav. Org.* **2010**, *73*, 433–438.

Arakawa, K.; Kawai, Y.; Iioka, H.; Tanioka, M.; Nishimura, J.; Kitazawa, H.; Tsurumi, K.; Saito, T. Effects of Gassericins A and T, Bacteriocins Produced by *Lactobacillus gasseri*, with Glycine on Custard Cream Preservation. *J. Dairy Sci.* **2008**, *92*, 2365–2372.

Argudin, M. A.; Mendoza, M. C.; Rodicio, M. R. Food Poisoning and *Staphylococcus aureus* Enterotoxins. *Toxins (Basel).* **2010**, *2*, 1751–1773.

Arques, J. L.; Rodríguez, E.; Nunez, M.; Medina, M. Combined Effect of Reuterin and Lactic Acid Bacteria Bacteriocins on the Inactivation of Food-borne Pathogens in Milk. *Food Control.* **2011**, *22*(3), 457–461.

Aziz, T. Thermal Processing of Dahi to Improve its Keeping Quality. *Indian J. Nutr. Diet.* **1985**, *22*, 80–87.

Baines, D.; Seal, R. *Natural Food Additives, Ingredients and Flavourings.* Woodhead Publishing: Cambridge, UK, 2012.

Barak, S.; Mudgil, D. Locust Bean Gum: Processing, Properties and Food Applications—A Review. *Int. J. Biol. Macromol.* **2014**, *66*, 74–80.

Barash, J. R.; Hsia, J. K.; Arnon, S. S. Presence of Soil-Dwelling Clostridia in Commercial Powdered Infant Formulas. *J. Pediatr.* **2010**, *156*(3), 402–408.

Baruzzi, F.; Lagonigro, R.; Quintieri, L.; Morea, M.; Caputo, L. Occurrence of Non-Lactic Acid Bacteria Populations Involved in Protein Hydrolysis of Coldstored High Moisture Mozzarella Cheese. *Food Microbiol.* **2012**, *30*, 37–44.

Batt, C. A.; Tortorello, M. *Encyclopedia of Food Microbiology* (2nd ed.). Elsevier: London, UK, 2014.

Bennett, S. D.; Walsh, K. A.; Gould, L. H. Foodborne Disease Outbreaks Caused by *Bacillus cereus*, *Clostridium perfringens*, and *Stayphylococcus aureus* – United States, 1998–2008. *Clin. Infect. Dis.* **2013**, *57*, 425–433.

Besser, R. E.; Lett, S. M.; Weber, L. T.; Doyle, M. P.; Barrett, T. J.; Wells, J. G.; Griffin, P. M. An Outbreak of Diarrhoea and Hemolytic Uremic Syndrome from *Escherichia coli* O157:H7 in Fresh-Pressed Apple Cider. *J. Am. Med. Assoc.* **1993**, *269*, 2217–2220.

Bilandzic, N.; Bozic, D.; Dokic, M.; Sedak, M.; Solomun Kolanovic, B.; Varenina, I. Assessment of Aflatoxin M1 Contamination in Milk of Four Dairy Species in Croatia. *Food Control.* **2014**, *43*, 18–21.

Blaser, M. J. How Safe is Our Food? Lessons from an Outbreak of Salmonellosis. *N. Engl. J. Med.* **1996**, *334*(20), 1324–1325.

Blocher, J.; Busta, F. Bacterial Spore Resistance to Acid. *Food Technology.* **1983**, 37.

Bottazzi, V.; Battistotti, B.; Dellaglio, F. Characterization of Streptococci present in Fontina cheese. *Scienza e Tecnica Lattiero–Casearia.* **1976** 27, 476–479.

Bouman, S.; Lund, D. B.; Driessen, F. M.; Schmidt, D. G. Growth of Thermoresistant *Streptococci* and Deposition of Milk Constituents on Plates of Heat Exchangers during Long Operating Times. *J. Food Prot.* **1982**, *45*, 806–812.

Brayden, W. L. Mycotoxin Contamination of the Feed Supply Chain: Implications for Animal Productivity and Feed Security. *Anim. Feed Sci. Technol.* **2012**, 173, 134–158.

Brennan, N. M.; Ward, A. C.; Beresford, T. P.; Fox, P. F.; Goodfellow, M.; Cogan, T. M. Biodiversity of the Bacterial Flora on the Surface of a Smear Cheese. *Appl. Environ. Microbiol.* **2002**, *68*, 820–830.

Buchanan, R. L. Predictive Food microbiology. *Trends Food Sci. Technol.* **1993**, *4*, 6–11.

Buening-Pfaue, H. Microbiological Evaluation of Baked Confectionery Products. *Lebensm u Gerich Chern.* **1978**, *32*, 14–21.

Bunning, V. K.; Crawford, R. G.; Tierney, J. T.; Peeler, J. T. Thermotolerance of Listeria Monocytogenes and *Salmonella typhimurium* after Sublethal Heat Shock. *Appl. Environ. Microbiol.* **1990**, *56*, 3216–3219.

Burt, S. Essential Oils: Their Antibacterial Properties and Potential Applications in Foods E A Review. *Int. J. Food Microbiol.* **2004**, *94*, 223–253.

Busscher, H. J.; Bruinsma, G.; van Weissenbruch, R.; Leunisse, C.; van der Mei, H. C.; Dijk, F. The Effect of Buttermilk Consumption on Biofilm Formation on Silicone

Rubber Voice Prostheses in an Artificial Throat. *Eur. Arch. Otorhinolaryngol.* **1998**, *255*, 410–413.

Cabrini, A.; Neviani, E. Pseudomonas as a Cause of the Bitter Flavour and Putrefied Smell on the Surface of Mozzarella Cheese. *Food Microbiol. Toxicol.* **1983**, *8*, 90–93.

Cantoni, C.; Bersani, C. Mozzarelle Blu: Cause Ed Ipotesi. *Ind. Aaliment.* **2010**, *49*, 27–30.

Cantoni, C.; Soncini, G.; Milesi, S.; Cocolin, L.; Iacumin, L. Colorazioni Anomale E Rigonfiamento Di Formaggi Fusi E Mozzarelle. *Ind. Aliment.* **2006**, *45*, 276–281.

Cantoni, C.; Soncini, G.; Milesi, S.; Cocolin, L.; Iacumin, L.; Comi, G. Additional Data about Some Defects of Cheeses: Discoloration and Blowing. *Ind. Aliment.* **2006**, *45*, 276–281.

Cantoni, C.; Stella, S.; Cozzi, M. Blue Colouring of Mozzarella Cheese. *Ind. Aliment.* **2003**, *42*, 840–843.

Cantoni, C.; Stella, S.; Cozzi, M.; Iacumin, L.; Comi, G. Blue Colour of Mozzarella Cheese. *Ind. Aliment.* **2003**, *42*, 840–843.

Carpentier, B.; Cerf, O. Biofilms and Their Consequences, with Particular Reference to Hygiene in the Food Industry. *J. Appl. Bacteriol.* **1993**, *75*, 499–511.

Cava, R.; Nowak, E.; Taboada, A.; Marin-Iniesta, F. Antimicrobial Activity of Clove and Cinnamon Essential Oils Against Listeria Monocytogenes in Pasteurized Milk. *J. Food Prot.* **2007**, *70*(12), 2757–2763.

Cegielska-Radziejewska, R.; Lesnierowski, G.; Kijowski, J. Antibacterial Activity of Hen Egg White Lysozyme Modified by Thermochemical Technique. *Eur. Food Res. Technol.* **2009**, *228*(5), 841–845.

Centers for Disease Control and Prevention. Outbreak of Listeria Monocytogenes Infections Associated with Pasteurized Milk from a Local Dairy–Massachusetts, 2007. *MMWR Morb. Mortal. Wkly. Rep.* **2008**, *57*, 1097–1100.

Centers for Disease Control and Prevention. Diarrheagenic *E. coli* (non shigatoxin producing *E. coli*). 2011. Available at http://www.cdc.gov/nczved/divisions/dfbmd/diseases/diarrheagenic ecoli/technical.html.

Champagne, C. P.; Laing, R. R.; Roy, D.; Mafu, A. A.; Griffiths, M. W. Psychrotrophs in Dairy Products: Their Effects and Their Control. *Crit. Rev. Food Sci. Nutr.* **1994**, *34*, 1–30.

Chang, C. M.; Chiang, M. L.; Chou, C. C. Response of Heat-Shocked Vibrio Parahaemolyticus to Subsequent Physical and Chemical Stresses. *J. Food Prot.* **2004**, *67*, 2183–2188.

Chavez, B. E.; Ledeboer, A. M. Drying of Probiotics: Optimization of Formulation and Process to Enhance Storage Survival. *Drying Technol.* **2007**, *25*(7–9), 1193–1201.

Christiansen, K. F.; Vegarud, G.; Langsrud, T.; Ellekjaer, M. R.; Bjorg Egelandsdal, B. Hydrolyzed Whey Proteins as Emulsifiers and Stabilizers in High-Pressure Processed Dressings. *Food Hydrocoll.* **2004**, *18*, 757–767.

Cocolin, L.; Aggio, D.; Manzano, M.; Cantoni, C.; Comi, G. An Application of PCR-DGGE Analysis to Profile the Yeast Populations in Raw Milk. *Int. Dairy J.* **2002**, *12*, 407–411.

Cocolin, L.; Innocente, N.; Biasutti, M.; Comi, G. The Late Blowing in Cheese: a New Molecular Approach Based on PCR and DGGE to Study the Microbial Ecology of the Alteration Process. *Int. J. Food Microbiol.* **2004**, *90*, 83–91.

Codex. CODEX STAN 243-2003: Standard for fermented milks (revised in 2010 ed.; Vol. 2012), 2003. http://www.fao.org/docrep/015/i2085e/i2085e00.pdf.

Cole, R. J.; Cox, R. H. The Aflatoxins. *Hand Book of Toxic Fungal Metabolites.* Academic Press: (London) LDT, 1981, pp 1–66.

Coppola, S.; Blaiotta, G.; Ercolini, D. Dairy Products. In *Molecular Techniques in the Microbial Ecology of Fermented Foods;* Cocolin, L., Ercolini, D., Eds.; Springer ScienceþBussiness Media: New York, 2008.

Corredig, M.; Roesch, R. R.; Dalgleish, D. G. Production of a Novel Ingredient from Butter-milk. *J Dairy Sci.* **2003**, *86*, 2744–2750.

Courtin, P.; Rul, F. Interactions between Microorganisms in a Simple Ecosystem: Yogurt Bacteria as a Study Model. *Lait.* **2004**, *84*, 125–134.

Coveney, H. M.; Fitzgerald, G. F.; Daly, C. A Study of the Microbiological Status of Irish Farmhouse Cheeses with Emphasis on Selected Pathogenic and Spoilage Microorganisms. *J. Appl. Bacteriol.* **1994**, *77*, 621–630.

Cowden, J. M.; Chisholm, D.; O'Mahony, M.; Lynch, D.; Mawer, S. L.; Spain, G. E. Two Outbreaks of *Salmonella enteritidis* Phage Type 4 Infection Associated with the Consump-tion of Fresh Shell–Egg Products. *Epidemiol. Infect.* **1989a**, *103*(1), 47–52.

Cowden, J. M.; Lynch, D.; Joseph, C. A.; O'Mahony, M.; Mawer, S. L.; Rowe, B. Case-Control Study of Infections with *Salmonella enteritidis* Phage Type 4 in England. *BMJ.* **1989b**, *299*(6702), 771–773.

Cox, W. Subject: Bitty Cream and Related Problems: Problems Associated with Bacterial Spores in Heat-Treated Milk and Dairy Products. *Int. J. Dairy Technol.* **1975**, *28*(2), 59–68.

Critchley, E.; Hayes, P.; Isaacs, P. Outbreak of Botulism in North West England and Wales, June, 1989. *Lancet.* **1989,** *334*(8667), 849–853.

Daniels, N. A.; MacKinnon, L.; Rowe, S. M.; Bean, N. H.; Griffin, P. M.; Mead, P. S. Foodborne Disease Outbreaks in United States schools. *Pediatr. Infect. Dis. J.* **2002**, *21*(7), 623–628.

De Buyser, M. L.; Dufour, B.; Maire, M.; Lafarge, V. Implication of Milk and Milk Products in Food-Borne Diseases in France and in Different Industrialised Countries. *Int. J. Food Microbiol.* **2001**, *67*(1–2), 1–17.

de Jong, P.; te Giffel, M. C.; Kiezebrink, E. A. Prediction of the Adherence, Growth and Release of Micro-Organisms in Production Chains. *Int. J. Food Microbiol.* **2002**, *74*, 13–25.

deWijk, R. A.; van Gemert, L. J.; Terpstra, M. E. J.; Wilkinson, C. L. Texture of Semi-solids: Sensory and Instrumental Measurements on Vanilla Custard Desserts. *Food Quality Pref.* **2003**, *14*, 305–317.

Decastelli, L.; Lai, J.; Gramaglia, M.; Monaco, A.; Nachtmann, C.; Oldano, F. Aflatoxins Occurrence in Milk and Feed in Northern Italy During 2004–2005. *Food Control.* **2007**, *18*, 1263–1266.

Di Pietro, S.; Haritchabalet, K.; Cantoni, G.; Iglesias, L.; Mancini, S.; Temperoni, A. Surveillance of Foodborne Diseases in the Province of Rio Negro, Argentina, 1993–2001. *Medicina (B Aires).* **2004**, *64*(2), 120–124.

Dickinson, E. Hydrocolloids at Interfaces and the Influence on the Properties of Dispersed Systems. *Food Hydrocoll.* **2003**, *17*, 25–39.

Dieu, B.; CuQ, J. Process for Producing a Sweet Custard Foodstuff With a Long Term Shelf Life Based on Milk and Eggs. US Patent 4,877,625, 1989.

Dinesh, P.; Boghra, V.; Sharma, R. Effect of Antioxidant Principles Isolated from Mango (*Mangifera indica* L.) Seed Kernels on Oxidative Stability of Ghee (butter fat). *J. Food Sci. Technol.* **2000**, *37*(1), 6–10.

Dobbs, C. M.; Bell, L. N. Storage Stability of Tagatose in Buffer Solutions of Various Compo-sitions. *Food Res. Int.* **2010**, *43*, 382–3869.

Dolci, P.; Barmaz, A.; Zenato, S.; Pramotton, R.; Alessandria, V.; Cocolin, L.; Rantsiou, K.; Ambrosoli, R. Maturing Dynamics of Surface Microflora in Fontina PDO Cheese Studied By Culture-Dependent and -Independent Methods. *J. Appl. Microbiol.* **2009**, *106*, 278–287.

Doyle, M. P.; Marth, E. H. Thermal Inactivation of Conidia from *Aspergillus flavus* and *Aspergillus parasiticus*. I. Effects of Moist Heat, Age of Conidia and Sporulation Medium. *J. Milk Food Technol.* **1975**, 678–682.

Driehuis, F.; Spanjer, M. C.; Schoten, J. M.; Giffel, M. C. Occurrence of Mycotoxins in Feedstuffs of Dairy Cows and Estimation of Total Dietary Intakes. *J Dairy Sci.* **2008**, *91*, 4261–4271.

Duarte, S. C.; Almeida, A. M.; Teixeira, A. S.; Pereira, A. L.; Falcao, A. C.; Pena, A. Aflatoxin M1 in Marketed Milk in Portugal: Assessment of Human and Animal Exposure. *Food Control.* **2013**, *30*, 411–417.

Duartea, S. C.; Almeida, A. M.; Teixeira, A. S.; Pereira, A. L.; Falcão, A. C.; Pena, A. Lino, C. M. Aflatoxin M1 in Marketed Milk in Portugal: Assessment of Human and Animal Exposure. *Food Control.* **2013**, *30*(2), 411–417.

Duboc, P.; Mollet, B. Applications of Exopolysaccharides in the Dairy Industry. *Int. Dairy J.* **2001**, *11*, 759–768.

Ehling-Schulz, M.; Fricker, M.; Scherer, S.; *Bacillus cereus*, the Causative Agent of an Emetic Type of Food-Borne Illness. *Mol. Nutr. Food Res.* **2004**, *48*, 479–487.

Ehling-Schulz, M.; Messelhäusser, U.; Granum, P. E. Bacillus Cereus in Milk and Dairy Production. In *Rapid Detection, Characterization and Enumeration of Food-borne Pathogens*; Hoorfar, J. Ed.; ASM Press: Washington, DC, 2011, pp. 275–289.

El-Ziney, M.; Van Den Tempel, T.; Debevere, J.; Jakobsen, M. Application of Reuterin Produced by *Lactobacillus reuteri* 12002 for Meat Decontamination and Preservation. *J. Food Prot.* **1999**, *62*(3), 257–261.

Everett, D. W.; McLeod, R. E. Interactions of Polysaccharide Stabilisers with Casein Aggregates in Stirred Skim-milk Yoghurt. *Int. Dairy J.* **2005**, *15*, 1175–1183.

Fallah, A. A. Aflatoxin M1 Contamination in Dairy Products Marketed in Iran During Winter and Summer. *Food Control.* **2010**, *21*, 1478–1481.

Farber, J. M.; Brown, B. E. Effect of Prior Heat Shock on the Heat Resistance of *Listeria monocytogenes* in meat. *Appl. Environ. Microbiol.* **1990**, *56*, 1584–1587.

Fauquant, J.; Beaucher, E.; Sinet, C.; Robert, B.; Lopez, C. Combination of Homogenization and Cross-Flow Microfiltration to Remove Microorganisms from Industrial Buttermilks With an Efficient Permeation of Proteins and Lipids. *Innov. Food Sci. Emerg. Technol.* **2014**, *21*, 131–141.

FDA. Direct Food Substances Affirmed as Generally Recognized as Safe: Egg White Lysozyme. *Fed. Regist.* **1998**, *63*(4), 12421–12426.

Ferron, P.; Michard, I. Distribution of *Listeria* spp. In Confectioners' Pastries from Western France: Comparison of Enrichment Methods. *Int. J. Food Microbiol.* **1993**, *18*, 289–303.

Fink-Gremmels, J. Mycotoxins in Cattle Feeds and Carry-over to Dairy Milk: A Review. *Food Addit. Contam.* A. **2008**, *25*, 172–180.

Flint, S. H.; van der Elzen, H.; Brooks, J. D.; Bremer, P. J. Removal and Inactivation of Thermo-Resistant *Streptococci Colonising* Stainless Steel. *Int. Dairy J.* **1999**, *9*, 429–436.

Forsythe, S. J. *Enterobacter sakazakii* and Other Bacteria in Powdered Infant Milk Formula. *Mater. Child Nutr.* **2005**, *1*, 44–50.

Franzetti, L.; Scarpellini, M. Characterisation of *Pseudomonas* spp. Isolated from foods. *Ann. Microbiol.* **2007**, 57 (1), 39–47.

Garde, S.; Arias, R.; Gaya, P.; Nuñez, M. Occurrence of *Clostridium* spp. in Ovine Milk and Manchego Cheese with Late Blowing Defect: Identification and Characterization of Isolates. *Int. Dairy J.* **2011**, *21*, 272–278.

Garde, S.; Avila, M.; Gaya, P.; Arias, R.; Nunez, M. Sugars and Organic Acids in Raw and Pasteurized Milk Manchego Cheeses with Different Degrees of Late Blowing Defect. *Int. Dairy J.* **2012**, *25*, 87–91.

Garrote, G. I.; Abraham, A. G.; De Antoni, G. L. Inhibitory Power of Kefir: the Role of Organic Acids. *J. Food Prot.* **2000**, *63*, 364–369.

Garti, N.; Reichman, D. Hydrocolloids as Food Emulsifiers and Stabilizers. *Food Microstruct.* **1993**, *12*, 411–426.

Ghose, D. N.; Chattoraj, S. B. *Staphylococcal* Infection from Dahi. *Indian J. Public Health.* **1963**, *7*, 1–4.

Giteru, S. G.; Coorey, R.; Bertolatti, D.; Watkin, E.; Johnson, S.; Fang, Z. Physicochemical and Antimicrobial Properties of Citral and Quercetin Incorporated Kafirin-Based Bioactive Films. *Food Chem.* **2015**, *168*, 341–347.

Gleeson, D.; O'Connell, A.; Jordan, K. Review of Potential Sources and Control of Thermoduric Bacteria in Bulk-Tank Milk. *Ir. J. Agric. Food Resarch.* **2013**, *52*(2), 217–227.

Gomez-Torres, N.; Avila, M.; Gaya, P.; Garde, S. Prevention of Late Blowing Defect by Reuterin Produced in Cheese by a *Lactobacillus reuteri* Adjunct. *Food Microbiol.* **2014**, *42*, 82–88.

Gould, L. H.; Mungai, E.; Behravesh, C. B. Outbreaks Attributed to Cheese: Differences Between Outbreaks Caused by Unpasteurized and Pasteurized Dairy Products, United States, 1998–2011. *Foodborne Pathog. Dis.* **2014**, *11*, 545–551.

Griffiths, M.; Phillips, J.; West, I.; Muir, D. The Effect of Extended Low-Temperature Storage of Raw Milk on the Quality of Pasteurized and UHT Milk. *Food Microbiol.* **1988**, *5*(2), 75–87.

Gulmez, M.; Guven, A. Survival of *Escherichia coli* 0157:H7, *Listeria monocytogenes* 4b and *Yersinia enterocolitica* O3 in Different Yogurt and Kefir Combinations as Pre Fermentation Contaminant. *J. Appl. Microbiol.* **2003**, *95*, 631–636.

Halder, A.; Black, D. G.; Davidson, P. M.; Datta, A. Development of Associations and Kinetic Models for Microbiological Data to be Used in Comprehensive Food Safety Prediction Software. *J. Food Sci.* **2010**, *75*, R107–R120.

Hanson, M.; Wendorff, W.; Houck, K. Effect of Heat Treatment of Milk on Activation of *Bacillus spores*. *J. Food Prot.* **2005**, *68*(7), 1484–1486.

Hantsis-Zacharov, E.; Halpern, M. *Culturable psychrotrophic* Bacterial Communities in Raw Milk and Their Proteolytic and Lipolytic Traits. *Appl. Environ. Microbiol.* **2007**, *73*, 7162–7168.

Hassan, A. N.; Frank, J. F. Microorganisms Associated with Milk. In *Encyclopedia of Dairy Sciences* (2nd ed.); Fuquay, J. W.; Fox, P. F; McSweeney, P. L. H, Eds.; Academic Press: San Diego, CA, USA, 2011, pp. 447–457.

Hassan, L. K.; Haggag, H. F.; ElKalyoubi, M. H.; EL-Aziz, M.; El-Sayed, M. M.; Sayed, A. F. Physico-Chemical Properties of Yoghurt Containing Cress Seed Mucilage or Guar Gum. *Ann. Agric. Sci.* **2015**, *60*(1), 21–28.

Hendry, G. A. F.; Houghton, J. D. Natural Food Colorants. Springer Science Business Media: London, UK, 1996.

Hennekinne, J. A.; De Buyser, M. L.; Dragacci, S. *Staphylococcus aureus* and its Food Poisoning Toxins: Characterization and Outbreak Investigation. *FEMS Microbiol. Rev.* **2012**, *36*(4), 815–836.

Hennessy, T. W.; Hedberg, C. W.; Slutsker, L.; White, K. E.; Besser-Wiek, J. M.; Moen, M. E. A National Outbreak of *Salmonella enteritidis* Infections from Ice Cream. The Investigation Team. *N. Engl. J. Med.* **1996**, *334*(20), 1281–1286.

Hup, G.; Bangma, A.; Stadhouders, J.; Bouman, S. Growth of Thermoresistant *Streptococci* In Cheese–Milk Pasteurisers: I. Some Observations in Cheese Factories. *North Eur. Dairy J.* **1980**, *46*, 245–251.

Irlinger, F.; Mounier, J. Microbial Interactions in Cheese: Implications for Cheese Quality and Safety. *Curr. Opin. Biotechnol.* **2009**, *20*, 142–148.

Jay, J. M. *Modern Food Microbiology.* Van Nostrand Reinhold: New York, USA, 1992.

Jorgensen, F.; Hansen, T. B.; Knochel, S. Heat Shock Induced Thermotolerance in *Listeria monocytogenes* 13–249 is Dependent on Growth Phase, pH and Lactic Acid. *Food Microbiol.* **1999**, *16*, 185–194.

Kim, B. Y.; Lee, J. Y.; Ha, S. D. Growth Characteristics and Development of a Predictive Model for Bacillus Cereus in Fresh Wet Noodles with Added Ethanol and Thiamine. *J. Food Prot.* **2011**, *74*, 658–664.

Klijn, N.; Nieuwenhof, F. F. J.; Hollwerf, J. D.; Vanderwaals, C. B.; Weerkamp, A. H. Identification of *Clostridium tyrobutyricum* as the Causative Agent of Late Blowing in Cheese by Species-Specific PCR Amplification. *Appl. Environ. Microbiol.* **1995**, *61*, 2919–2924.

Knight, G. C.; Nicol, R. S.; McMeekin, T. A. Temperature Step Changes: A Novel Approach To Control Biofilms of *Streptococcus thermophilus* in a Pilot Plant-Scale Cheese–Milk Pasteurisation Plant. *Int. J. Food Microbiol.* **2004**, *93*, 305–318.

Kober, A. K. M. H.; Mannan, M. A.; Debnath, G. K.; Akter, S. Quality Evaluation of Dahi/ Yogurt Sold in Chittagong Metropolitan Area in Bangladesh. *Int. J. Sustainable Agric. Technol.* **2007**, *3*, 41–46.

Kumar, C. G.; Anand, S. K. Significance of Microbial Biofilms in Food Industry: A Review. *Int. J. Food Microbiol.* **1998**, *42*, 9–27.

Lacroix, C. *Protective Cultures, Antimicrobial Metabolites and Bacteriophages for Food and Beverage Biopreservation.* Woodhead Publishing: Cambridge, UK, 2011.

Langa, S.; Landete, J. M.; Martin-Cabrejas, I.; Rodriguez, E.; Arques, J. L.; Medina, M. In situ Reuterin Production by *Lactobacillus reuteri* in Dairy Products. *Food Control.* **2013**, *33*, 200–206.

Le Bourhis, A. G.; Dore, J.; Carlier, J. P.; Chamba, J. F.; Popoff, M. R.; Tholozan, J. L. Contribution of *C. beijerinckii* and *C. sporogenes* in Association with *C. tyrobutyricum* to the Butyric Fermentation in Emmental Type Cheese. *Int. J. Food Microbiol.* **2007**, 113, 154–163.

Le Bourhis, A. G.; Saunier, K.; Dore, J.; Carlier, J. P.; Chamba, J. F.; Popoff, M. R. Development and Validation of PCR Primers to Assess the Diversity of *Clostridium* spp. in Cheese by Temporal Temperature Gradient Gel Electrophoresis. *Appl. Environ. Microbiol.* **2005**, *7*, 29–38.

Le Loir, Y.; Baron, F.; Gautier, M. *Staphylococcus aureus* and Food Poisoning. *Genet. Mol. Res.* **2003**, *2*, 63–76.

Leclercq, A.; Wanegue, C.; Baylac, P. Comparison of Fecal Coliform Agar and Violet Red Bile Lactose Agar for Fecal Coliform Enumeration in Foods. *Appl. Environ. Microbiol.* **2002**, *68*, 1631–1638.

Leonnerdal, B. Biological Effects of Novel Bovine Milk Fractions. Nestle Nutrition Workshop Series. *Paediatr. Prog.* **2011**, *67*, 41–54.

Lin, Y. D.; Chou, C. C. Effect of Heat Shock on Thermal Tolerance and Susceptibility of *Listeria monocytogenes* to Other Environmental Stresses. *Food Microbiol.* **2004**, *21*, 605–610.

Linton, R. H.,Webster, J. B.; Pierson, M. D.; Hanckney, C. R. The Effect of Sublethal Heat Shock and Growth Atmosphere on the Heat Resistance of *Listeria monocytogenes* Scott A. *J. Food Prot.* **1992**, *55*, 84–87.

Lucey, J. A. Formation and Physical Properties of Milk Protein Gels. *J Dairy Sci.* **2002**, *85*, 281–294.

MacDougall, D. B. *Colour in Food*. Woodhead Publishing Limited: Cambridge, UK, 2002.

Mackey, B. M.; Derrick, C. Heat Shock Protein Synthesis and Thermotolerance in *Salmonella typhimurium*. *J. Appl. Bacteriol.* **1990**, *69*, 373–383.

Manfreda, G.; De Cesare, A. The Challenge of Defining Risk-Based Metrics to Improve Food Safety: Inputs from the BASELINE Project. *Int. J. Food Microbiol.* **2014**, 184, 2–7.

Massa, S.; Altieri, C.; Quaranta, V.; De Pace, R. Survival of *Escherichia coli* O157:H7 in Yoghurt During Preparation and Storage at 4 1C. *Lett. Appl. Microbiol.* **1997**, *24*, 347–350.

Massa, S.; Gardini, F.; Sinigaglia, M.; Guerzoni, M. E. *Klebsiella pneumoniae* as a Spoilage Organism in Mozzarella Cheese. *J Dairy Sci.* **1992**, *75*, 1411–1414.

Mauro, A.; Delia, S.; Laganà, P. Study about the Microbiological Variations of Refrigerated Mozzarella Cheese Samples. *Industrie Alimentari.* **2005**, *44*(443), 18–21.

Mayenobe, D.; Didienne, R.; Pradel, G. Caracterisation des gonflements tardifs dans les fromages de St-Nectaire et certaines pates pressees. *Lait.* **1983**, *63*, 15–24.

McMeekin, T. A.; Baranyi, J.; Bowman, J.; Dalgaard, P.; Kirk, M.; Ross, T.; Schmid, S.; Zwietering, M. H. Information Systems in Food Safety Management. *Int. J. Food Microbiol.* **2006**, 112, 181–194.

McMeekin, T. A.; Olley, J.; Ratkowsky, D. A.; Ross, T. Predictive Microbiology: Towards the Interface and Beyond. *Int. J. Food Microbiol.* **2002**, *73*, 395–407.

Mead, P. S.; Slutsker, L.; Dietz, V.; McCaig, L. F.; Bresse, J. S.; Shapiro, C. Food-Related Illness and Death in the United States. *Emerg. Infect. Dis.* **1999**, *5*, 607–625.

Meyrand, A.; Boutrand-Loei, S.; Ray-Gueniot, S.; Mazuy, C.; Gaspard, C. E.; Jaubert, G. Growth and Enterotoxin Production of *Staphylococcus aureus* During the Manufacture and Ripening of Camembert-Type Cheeses From Goats' Milk. *J. Appl. Microbiol.* **1998**, *85*, 537–544.

Michard, J.; Jardy, N.; Gey, I. L. Coliformes and coliformes fecaux dans Ie patisseries. *Microbiol. Aliment Nutr.* **1986**, *4*, 205–216.

Minguella, J. A. Environmental Contamination of Biological Origin in Patisserie Products. *Pasticceria Internazionale.* **1981**, *4*, 94–97.

Montel, M. C.; Buchin, S.; Mallet, A.; Delbes-Paus, C.; Vuitton, D. A.; Desmasures, N.; Berthier, F. Traditional Cheeses: Rich and Diversemicrobiotawith Associated Benefits. *Int. J. Food Microbiol.* **2014**, *177C*, 136–154.

Morgan, D.; Newman, C. P.; Hutchinson, A. M.; Walker, M.; Rowe, B.; Maijd, F. Verotoxin-Producing *Escherichia coli* 0157 Infection Associated with Consumption of Yoghurt. *Epidemiol. Infect.* **1993**, 111, 181–187.

Mounier, J.; Gelsomino, R.; Goerges, S.; Vancanneyt, M.; Vandemeulebroecke, K.; Hoste, B.; Scherer, S.; Swings, J.; Fitzgerald, G. F.; Cogan, T. M. Surface Microflora of Four Smear-Ripened Cheeses. *Appl. Environ. Microbiol.* **2005**, *71*, 6489–6500.

Munsch-Alatossava, P.; Alatossava, T. Phenotypic Characterization of Raw Milk-Associated Psychrotrophic Bacteria. *Microbiol. Res.* **2006**, 161, 334–346.

Murano, E. A.; Pierson, M. D. Effect of Heat Shock and Growth Atmosphere on the Heat Resistance of *Escherichia coli* O157:H7. *J. Food Prot.* **1992**, *55*, 171–175.

Murdock, C.; Matthews, K. Antibacterial Activity of Pepsin-Digested Lactoferrin on Food-borne Pathogens in Buffered Broth Systems and Ultra-High Temperature Milk with EDTA. *J. Appl. Microbiol.* **2002**, *93*(5), 850–856.

Nauta, M. J. Modelling Bacterial Growth in Quantitative Microbiological Risk Assessment: Is It Possible? *Int. J. Food Microbiol.* **2002**, *73*, 297–304.

Newkirk, R.; Hedberg, C.; Bender, J. Establishing a Milkborne Disease Outbreak Profile: Potential Food Defense Implications. *Foodborne Pathog. Dis.* **2011**, *8*(3), 433–437.

Nikoofar, E.; Hojjatoleslami, M.; Shariaty, M. A. Surveying the Effect of Quince Seed Muci-lage as a Fat Replacer on Texture and Physicochemical Properties of Semi Fat Set Yoghurt. *Int. J. Farm. Allied Sci.* **2013**, *2*, 861–865.

Northolt, M. D.; Van Egmond, H. P.; Soentoro, P.; Deijll, E. Fungal Growth and the Presence of Sterigmatocystin in Hard Cheese. *J. Assoc. Off. Anal. Chem.* **1980**, *63*(1), 115.

Notermans, S.; Dormans, J. A. M. A.; Mead, G. C. Contribution of Surface Attachment to the Establishment of Microorganisms in Food Processing Plants: A Review. *Biofouling.* **1991** *5*, 21–36.

O'Brien-Nabors, L. *Alternative Sweeteners.* Marcel Dekker: New York, USA, 2001.

Oliveira, T. M.; Soares, N. F. F.; Pereira, R. M.; Fraga, K. F. Development and Evaluation of Antimicrobial Natamycin-Incorporated Film in Gorgonzola Cheese Conservation. *Packag. Technol. Sci.* **2007**, *20*, 147–153.

Oliver, S. P.; Jayarao, B. M.; Almeida, R. A. Foodborne Pathogens in Milk and the Dairy Farm Environment: Food Safety and Public Health Implications. *Foodborne Pathog. Dis.* **2005**, *2*, 115–129.

Parmar, S.; Sharma, R. Effect of Mango (*Mangifera indica* L.) Seed Kernels Pre-extract on the Oxidative Stability of Ghee. *Food Chem.* **1990**, *35*(2), 99–107.

Parvar, M. B.; Tehrani, M. M.; Razavi, S. M. A. Effects of a Novel Stabilizer Blend and Presence of κ-Carrageenan on Some Properties of Vanilla Ice Cream During Storage. *Food Biosci.* **2013**, *3*, 10–18.

Phelan, M.; Aherne, A.; FitzGerald, R. J.; O'Brien, N. M. Casein-Derived Bioactive Peptides: Biological Effects, Industrial Uses, Safety Aspects and Regulatory Status. *Int. Dairy J.* **2009**, *19*(11), 643–654.

Pietri, A.; Bertuzzi, T.; Bertuzzi, P.; Piva, G. Aflatoxin M1. Occurrence in Samples of Grana Padano. *Food Addit. Contam.* **1997**, *14*, 341–344.

Pitt, J. I.; Hocking, A. D. *Fungi and Food Spoilage* (3rd ed.). Springer: New York, 2009.

Prandini, A.; Tansini, G.; Sigolo, S.; Filippi, L.; Laporta, M.; Piva, G. On the Occurrence of Aflatoxin M1 in Milk and Dairy Products. *Food Chem. Toxicol.* **2009**, *47*, 984–991.

Princewell, T.; Agba, M. Examination of Bovine Faeces for the Isolation and Identification of *Clostridium* species. *J. Appl. Microbiol.* **1982**, *52*(1), 97–102.

Puravankara, D.; Boghra, V.; Sharma, R. S. Effect of Antioxidant Principles Isolated from Mango (*Mangifera indica* L) seed Kernels on Oxidative Stability of Buffalo Ghee (Butter-Fat). *J. Sci. Food Agric.* **2000**, *80*(4), 522–526.

Quigley, L.; O'Sullivan, O.; Stanton, C.; Beresford, T. P.; Ross, R. P.; Fitzgerald, G. F. The Complex Microbiota of Raw Milk. *FEMS Microbiol. Rev.* **2013**, *37*, 664–698.

Rahimi, E.; Bonyadian, M.; Rafei, M.; Kazemeini, H. R. Occurrence of Aflatoxin M1 in Raw Milk of Five Dairy Species in Ahvaz, Iran. *Food Chem. Toxicol.* **2010**, *48*, 129–131.

Restaino, L.; Frampton, E. W.; Lionberg, W. C.; Becker, R. J. A Chromogenic Plating Medium for the Isolation and Identification of *Enterobacter sakazakii* from Foods, Food Ingredi-ents, and Environmental Sources. *J. Food Prot.* **2006**, *69*, 315–322.

Roberts, T.; Hitchins, A. Resistance of Spores. In *The Bacterial Spore*; Gould, G. W., Hurst, A., Eds.; Academic Press: London, 1969, pp 611–670.

Rombaut, R.; Van Camp, J.; Dewettinck, K. Phospho- and Sphingolipid Distribution During Processing of Milk, Butter and Whey. *Int. J. Food Sci. Technol.* **2006**, *41*, 435–443.

Ruangwises, N.; Ruangwises, S. Aflatoxin M1 Contamination in Raw Milk Within the Central Region of Thailand. *Bull. Environ. Contam. Toxicol.* **2010**, *85*, 195–198.

Russell, A. Bacterial Spores and Chemical Sporicidal Agents. *Clin. Microbiol. Rev.* **1990**, *3*(2), 99–119.

SADHA (South Australian Department for Health and Ageing). The State of Public and Environmental Health Report for South Australia 2010e2011, 2012. Available at http://www.sahealth.sa.gov.au/wps/wcm/connect/a7bd2d004aafc0168918ad1be4847105/PublicEnvironHealthReport10-11-PHCS20120329.pdf?MOD¼AJPERES&CACHEID¼a7bd2d004aafc0168918ad1be4847105&CACHE¼NONE.

Saltmarsh, M. *Essential Guide to Food Additives* (4th ed.). RSC Publishing: Cambridge, UK, 2013.

Samant, S. K.; Singhal, R. S.; Kulkarni, P. R.; Rege, D. V. Protein-Polysaccharide Interactions: A New Approach in Food Formulations. *Int. J. Food Sci. Technol.* **2007**, *28*, 547–562.

Santivarangkna, C.; Kulozik, U.; Foerst, P. Inactivation Mechanisms of Lactic Acid Starter Cultures Preserved by Drying Processes. *J. Appl. Microbiol.* **2008**, *105*(1), 1–13.

Schanbacher, F.; Talhouk, R.; Murray, F.; Gherman, L.; Willett, L. Milk-borne Bioactive Peptides. *Int. Dairy J.* **1998**, *8*(5–6), 393–403.

Schornsteiner, E.; Mann, E.; Bereuter, O.; Wagner, M.; Stephan Schmitz-Esser, S. Cultivation-Independent Analysis of Microbial Communities on Austrian Raw Milk Hard Cheese Rinds. *Int. J. Food Microbiol.* **2014**, 180, 88–97.

Scotter, M. The Chemistry and Analysis of Annatto Food Colouring: A Review. *Food Addit. Contam.: A.* **2009**, *26*, 1123–1145.

Senini, L.; Cappa, F.; Cocconcelli, P. S. Use of rRNA-Targeted Oligonucleotide Probes for the Characterization of the Microflora from Fermentation of Fontina Cheese. *Food Microbiol.* **1997**, 14, 469–476.

Shaker, R. R.; Osaili, T. M.; Ayyash, M. Effect of Thermophilic Lactic Acid Bacteria on the Fate of Enterobacter Sakazakii During Processing and Storage of Plain Yogurt. *J. Food Saf.* **2008**, *28*, 170–182.

Shankaran, R.; Leela, R. K. Prevalence of Enterotoxigenic *Staphylococci* in Bakery Products. 1. *Food Prot.* **1983**, *46*, 95–97.

Sharp, J. C. Infections Associated with Milk and Dairy Products in Europe and North America, 1980–85. *Bull. World Health Org.* **1987**, *65*(3), 397–406.

Shimazaki, K. Lactoferrin: a Marvellous Protein in Milk? *Anim. Sci. J.* **2000**, *71*(4), 329–347.

Sibanda, L.; Saeger, S. D.; Peteghem, C. V. Development of a Portable Field Immunoassay for the Detection of Aflatoxin M1 in Milk. *Int. J. Food Microbiol.* **1999**, *48*, 203–209.

Smith, J.; Hong-Shum, L. *Food Additives Databook*. Blackwell Publishing: Oxford, UK, 2011.

Smith, V. J.; Desbois, A. P.; Dyrynda, E. A. Conventional and Unconventional Antimicrobials from Fish, Marine Invertebrates and Micro-Algae. *Mar. Drugs.* **2010**, *8*(4), 1213–1262.

Sodini, I.; Morin, P.; Olabi, A.; Jimenez-Flores, R. Compositional and Functional Properties of Buttermilk: A Comparison Between Sweet, Sour, and Whey Buttermilk. *J Dairy Sci.* **2006**, *89*, 525–536.

Soncini, G.; Marchisio, E.; Cantoni, C. Causes of Chromatic Alterations in Mozzarella Cheese. *Ind. Aliment.* **1998**, *37*, 850–855.

Stenfors Arnesen, L. P.; Fagerlund, A.; Granum, P. E. From Soil to Gut: *Bacillus cereus* and its Food Poisoning Toxins. *FEMS Microbiol. Rev.* **2008**, *32*, 579–606.

Supavititpatana, P.; Wirjantoro, T. I.; Arunee Apichartsrangkoon, A.; Raviyan, P. Addition of Gelatin Enhanced Gelation of Corn–Milk Yogurt. *Food Chem.* **2008**, 106, 211–216.

Swaminathan, B.; Gerner-Smidt, P. The Epidemiology of Human Listeriosis. *Microb. Infect.* **2007**, *9*, 1236–1243.

Tamine, A. Y.; Robinson, R. K. *Yoghurt. Science and Technology*. Pergamon Press: Oxford, UK, 1985.

Todds, E. C. D.; Jarvis, G. A.; Weiss, K. E, Riedel, G. W, Charbonneau, S. Microbiological Quality of Frozen Cream-Type Pies Sold in Canada. *J. Food Prot.* **1993**, *46*, 34–40.

Tsai, G. J.; Wu, Z. Y.; Su, W. H. Antibacterial Activity of a Chitooligosaccharide Mixture Prepared by Cellulase Digestion of Shrimp Chitosan and its Application to Milk Preservation. *J. Food Prot.* **2000**, *63*(6), 747–752.

Turconi, G.; Guarcello, M.; Livieri, C.; Comizzoli, S.; Maccarini, L.; Castellazzi, A. M. Evaluation of Xenobiotics in Human Milk and Ingestion by the Newborn E Epidemiological Survey in Lombardy (Northern Italy). *Eur. J. Nutr.* **2004**, *43*, 191–197.

Vissers, M.; Driehuis, F.; Te Giffel, M.; De Jong, P.; Lankveld, J. Concentrations of Butyric Acid Bacteria Spores in Silage and Relationships with Aerobic Deterioration. *J. Dairy Sci.* **2007a**, *90*(2), 928–936.

Vissers, M.; Driehuis, F.; Te Giffel, M.; De Jong, P.; Lankveld, J. Minimizing the Level of Butyric Acid Bacteria Spores in Farm Tank Milk. *J. Dairy Sci.* **2007b**, *90*(7), 3278–3285.

Vissers, M.; Te Giffel, M.; Driehuis, F.; De Jong, P.; Lankveld, J. Minimizing the Level of *Bacillus cereus* Spores in Farm Tank Milk. *J Dairy Sci.* **2007c**, *90*(7), 3286–3293.

Wang, M.; Ran, L.; Wang, Z.; Li, Z. Study on National Active Monitoring for Food Borne Pathogens and Antimicrobial Resistance in China 2001. *Wei Sheng Yan Jiu.* **2004**, *33*(1), 49–54.

Weagant, S. D.; Bryant, M. L.; Park, D. H. Survival of *Escherichia coli* O157:H7 in Mayonnaise and Mayonnaise-Based Sauces at Room and Refrigerated Temperatures. *J. Food Prot.* **1994**, *57*, 629–631.

Wong, A. C. L.; Cerf, O. Biofilms: Implications for Hygiene Monitoring of Dairy Plant Surfaces. *Bull. IDF.* **1995**, 302, 40–44.

Wong, H. C.; Chang, M. H.; Fan, J. Y. Incidence and Characterization of *Bacillus cereus* Isolates Contaminating Dairy Products. *Appl. Environ. Microbiol.* **1988**, *54*, 699–702.

Xue, J.; Michael Davidson, P.; Zhong, Q. Thymol Nanoemulsified by Whey Protein-Maltodextrin Conjugates: The Enhanced Emulsifying Capacity and Antilisterial Properties in Milk by Propylene Glycol. *J. Agric. Food Chem.* **2013**, *61*, 12720−12726.

Yadav, H.; Shalini, J.; Sinha, P. R. Effect of Dahi Containing *Lactococcus lactis* on the Progression of Diabetes Induced by a High Fructose Diet in Rats. *Biosci. Biotechnol. Biochem.* **2006**, *70*, 1255–1258.

Yadav, H.; Shalini, J.; Sinha, P. R. Formation of Oligosaccharides in Skim Milk Fermented with Mixed Dahi Cultures, *Lactococcus lactis* subsp. Diacetylactis and Probiotic Strains of Lactobacilli. *J. Dairy Res.* **2007**, *74*, 154–159.

Zhao, Q.; Zhao, M.; Yang, B.; Cui, C. Effect of Xanthan Gum on the Physical Properties and Textural Characteristics of Whipped Cream. *Food Chem.* **2009**, 116, 624–628.

Zottola, E. A.; Sasahara, K. C. Microbial Biofilms in the Food Industry—Should They be a Concern? *Int. J. Food Microbiol.* **1994**, *23*, 125–148.

Zou, Y.; Lee, H. Y.; Seo, Y. C.; Ahn, J. Enhanced Antimicrobial Activity of Nisin-Loaded Liposomal Nanoparticles Against Foodborne Pathogens. *J. Food Sci.* **2012**, *77*(3), M165–M170.

CHAPTER 6

LACTIC ACID BACTERIA (LAB) BACTERIOCINS: AN ECOLOGICAL AND SUSTAINABLE BIOPRESERVATIVE APPROACH TO IMPROVE THE SAFETY AND SHELF LIFE OF FOODS

AMI PATEL[1], NIHIR SHAH[1], and DEEPAK KUMAR VERMA[2*]

[1]*Division of Dairy and Food Microbiology, Mansinhbhai Institute of Dairy & Food Technology, MIDFT, Dudhsagar Dairy Campus, Mehsana 384002, Gujarat, India*

[2]*Department of Agricultural and Food Engineering, Indian Institute of Technology, Kharagpur 721302, West Bengal, India*

*Corresponding author. E-mail: deepak.verma@agfe.iitkgp.ernet.in; rajadkv@rediffmail.com

CONTENTS

ABSTRACT

This chapter deals with the application of bacteriocins as a biopreservative agent in various foods, such as milk and milk products; vegetable and fruit products; cereal and legume products; eggs, meat, and meat products; and fish and other seafood products to improve their safety and keeping quality. It also covers different aspects associated with bacteriocins such as classification, mode of action, sustainability in food products, and factors affecting biosynthesis and activity. The outcomes of various studies associated with the bacteriocin application, either used alone or in conjunction with other antimicrobials against spoilage bacteria and/or foodborne pathogens have been examined. It also highlights the combinations of bacteriocin with novel food packaging and food-processing methods, such as pulsed electric field, high-pressure homogenization, high hydrostatic pressure, irradiation, etc., and the last section talks about the regulatory standards and other factors challenged to use of bacteriocins in brief.

6.1　INTRODUCTION

Milk, vegetable, fruit, cereal, meat, and fish are nutritionally rich foods, consumed almost during every meal in various forms, that is, raw, processed or fermented products. To offer safe and quality foods with preferably longer shelf life, most of the dairy and food industries make use of chemical preservatives to protect such perishable food products from microbial attack. The awareness and need for chemical preservatives free foods had enforced dairy and food industry to incorporate compounds which are safe, natural, and have no side effects. Biopreservation is the method that utilizes natural or controlled microorganisms or their antimicrobial metabolites to preserve and enhance the shelf life of food. Such beneficial microorganisms or their fermentation end-products usually controls spoilage causing bacteria and render pathogens inactive in food products.

6.2　BIOPRESERVATIVE APPROACH FOR FOODS

Usually food products are preserved by making use of food grade salt, sugar, nitrite, sulphite, and organic acids like acetate, lactate, propionate, benzoate, and sorbic acid, singly or in combinations, together with or without heat treatment. Considerable economic loss together with higher morbidity rates in developed as well as developing nations has been reported for foodborne illnesses. This has brought the microbial community, especially bacteria, in focus; those have an inherent attribute to produce antimicrobial compounds and exert antagonistic activity against competing microflora. Such compounds mainly comprise organic acids, hydrogen peroxide (H_2O_2), antibiotics, lysozyme, bacteriocins, and antimicrobial peptides. Among these, bacteriocins from Gram-positive and Gram-negative bacteria have gained a great attention worldwide over the last few decades. Bacteriocins may be defined as low molecular weight peptides or proteins (usually having 30–60 amino acids) that are ribosomally synthesized, extracellulary released, and have a bactericidal or bacteriostatic effect on closely related species (Klaenhammer, 1993; Settanni and Corsetti, 2008). Bacteriocins are reported to degrade by proteases (like trypsin and pancreatin) present in gastrointestinal tract without affecting intestinal microflora which makes it safe for human consumption (Parada et al., 2007). Increasing scientific evidences for the isolation and characterization of bacteriocin producing strains as well as bacteriocins in last few years indicate higher concern and emphasis given to biological preservatives at industrial level. Further, bacteriocins have been widely explored to prove their potential efficacy for health applications of humans and animals in pharmaceuticals and biomedical laboratories (Dicks et al., 2011).

6.3 LACTIC ACID BACTERIA (LAB) BACTERIOCINS: AN ECOLOGICAL AND SUSTAINABLE BIOPRESERVATIVE

Lactic acid bacteria (LAB) have apparent significance in food fermentation, development of typical organoleptic taste and flavour within food matrixes, and governed GRAS (Generally Regarded as Safe) status. These food-grade bacteria produce diverse antimicrobial substances especially bacteriocins which enhance the microbiological food safety with subsequent reduction in the risk and growth of foodborne pathogens as well as spoilage microorganisms without altering sensory characteristics of final product. At commercial level, nisin is the first bacteriocin applied in various food products to prolong their keeping quality. It is synthesized by *Lactococcus lactis* subsp. *lactis* and has a wide spectrum activity against Gram-positive bacteria and sporeformers, but it cannot inhibit Gram-negative bacteria, yeast, and molds. After nisin, pediocin PA-1/AcH obtained from *Pediococcus* spp., lactacin 3147 from Lactococci and enterocin AS-48 from *Enterococcus faecalis* are most likely bacteriocin to be used as biopreservative in near times. Strains of LAB reported to produce diverse bacteriocins such as plantaricin, helveticin, acidophilin, sakacin, lactacin B, and bulgaricin which are yet to be utilized commercially. A nonprotein bacteriocin reuterin (chemically 3-hydroxy propionaldehyde) synthesized by *Lb. reuteri* also possess a broad inhibitory action against Gram-positive and Gram-negative bacteria, yeasts, and molds. Comparatively, it is much stable over a wide pH range. Propionicins is another antimicrobial peptide obtained from *Propionibacterium freudenreichii* subsp. *shermanii,* which revealed efficient inhibition of contaminating microflora.

Generally, bacteriocins are applied individually in majority of food products within the permissible concentration of specific country. However, a new concept has been developed in that bacteriocins are employed as an integral part of hurdle technology for food preservations. The term "hurdle" involves properties like storage at low temperature, processing at high temperature, low pH, water activity (a_w), and/or change in redox potential (Eh) together with the presence of biopreservatives or other preservative substances in food products. The intensity of the hurdle can be determined and controlled according to the spoilage microorganism(s) type and regulated as per consumer wellbeing and preference without compromising the quality and appearance of the final food product (Alasalvar, 2010; Yang et al., 2012). Furthermore, their use in combination with other chemical and physical methods like heat treatment, irradiation, or involving chemical agents efficiently controls spoilage bacteria and pathogens. The novel mild nonthermal techniques like high hydrostatic pressure (HHP), pulsed electric field (PEF), and vacuum or modified atmosphere packaging (MAP) increases the permeability of cytoplasmic membranes and thus, positively influences the action of numerous bacteriocins. Remarkably, collective treatments of bacteriocin with particular hurdles reported to affect the outer-membrane permeability of Gram-negative bacteria and increasing the efficiency of some LAB-derived bacteriocins (Galvez et al., 2007; Dortu and Thonart, 2009).

Several possible strategies used to apply bacteriocins in the food preservation may involve: (1) in situ production of bacteriocin within the food matrixes by making the use of bacteria either in form of starter culture or as protective cultures, (2) incorporation of bacteriocin preparation (purified or partially purified) in food products as biopreservative agent, and (3) use of bacteriocin containing substrate (or foodstuff) as an ingredient in food formulation, in which the product would have previously fermented with a bacteriocinogenic strain. Among these, a particular approach may be employed depending on the food product type, environmental conditions, as well as different intrinsic and extrinsic factors occurred while product handing, processing, storage, and distribution. In situ bacteriocin producing starter cultures find good application in fermented foods, although incorporation of fully or partially purified bacteriocins provide a more controllable preservative tool in refrigerated foods, such as vegetable salads or minimally processed meat products. An advance more effective approach is to incorporate the bacteriocin into the packaging material; such practice is termed as active packaging or antimicrobial packaging. The normal protective function of packaging materials is found to enhance through this technique. The efficacy of bacteriocins activity could be conserved up till consumer procure the food product or even beyond via control released of bacteriocin or any other suitable antimicrobial agent. Usually, packaging materials are made up of polymers like cellulose-based packaging or polypropylene in which bacteriocins is either coated or incorporated within the films.

6.3.1 CLASSIFICATION OF LAB BACTERIOCINS

During last few decades, bacteriocin research has progressed exponentially in the field of bacteriocin, and diverse kinds of antimicrobial peptides have been reported from various food grade bacteria. It is quite difficult to format the lasting natural classification scheme that covers all of the existing bacteriocins and antimicrobial peptides. Based on recent advances in bacteriocin research, several revised classification schemes has been proposed which mainly focused on reforming subclasses of class II, modification of classes III and IV (Cotter et al., 2005). However, the classification format is still under debate with new groups being proposed based on the huge number of new bacteriocins that have been extracted, identified, as well as characterized and the evolving definition for these antimicrobial peptides.

Recent bacteriocin classification schemes offer concise and appealing grouping of these proteins, but the diversity and insufficient structural information of several bacteriocins render these systems imperfect (Snyder and Worobo, 2013). Different heterogeneous bacteriocins have been classified into four classes viz., Class I bacteriocins, called lantibiotics (containing lanthionine groups) are small peptides (<5 kDa) and that undergo extensive posttranslational modifications and involves two additional subclasses (Figure 6.1); Class II bacteriocins are small (<10 kDa), heat-stable, nonlantibiotics which do not go through extensive post-translational modifications (Cotter et al., 2005; Perez et al., 2014) and further divided into four subclasses. The Class III bacteriocins are large in size (>30 kDa), heat-labile nonlantibiotics while other complex bacteriocins containing glycol and/or lipid moieties are kept under Class IV. Conversely, the Class III group bacteriocins have been proposed to be reclassified as bacteriolysins as they are lytic enzymes rather than peptides (Heng et al., 2007).

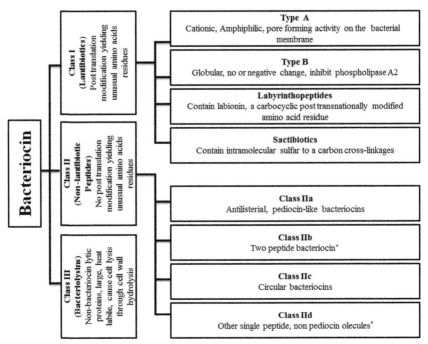

*Some of these are synthesized without leader peptide and could be included in a separate class of leaderless bacteriocins

FIGURE 6.1   Classification scheme of LAB-derived bacteriocins.

6.3.2   MECHANISM OF ACTION

Nisin possesses a wide spectrum inhibitory activity against Gram-positive bacteria and also found to inhibit germination of *Bacillus* sp. and *Clostridium* sp. spores (de Arauz et al., 2009). On the other hand, lactococcin A has a narrow spectrum of activity and inhibit other lactococci only. LAB derived bacteriocins antagonize sensitive bacteria through diverse and distinct mechanisms. The structure–function interactions have been established for only few bacteriocins. Most of the bacteriocins reported to target cell membrane, cell wall, enzymes, or nucleic acids of microbial cell (Snyder and Worobo, 2013). Initial electrostatic attraction between the antimicrobial peptide and target cell membrane is considered as the driving force for successive events.

The mechanism of action of nisin on target microbial cell has been studied in much detail compare to other bacteriocins. It can act as a surface active agent that imparts its binding to various cell components like fatty acids of phospholipids, and this trait favors the adsorption of bacteriocin to solid surfaces for destroying cells that adhere consequently. Nisin induces pore formation in the sensitive bacteria that interrupt the proton motive force as well as pH equilibrium leading leakage of ions, ATP hydrolysis, and finally results in cell death (Benz et al., 1991). Other lantibiotics such as subtilin, epidermin, Pep5, and lacticin 3147 also proposed to form pores in target cell (Brotz et al., 1998; Schuller et al., 1989).

In some cases, the interaction between a target cell and a bacteriocin may involve attractive but nonspecific forces, with a negatively charged bacterial cell membrane and positively charged bacteriocin; nevertheless, it is assumed that major bacteriocins bind to particular receptors on the surface of sensitive cell (Hassan et al., 2012). For instance, nisin is able to bind with lipid II molecule (a peptidoglycan precursor), and thereby, it interferes with the cell wall biosynthesis in sensitive cell (Linnett and Strominger, 1973). It is observed that most of Class II bacteriocins predominantly forms pores in the target cell, disturbing cytoplasmic membrane integrity and function through leakage of amino acids and ions and depletion of intracellular ATP. According to Thomas et al. (2000), nisin inactivates endospore by means of preventing postgermination swelling and subsequent spore outgrowth. Bacteriocins are highly specific in their action and act in concentration dependent manner, that is, bacteriocin is more specific in action at low concentration, whereas at high concentration, it is often much nonspecific activity (Gabrielsen et al., 2014). This specific activity at lower concentrations make bacteriocins extreme potent, effective at pico- to nanomolar concentrations, whereas eukaryotic antimicrobial peptides necessitated micromolar concentrations for the inhibitory effect (Hassan et al., 2012).

6.4 BACTERIOCINS SUSTAINABILITY IN FOOD SYSTEMS

Bacteriocin applications in food system have been extensively investigated. In food products, many factors showed to influence bacteriocin biosynthesis and activity. Several factors are responsible to affect the biosynthesis of bacteriocin, viz. (a) inadequate physical conditions and chemical composition of food such as temperature, osmotic pressure, water activity, pH, nutrient content, etc., for instance, because of higher solubility at low pH, nisin is more efficient at pH 2 than at pH 8 (Cleveland et al., 2001; Daeschel et al., 1992), whereas pediocin is much stable for broader pH range, (b) spontaneous loss in production capacity, (c) inactivation of the producing strain by bacteriophage, and (d) antagonistic behavior by other microflora of foods. It is linked with the competition of other microbiota for nutrients, which may suppress the growth of bacteriocinogenic strain and ultimately bacteriocin production.

On the other hand, efficiency of bacteriocin synthesized in food system is negatively influenced by some other factors that includes, (1) development of resistance by pathogens to the bacteriocin, (2) interference of several food components like phospholipids, fat, and protein with bacteriocin molecules, for instance, Nisin has limited efficacy against foodborne pathogen *L. mono-cytogenes* if incorporated meat prior to thermal processing because undenatured meat proteins bind bacteriocin and thus inhibit its activity, (3) inappropriate environmental conditions mandatory for the biological activity of bacteriocin, (4) inactivation of bacteriocin by other additives, (5) low solubility and diffusion and/or irregular distribution of bacteriocins in food matrix. Moreover, activity of nisin may diminished by addition of divalent cations such as Ca^{2+} or Mg^{2+} as they are bounded to anionic phospholipids in the cell membrane. It provides more rigidity to cytoplasmic membrane and thereby decreases the affinity of the bacteriocin (Cleveland et al., 2001; Galvez et al., 2007; Schillinger et al., 1996). Further, in the presence of an enzyme glutathione *S*-transferase, nisin is inactivated by glutathione which is found in raw meats. Thus, in meat products, such reactions greatly influence the activity of bacteriocin (Cleveland et al., 2001). On the other hand, nisin found to inhibit resistant Gram-positive and Grams-negative bacteria when used in conjunction with surfactants, chelators (EDTA), and adjuvants.

It is advisable to isolate and subsequently employ the bacteriocin from the inherent bacteriocinogenic strains of specific food substrate, only, for the preservation, that is, if the bacteriocin is obtained from strains isolated from vegetables, then it should be utilized for preservation of specific vegetables only. It would help to resolve the hurdles associated with environmental adaptations and inherent factors affecting the activity and stability of bacteriocin as well as bacteriocin producing strain, too within the food matrixes.

6.5 APPLICATIONS OF LAB BACTERIOCINS IN FOOD

6.5.1 IN MILK AND DAIRY PRODUCTS

Numerous studies had reported the antimicrobial activity of purified bacteriocins and/or bacteriocinogenic strains for preservation of milks and milk products either to inactivate foodborne pathogens, spoilage organisms, and/or both. In different industrialized countries, milk and dairy products implicated 1–5% of the total bacterial foodborne outbreaks in that 53.1% attributed to cheese followed by 39.1% to milk, and 7.8% to other dairy products as acknowledged from foodborne disease reports (De Buyser et al., 2001; Claeys et al., 2013). Several studies demonstrating the application of bacteriocins on milk and dairy products, either alone or together with other physical or chemical treatments, have been compiled in Table 6.1.

6.5.1.1   USE OF PURIFIED OR SEMIPURIFIED NISIN, SINGLE, OR IN COMBINATION WITH OTHER TREATMENTS

To overcome the shelf-life problems coupled with warm weather, transport to long-distance, and insufficient cooling or refrigeration systems, addition of bacteriocin in milk is permitted in several countries (Thomas et al., 2000). In several cases, even under poor refrigeration conditions, nisin together with heat treatments found to extend the keeping quality of raw milk (Davies and Delves-Broughton 1999; Penna and Moraes, 2002). Different researchers from worldwide used nisin to increase the keeping quality of milk and traditional dairy products as listed in Table 6.1, and it can be indicated that very limited work is published from India. In a recent approach, biopreservative effect of nisin was investigated at various concentrations such as 50, 100, 200, and 300 IU/ml to increase the shelf life of standardized, pasteurized milk. The keeping quality was extended to a minimum of 1 week at 4 °C, and increasing concentrations of nisin did not show additional effect, instead significantly affected the sensory scores of milk (Radha, 2014). Nisin in combination with heat treatment (80–100 °C) found to reduce the D-value of *B. cereus* (Penna and Moraes, 2002). Nisin is widely applied to avoid the gas blowing defect in cheese produced by *Clostridium tyrobutyricum* and, further, to restrict the growth of postprocessing contaminants (Thomas et al., 2000; Deegan et al., 2006). Bacteriocin is also effective in many other heat treated milk products, that is, clotted cream, canned evaporated milks, flavoured milk, paneer, or chilled desserts, and much work is done on different types of cheese varieties (prepared with or without starters). Earlier, Kumar et al. (1998) analyzed effect of nisin on the quality of traditional Indian fermented milk product—*dahi*, whereas in another approach, authors analyzed the biopreservative action of nisin in *lassi*-fermented milk

TABLE 6.1 Studies on Application of LAB Bacteriocins in Milk and Milk Products.

Bacteriocin in single or with other treatments	Product	Activity observed	References
Nisin	Cheese	Concentration of 400 IU/g was adequate to prevent spoilage in cheese due to *C. sporogenes* during 90 days of storage period	Roberts and Zottola (1993)
Nisin	Processed cheese	Total plate count and anaerobic spore counts were suppressed at the concentration of 100–500 ppm in processed cheese during storage (5 or 21 °C), whereas the growth of *B. cereus*, *G. stearothermophilus* and *B. subtilis* were prevented at very low concentration (5 ppm)	Plockova et al. (1996)
Nisin	Cheese	2.5 ppm was sufficient to reduce *L. monocytogenes* growth for more than 8 weeks in Ricotta-cheese compared to control sample	Davies et al. (1997)
Nisin	Cheese	Soybean phosphatidylcholine encapsulated nisin indicated superior inhibition of *L. monocytogenes* in Minas frescal cheese than added free nisin at 7 °C/21 days	Malheiros et al. (2012)
Nisin	Pasteurized milk	Shelf life was extended to minimum of 1 week at 4 °C with increasing concentration of nisin	Radha (2014)
Bacteriocin from *Lb. acidophilus*	Pasteurized milk	Successfully retarded the growth of contaminants, i.e., total bacteria, aerobic spore-formers, and psychrotrophs, and during refrigerator storage, extended the keeping quality of pasteurized milk up to 12 days	Ibrahim and Elbarbary (2012)
Nisin and Heat treatment	Raw milk	Shown decrease in *D*-values for *B. cereus* and *B. stearothermophilus* as well as native microflora of raw milk	Davies and Delves-Broughton (1999), Thomas et al. (2000)
Nisin and heat treatment in the range of 80–100 °C	Milk	Up to 40% reduction in *D*-value of *B. cereus*	Penna and Moraes (2002)

TABLE 6.1 *(Continued)*

Bacteriocin in single or with other treatments	Product	Activity observed	References
Nisin (500 IU/mg) and heat treatment at 121 °C/5 min	Milk model system	Highest inactivation of *B. stearothermophilus* into acidified concentrated buffalo milk model system	Rao and Mathur (1996)
Nisin (40 IU/ml) and heat treatment at 72 °C/15 s	Milk	Increased the shelf life up to 7 days at 10 °C with significantly lower count of *Lactobacillus*	Wirjantoro and Lewis (1996)
Nisin and *B. licheniformis* ZJU12 (cell-free supernatant)	—	Inhibited the growth *S. aureus*, *B. cereus*, and *M. flavus*—foodborne pathogens	He and Chen (2006)
Nisin and Monolaurin	Milk	Seems to be effective against bacilli in milk especially inhibited *B. licheniformis* with increase in pH when added simultaneously to milk	Mansour et al. (1999)
Nisin and Monolaurin	Skim milk	Exerted bactericidal effect against *Bacillus* spp. in skim milk	Mansour and Milliere (2001)
Nisin and LPS	Skim milk	Through synergistic effect retarded *L. monocytogenes* proliferation in reconstituted skim milk for longer time period	Boussouel et al. (1999)
Nisin and Lysozyme	Milk	Restricted blowing of cheese milk as well as activity of *Bacillus* spp. and growth of different strain of *Lactobacilli*. No adverse effect on bacteriocin producing starter observed	Kozakova et al. (2005)
Nisin and Reuterin	Milk	Inhibited *L. monocytogenes* and *S. aureus* growth in milk	Arques et al. (2004)
Nisin and HHP	Cheese	Sharp decrease in the count of *S. aureus* when nisin-containing cheese subjected to HHP treatment	Arques et al. (2005a)
Nisin and HHP	Cheese	More than 2 log cycle reduction in aerobic mesophilic bacteria	Capellas et al. (2000)
Nisin (500 IU/ml) and HHP (500 Mpa/5 min)	Milk	Complete inactivation of *P. fluorescence* and *E. coli* observed while decrease in cell count of *L. innocua* was more than 8.3 log cycle	Black et al. (2005)

TABLE 6.1 *(Continued)*

Bacteriocin in single or with other treatments	Product	Activity observed	References
Nisin (100 IU/ml) and PEF (32 pulses at 50 kV/cm)	Skim milk	Inactivated *L. innocua* population up to 3.8 log in skim milk	Calderon-Miranda et al. (1999)
Nisin (38 IU/ml), PEF (50 pulses at 80 kV/cm, 52 °C) and lysozyme (1638 IU/ml)	Raw milk	Observed 7.0 log reduction of the milk endogenous microbiota	Smith et al. (2002)
Nisin, PEF and carvacrol	Milk	Enhanced inactivation of vegetative cells of foodborne pathogen *B. cereus*	Pol et al. (2001)
Lacticin 3147	Yoghurt, Cheese	Rapid inactivation of *L. monocytogenes* in yoghurt within 60 min, whereas 85% reduction in viable count of *S. aureus* in cottage cheese within 120 min	Morgan et al. (2001)
Enterocin AS-48 (20–35 μg/g)	Milk	Reduced the count of *B. cereus*—both viable cells or endospores inoculated in boiled rice and commercial rice-based formula prepared in whole milk below detection level during storage of 15 days between 6 °C and 37 °C and prevented production of enterotoxin, too	Grande et al. (2006a)
Enterocin AS-48 (20–35 μg/g)	Milk	Addition of sodium lactate reduced the requiring concentration of AS-48 (8–16 μg/g). Combination of heat treatment and AS-48 decreased the *D*-value for bacteriocin resistant endospores of *B. cereus*	Grande et al. (2006b)
Enterocin AS-48	—	Resulted into complete inhibition of *L. monocytogenes* and *B. cereus* cells depending on the nature of substrate. Higher conc. (50 μg/g) required for inactivation of *S. aureus*	Martinez Viedma et al. (2009a,b)
Nisin and Potassium sorbate	*Paner*	Improved the keeping quality of *paneer*	Thakral et al. (1990)

TABLE 6.1 *(Continued)*

Bacteriocin in single or with other treatments	Product	Activity observed	References
Bacteriocin from *Enterococcus faecium* BS 13	*Paner & Khoya*	Extended the shelf life of *paneer* and *khoya* under refrigeration conditions as compared to control, bacteriocin-supplemented product was more acceptable as compared control	Bali et al. (2013)
Microgard™-100 and pediocin	*Doda burfi*	Total bacterial count increased from "nil" count to 3.63 log cfu/g after 27 days of storage in treated samples relative to control sample that showed 3.36 log cfu/g after 12 days	Chawla et al. (2015)

product during storage at different temperatures (Kumar and Prasad, 1996). The effective concentration of nisin varied for different types of dairy products, for example, 200 IU for *kheer* (Singh et al., 1987), 100–200 IU/g for *khoa*, and 85–100 IU/g for yoghurt (Gupta et al., 1989). In another study, Rao (1990) mentioned that at least 500 IU/g of initial nisin concentration is essential for attaining a residual value of about 50 IU/g following thermal processing of in-package sterilized *paneer* like product at 118 °C for 5 min.

The application of encapsulation and immobilization techniques for nisin and other bacteriocin during the preservation of dairy products had also given promising outcomes. Malheiros et al. (2012) suggested that encapsulation of bacteriocin in liposomes of partially purified soybean phosphatidylcholine could serve as a potential platform to rule out foodborne pathogens from cheeses. Cao-Hoang et al. (2010) applied nisin immobilized in sodium alginate on the surface and in depth of mini red Babybel cheese to reduce the growth of *Listeria innocua* which already have been spreaded over the surface. They noticed 1.1 log cfu/g reductions on the surface after storage at 4 °C for 1 week to that of control. Nisin in combination with nonthermal treatments, like HHP, PEF, high intensity pulsed-electric field (HIPEF), monolaurin, and lactoperoxidase system (LPS), could enhance the shelf life of milk and dairy products even under poor refrigeration. Monolaurin is a monoester of lauric acid which shows antimicrobial activity that can be further intensified after being mixed with nisin. Most of these treatments either in combination with nisin or other bacteriocin may provide a suitable alternative to traditional thermal treatment and minimal effect on the sensory or organoleptic qualities of milk or related dairy products. The combination of LPS and nisin improves the keeping quality of raw milk at room temperature (Boussouel et al., 1999). Authors observed synergistic and stable inhibitory effect toward *L. monocytogenes* in reconstituted skim milk. The synergism between LPS and nisin is explained through their mode of action, that is, both can affect the cytoplasmic membrane. Hypothiocyanite, the first reaction product of LPS, interacts with the thiol group of proteins, and thus, inactivate essential enzymes and proteins of microbial cell (Boots and Floris, 2006) while nisin forms wedge-like pore through binding to phospholipid molecules of the cytoplasmic membrane (Moll et al., 1997).

Application of nisin (100 IU/ml) in combination with PEF (32 pulses at 50 kV/cm) led to inhibit *Listeria innocua* in skim milk synergistically (Calderón-Miranda et al., 1999). The normal microflora of raw milk was declined as per the cell count result (Smith et al., 2002) and in skim milk inoculated with *S. aureus* using PEF and HIPEF treatments, respectively, in combination with nisin (Sobrino-López et al., 2006). In both the studies, the inhibitory effect was more prominent in conjunction with third hurdle, that is, the addition of certain antimicrobials like carvacrol, lysozyme, or mild heat treatment (Smith et al., 2002; Sobrino-López and Martín-Belloso, 2008). According to Alpas and Bozoglu (2000), addition of bacteriocins powder (nisin and pediocin PA-1) in milk followed by HHP successively reduced *S. aureus* count and absence of cell growth for 30 days in samples during storage at 25 °C. Addition of lysozyme and nisin prior to HHP treatment enhanced the lethality of pressure resistant *E. coli* in skim milk (Garcia-Graells et al., 1999). Bacterial endospores are quite resistant to pressure, and thus, López-Pedemonte et al. (2003) successfully applied HHP cycles twice, first one to induce endospore germination, whereas the second one to eliminate the vegetative bacterial cells. Bacteriocin treatment together with HHP increases the efficiency of microbial inactivation as it causes sublethal injuries and sensitization in the specific microbial cell. It is believed that permeabilization of microbial cell and consequent sensitization to bacteriocin depends on certain variables, such as the treatment time and pressure. Exposure to low pressure with short time treatment causes temporary changes in cell membrane permeability of *E. coli* cells (Diels et al., 2005). On the other hand, HHP can cause permanent damage, and with this theory, Black et al., (2005) demonstrated in their study that significant cell damage sustained throughout the treatment and then after as well.

6.5.1.2   USE OF BACTERIOCINOGENIC STRAINS AND OTHER BACTERIOCINS TO CONTROL THE GROWTH OF SPOILAGE AND PATHOGENIC MICROFLORA

Nisin producing lactococci strains effectively restricted the growth of *S. aureus*, *L. monocytogenes*, and *Clostridium* sp. in various cheese types; however, poor technological performance (low acidification and proteolytic activity) recommends their use as adjunct cultures (Deegan et al., 2006; Galvez et al., 2008). Conversely, Dal Bello et al. (2011) performed a comparative study using four bacteriocin-producing *Lactococcus lactis* strains (two lacticin 481 producers, one nisin A producer and one nisin Z producer). Authors reported that the nisin A producing *Lc. lactis* 40FEL3, and up to certain level lacticin 481 producers (32FL1 and 32FL3), suppressed *L. monocytogenes* growth during Cottage cheese production and storage. *Lc. lactis* subsp. *lactis* 3147 has been recognized to produce lactacin. Several characteristics of lactacin 3147 are similar to nisin such as higher activity in acidic media and stability to mild thermal treatments. Although belongs to class-I bacteriocin, it is a hydrophobic in nature, two-component bacteriocin. Despite of an effective activity of lactacin 3147, failure of commercialization is attributed to its sensitivity to heat and less stability in foods. According to Morgan et al. (2001), more than 70% lactacin activity lost in concentrated powder product was observed within 5 months of room temperature storage, whereas there was no loss at low temperature during entire storage period. Lactacin 3147 combined with HHP treatment showed synergistic effect in controlling *S. aureus* and *L. monocytogenes* population and decreased the amount of bacteriocin required for the inhibition (Morgan et al., 2000). Modified starters producing lacticin 3147 successfully inhibited the growth of *L. monocytogenes* in various cheese types (O'Sullivan et al., 2006).

Majority of pediococci cannot ferment lactose, and thus, milk or dairy products are not a suitable substrate for the growth of pediococci. Only some strains like *Pd. acidilactici* NRRL-B-18925 can produce bacteriocin in milk-based medium. Pediocin production (about 3200–6400 units/ml after 8 h of incubation) in milk has been reported when such strain grown in coculture with yoghurt starter cultures as the excess sugar is released by the starters from hydrolysis of lactoses (Somkuti and Steinberg, 2010). The antilisterial activity and stability at broad pH range in aqueous solutions (even at ambient temperature) as well as during freezing show suitable appliance of pediocin PA-1/AcH in cottage cheese or cheese sauce (Nes et al., 1996; Rodríguez et al., 2002). Pediocin is available commercially as Alta™ product. Reviriego et al. (2005) isolated two strains of lactococci and transformed into pediocin producer which were identified as suitable candidates for food-grade bacteriocin producer. Enterococci occur naturally in cheeses, and several strains have been noticed to produce bacteriocin with marked antilisterial activity. Some bacteriocins like enterocin CCM 4231, enterocin CRL35, or enterocin AS-48 have been examined for dairy foods preservation and found to inhibit the *S. aureus*, *L. monocytogenes*, or *B. cereus* growth. The application of enterococcal bacteriocins in activated film coating greatly reduced the count of listeria on cheese surfaces (Iseppi et al., 2008). It could offer an effective barrier against crosscontamination of the cheeses.

Reuterin is synthesized by *Lb. reuteri*, which is reported to effectively diminish *L. monocytogenes* and *E. coli* count in milk and cottage cheese (El-Ziney and Debevere, 1998). The mixture of reuterin and nisin act synergistically to prevent the proliferation of pathogen when used in milk (Arques et al., 2004). Propionibacteria are associated with the fermentation of several milk products and may produce bacteriocin namely propionicins which have broad inhibitory spectra (Holo et al., 2002). Commercially available Microgard™ contains bacteriocin obtained from *Propionibacterium freudenreichii* subsp. *shermanii*. It is approved in some countries as an ingredient to be used in dairy products like yoghurt and cottage cheese. Apart from this, many bacteriocinogenic LAB are poorly adapted to milk environment unfortunately. Probiotic LAB of human intestinal origin showed little growth in milk (Avons et al., 2004). In some cases, milk is required to be supplemented with yeast extract, amino acids, or specific growth factors to encourage the growth of such bacteria.

6.5.2 IN VEGETABLE AND FRUIT FOOD PRODUCTS

Fresh vegetables and fruit products come into direct contact with soil which may serve as potential source of transmission or contamination of foodborne pathogens as well as spoilage causing organisms. Due to this, numbers of outbreaks have been reported (Lynch et al., 2009). Very few studies have been subjected in context to applying bacteriocins as biopreservative agent in vegetables compared to other food products, such as milk and dairy products, meat, and meat products as well as fish and seafood products. The growth of pathogens like *Escherichia coli*, *Salmonella*, *Listeria* sp., *Bacillus* sp., and other enterobacteria hampered when nisin, enterocin, and pediocin implicated on the surfaces of fresh produced. In general, bacteriocins can be applied on various kind of vegetables and fruit products such as (a) fresh raw products (cut or sliced), (b) semiprocessed and/or cooked ready to eat and canned products, and (c) fermented vegetable and fruit beverages. Several studies demonstrating applications of bacteriocins in vegetables or fruit food products have been compiled in Table 6.2.

6.5.2.1 FRESH RAW PRODUCTS (CUT OR SLICED PRODUCTS AND JUICES)

Fresh vegetables and fruits like perishable foods can easily get contaminated with various spoilage causing and pathogenic microorganisms belonging from soil, manure, water, insects, human handlers, and other processes. Nisin, pediocin, lactacin, and enterocin AS-48 have been applied to restrict and inactivate foodborne pathogens from the surfaces of freshly cut vegetables, fruits, and sprouted seeds; such studies have been comprised in Table 6.2. In regard to prevent transmission of microbes from the surface of vegetables and fruits and thus, to decrease the number of microflora on sliced vegetable and fruit surfaces during storage, treatment of bacteriocins has also been employed for sanitization of whole food product (Silveira et al.,

TABLE 6.2 Studies on Application of LAB Bacteriocins in Vegetable and Fruit Products.

Bacteriocin in single or/ with other treatments	Activity observed	References
Nisin	Addition to cabbage developed controlled fermentation process along with preventing the growth of homofermentative bacteria	Breidt et al. (1995)
Nisin	Prevented growth of *Lactobacilli* spp. responsible for over-ripening of *kimchi*	Choi and Park (2000)
Nisin	Restricted growth of aciduric and nonaciduric spore forming bacilli in canned products	Thomas et al. (2000, 2002)
Nisin and pediocin	Controlled the outgrowth of *B. subtilis* spores in sous vide mushrooms	Cabo et al. (2009)
Enterocin AS-48	*p*-Hydroxybenzoic acid and 2-nitropropanol together with bacteriocin showed synergistic effect against cocktail of *S. entrica* serovars in salads	Molinos et al. (2009b)
Enterocin AS-48	Endospore formers were inactivated in canned vegetables boiled rice and purees	Abriouel et al. (2010)
Enterocin AS-48 (10 mg/l)	Six vegetable products were preserved up to 30 days at different temperature	Grande et al. (2007b)
Enterocin AS-48 (7 µg/g)	Killed vegetative cells of *G. stearothermophilus* in canned corn and peas under simulated storage condition for 30 days	Martínez Viedma et al. (2010a)
Coated enterocin EJ97 in polythene film	Decreased the count of viable cells of *B. coagulans* in corn and peas stored at 4 °C together with EDTA	Martínez Viedma et al. (2010b)
Enterocin AS-48 (25 µg/ml) and polyphosphoric acid (0.1–2.0%)	Controlled the growth of *E. coli* O157:H7, *S. enteric*, *Shigella* sp., *Ent. aerogenes*, *Y. enterocloitica*, *A. hydrophilia*, and *Ps. fluorescens*. in stored (6 and 15 °C) sprout samples	Molinos et al. (2008)
Kimchicin	Prevented growth of pathogens in *kimchi*	Chang and Chang (2011)
Nisin	Decreased the viable *L. monocytogenes* count on apple and honeydew melon slices	Leverentz et al. (2003)

TABLE 6.2 (Continued)

Bacteriocin in single or/ with other treatments	Activity observed	References
Nisin (25–50 IU/ml)	Inhibited spore germination in orange and mixed fruit drinks	Yamazaki et al. (2000)
Nisin plus cinnamon	Inactivated cells of E. coli O157:H7 and S. typhimurium in apple juice	Yuste and Fung (2004)
Nisin and HHP	Reduction of aerobic mesophlic microbiota in cucumber juice	Zhao et al. (2014)
Nisin (100 U/ml) + PEF (80 kV/cm, 20 pulses, 50 °C)	Reduced total viable count in tomato juice by 4.4 log units when stored for 28 days at 4 °C	Nguyen and Mittal (2007)
Nisin or lysozyme and PEF	Viable bacterial count reduced up to 5.9 logs in red and white grape juice	Wu et al. (2005)
Nisin, lysozyme and/or PEF	Inactivated S. typhimurium from orange juice	Liang et al. (2002)
Enterocin AS-48	Individually or in conjunction with PEF technique, chelators, or heat treatment, inactivated both spoilage causers as well as pathogens in fruit juices	Abriouel et al. (2010)
Nisin (10 IU) and high pressure homogenization	Enhanced the bactericidal effect against L. innocua in apple juice and carrot juice	Pathanibul et al. (2009)
Enterocin AS-48	Prevented the growth of exopolysaccharide and 3-hydroxy propionaldehyde producing LAB strains in apple ciders and juice	Grande et al. (2006b); Martinez Viedma et al. (2008a)
Enterocin AS-48 (1.75 µg/ ml)	Completely suppressed the proliferation of G. stearothermophilus for minimum of 30 days of incubation at 45 °C in coconut juice and coconut milk	Martinez Viedma et al. (2009b)
Enterocin AS-48	In conjunction with essential oils, biological active substances, and chemical preservatives at the rate of 30 µg/g or 60 µg/g alone showed complete inhibition of L. monocytogenes	Molinos et al. (2009a)
Bacteriocin (5%) from Lb. fermentum UN01	Low number of colonies in bacteriocin treated sample in compared to control and improved the shelf life of apple juice	Udhayashree et al. (2012)
Thermophilin 110 from Stre. thermophilus	Suggested to inhibit Pediococci growth in brewing industry	Gilbreth and Somkuti (2005)

TABLE 6.2 *(Continued)*

Bacteriocin in single or/ with other treatments	Activity observed	References
Enterocins L50A and L50B from *E. faecium* L50	Bactericidal against beers spoilage *Lb. brevis* and *Pd. damnosus* in a dose and substrate dependent manner	Basanta et al. (2008)
Pediocin PD-1 from *Pd. pentosaceus*	Proposed to remove the biofilms of *O. oeni* from stainless steel surfaces as well as to control their proliferation in wine	Bauer et al. (2003)
Mundticin ATO6 from *Enterococcus mundtii* ATO6	Bacteriocin treatment (200 AU/ml) decreased *L. monocytogenes* count by 2 log cycles in mung bean sprouts, but it could not prevented the bacterial growth under a modified atmosphere at 8 °C during storage study	Bennik et al. (1999)

2008). Spraying, dipping, impregnation, and coating are the various modes for the application of bacteriocin to fresh-cut fruits and vegetables.

With compare to commercially available nisin, highest diminution in the *L. monocytogenes* count was achieved during 7 days of storage at 4 °C in minimally processed iceberg lettuce samples with bacteriocin RUC9 obtained from wild strain of *Lc. lactis* (Randazzo et al., 2009). The strain was originally recovered from minimally processed mixed salads. The individual or collective effect of nisin and pediocin with other preservatives like phytic acid, citric acid, sodium lactate, and potassium sorbate on sprouted mung beans, broccoli, and fresh-cut cabbage stored at 25 °C before treatment was studied against *L. monocytogenes*. About 2–4 log cycle reduction reported for individual treatment, but significant reduction was noticed for broccoli and fresh-cut cabbage with two mixtures; nisin-phytic acid and nisin-pediocin-phytic acid. However, pediocin showed potent inhibition than nisin, individually as well as in combination with organic acids (Bari et al., 2005). Molinos et al. (2005) immersed alfalfa, soybean sprouts, and green asparagus in three different concentrations (viz., 5, 12.5, and 25 µg/ml) of enterocin AS-48 for 5 min. Efficient reduction of *L. monocytogenes* (2–2.4 log cfu/g) was obtained in alfalfa and soybean sprouts with highest conc. (25 µg/ml) when stored at 6 and 15 °C right after treatment. These treatments failed for green asparagus, and instead, the mutual treatment of bacteriocin and chemicals like lactic acid, trisodium trimetaphosphate, peracetic acid, sodium hypochlorite, or *n*-propyl-*p*-hydroxybenzoate inhibited growth of *L. monocytogenes* more significantly. Ukuku and Fett (2004) demonstrated significant decrease in *Salmonella* that were directly inoculated onto fresh-cut cantaloupe, through combine treatments of nisin-sodium lactate, nisin-potassium sorbate, and nisin–sodium lactate–potassium sorbate. Similarly, nisin in conjunction with of citric acid, sodium lactate, or H_2O_2 as a decontaminator resulted in sanitization of whole cantaloupe and honeydew melon surfaces. Further, it also prohibited *L. monocytogenes* and *E. coli* transfer on fresh cut pieces (Ukuku et al., 2005).

Compared to cut food products, fruit and vegetable juices and related drinks enhance the functionality of bacteriocins since such fluid substrates increase diffusion, facilitate activity and solubility due to acidic pH, and reduce bacteriocin adsorption on fat (hydrophobic substance). Presence of organic matters together with other biological active molecules also increases bacteriocin potential. Several LAB and yeast could also grow in acidic substrates causing souring, ropiness, or gassiness. Usually low acidic pH of juices does not favor the growth of pathogens, but in some less acidic juices, they may proliferate. As illustrated in Table 6.2, potential bacteriocins demonstrated antimicrobial activity against several pathogens and spoilage bacteria like exopolysaccharide-producing lactobacilli, pediococci, and *Bacillus licheniformis* in juices. Bacteriocins also confer inactivation of endospore forming *G. stearothermophilus* and *Alicyclobacillus acidoterrestris*, and acrolein-producing bacteria. Addition of enterocins reduced viable count of foodborne pathogens *S. aureus*, *L. monocytogenes*, and *B. cereus* in soy milk, sport and energy drink, and lettuce juices at lower pH (Abriouel et al., 2010; Galvez et al., 2008).

The thermophilic endospore former *A. acidoterrestris* impart an unpleasant medicinal taste to fruit juices linked with the production of guaiacol. During food processing, this bacterium can withstand pasteurization and consequently spoil fresh and processed juices as well. Outcome of many investigations trigger effectiveness of nisin and enterocin AS-48 as possible hurdles against this bacterium (Komitopoulou et al., 1999; Yamazaki et al., 2000; Pena et al., 2009). Grande et al. (2005) studied the behavior of enterocin AS-48 against endospore former *A. acidoterrestris* in a range of fruit juice samples. Addition of bacteriocin at the rate of 2.5 μg/ml suppressed the growth of indicator organism in both natural (orange and apple juice) and commercial (orange, apple, pineapple, peach, and grapefruits) juices at three different incubation periods (4, 15, and 37 °C) up to 14 days and 2–3 months, respectively. From electron microscopy, authors revealed that bacteriocin treated vegetative cells established significant damage followed by cell lysis, whereas it inhibited endospore germination and disorganized the endospore structure. The nonspore former *Propionibacterium cyclohexanicum* is another major spoilage causing bacterium in juices. Walker and Philips (2008) determined the effect of nisin at the rate of 500 and 1000 IU/ml against this bacterium in orange juice. The spoilage was controlled till 15 days of storage; however, the regrowth could not be prevented during storage for longer time interval.

Combination of EDTA (20 nM) and nisin (300 IU) suggested to apply in freshly made apple cider to increase microbiological safety as it can suppressed the growth of Gram-negative foodborne pathogens, too like *Salmonella* and *E. coli* O157:H7 (Ukuku et al., 2009). Bacteriocins in combination with PEF and/or other compounds revealed improved bactericidal effects and reduced the risk for proliferation of survivors (Liang et al., 2002; Nguyen and Mittal, 2007; Wu et al., 2005; Zhao et al., 2014). Enterocin AS-48 (2 mg/l) in combination with PEF inhibited the growth of EPS producing spoilage bacterium *Lb. diolivorans* 29 (Martínez Viedma et al., 2009c). While Liang et al. (2006) noticed reduction of naturally occurring microbes especially fungi when treated with PEF and mixture of nisin and lysozyme. Nisin (10 IU) combined with high pressure homogenization

(HPH) reduced the intensity of the high pressure treatment to some extent for inactivation *L. innocua* (Pathanibul et al., 2009).

6.5.2.2   SEMIPROCESSED, COOKED READY TO EAT (RTE) AND CANNED PRODUCTS

Unhygienic practices during the processing, crosscontamination, personnel handling, improper storage, and no heat treatment prior to consumption attributed to microbiological safety of ready to eat foods like salads which are made up of cooked and/or uncooked vegetables with or without nonvegetarian foods. In salads, heat treatment offers background to pathogens for proliferation against low number of competitive microflora. Enterocin AS-48 and nisin, singly or in conjunction with other compounds (such as hydrocinnamic acid/carvacrol), have proven effective to prevent the proliferation of *Salmonella* sp., *S. aureus*, and *L. monocytogenes* (Grande et al., 2007a; Molinos et al., 2009a,b; Schillinger et al., 2001). In canned or heat-treated products, also addition of salt, sugar or chemical preservatives, and refrigeration or acidification is often demanded with regard to prevent growth of spore-formers. Results of certain studies suggest the practical application of bacteriocins (Thomas et al., 2000, 2002). Incorporation of pediocin, nisin, enterocin AS-48, or their combinations has been proposed to retard the outgrowth of endospore and synthesis of enterotoxins (*B. cereus* or *Cl. botulinum* toxins) in cooked as well as canned vegetables during storage (Galvez et al., 2008; Abriouel et al., 2010). Moreover, combining phenolic compounds like carvacrol, hydrocinnamic acid, eugenol, and geraniol together with bacteriocin enhanced antimicrobial action against bacilli (Grande et al., 2007b). Added bacteriocin also increased heat inactivation of *B. coagulans* endospores. Flat sour spoilage causing *Bacillus* sp. inhibited during storage in canned foods like tomato paste, peaches, and pineapple juice through added enterocin AS-48 for 15 days of storage at 37 °C (Lucas et al., 2006).

6.5.2.3   FERMENTED VEGETABLES AND FRUIT BEVERAGES

In the case of vegetables, generally spontaneous fermentation takes place, and microbial growth is affected by various internal and external factors. Incorporation of antibacterial peptides provides a suitable environment for selective group of microbes to bring about a desirable change in the finished fermented product. Lactic fermentation is the definite mechanism involve in the biological preservation of perishable vegetables and serve as an excellent source for the isolation of bacteriocin-producing LAB. Certain examples of antagonistic LAB isolates include *Pd. parvulus* from vegetables (Bennik et al., 1997); *Lb. sake* C2 from fermented cabbage (Gao et al., 2010); *Lb. fermentum* and *W. cibaria* strains from cucumber, cabbage, turmeric, and carrot (Patel et al., 2012); *P. pentosaceus* 05-10 from Sichuan pickle (Huang et al., 2009); *W. cibaria* from ripe mulberries (Chen et al., 2010); *Lb. pentosus* B96 from fermented green olives (Delgado et al., 2005); *P. pentosaceus* K23-2 from *kimchi* (Shin et al., 2008); and *Lc. lactis* 23 from fermented carrots (Uhlman et al., 1992). Such novel strains represent immense potential for exploration as starter or protective cultures during vegetable fermentations. Extensive research had taken place with regard to isolate antagonistic strains from vegetables, but only limited work has been done in the possible applications and further characterization of antimicrobial compound(s). Application of live cultures or bacteriocinogenic cultures is another alternate to restrict the growth of spoilage and pathogenic microbes in vegetables and fruits. Strains isolated from raw vegetables could perform better in the vegetable and easily get adapted under cold and moderate temperatures. However, the ecological factors affect the growth of culture as well as production of antimicrobials in situ.

Nisin was incorporated in *kimchi*-fermented vegetable product to suppress the proliferation of over-ripening responsible lactobacilli for of the product. A higher inhibition of *Lactobacillus* sp. was noticed in comparison to *Leuconostoc* sp. by Choi and Park (2000). Two most promising bacteriocinogenic strains, *E. faecium* BFE 900 and *Lb. plantarum* LPCO 10, were identified from table olive (Franz et al., 1996; Jimenez et al., 1993). Later on, in green olive fermentation, as a starter culture, the growth behavior of *Lb. plantarum* LPCO 10 was studied. Authors concluded that *Lb. plantarum* LPCO 10 produced highest lactic acid within short time period compared to its nonbacteriocin producing derivative. Further, olive juice broth acquired from green olives was found precious alternate of olive fermentation brine for Spanish style olive fermentation due to excellent bacteriocin production by *Lb. plantarum* LPCO 10.

Sauerkraut is a product containing shredded cabbage naturally fermented through the action of *Leu. mesenteroides* in the presence of brine solution or salt. In one of the earlier experiment, Stamer et al. (1971) illustrated the use of filter sterilized cabbage juice broth for development of paired starter culture system during manufacturing of sauerkraut. In one of the experiment, nisin-producing lactococcal strain and nisin-resistant leuconostoc strain were grown either in combination or separately for preparation of sauerkraut. Presence of lactococci did not affect the growth of leuconostoc, but presence of leuconostocs affected the growth of lactococci. Nisin level remained constant up to 12 days test period. Nisin addition helped to develop controlled fermentation process when nisin-resistant *Leu. mesenteroides* was inoculated in cabbage, and it also delayed the proliferation of homofermentative LAB (Breidt et al., 1995). Pickles made from cucumber (with salt and spices) were stored at 5°C for 3 weeks and afterward exposed to different storage temperature viz., 16, 25, and 30°C to elaborate LAB for their biopreservative potential (Reina et al., 2005). Authors identified 10 isolates as bacteriocin producer out of which 3 were having antilisterial activity.

In the case of fermented beverages, importance of bacteriocin addition has been investigated to avoid the spoilage in the final product. In some countries, addition of nisin in beer is permitted, but not in wine. Nisin could be applied at various stages during manufacturing of fermented beverages, such as for washing and cleaning of equipments; incorporation of fermenter tank to avoid growth of unwanted microbes and washing starters (yeast pitching) with bacteriocin to inhibit contaminating bacteria (Ogden and Tubb, 1985; Radler, 1990; Delves-Broughton et al., 1996; Thomas et al., 2000). Nisin treatment would also help to reduce pasteurization time-temperature regime and improve the keeping quality of uncontaminated beers. In wine industry, the quantity of sulfur dioxide used to kill LAB that could be minimized by addition of nisin and potassium metabisulphite together (Rojo-Bezares et al., 2007; Bartowsky, 2009). Bacteriocin often inhibits the growth of desire bacteria of fermentation. For instance, the growth of *Oenococcus oeni*, a bacterium responsible for the malolactic fermentation in wine found to get inhibit with nisin; thus in such case, bacteriocin resistant strain is developed to continue malolactic reactions. Similarly, nisin resistant lactic starters have also been applied in several studies to increase the microbial safety without affecting fermenting microbiota of different foods (Gálvez et al., 2008).

6.5.3 IN CEREAL AND LEGUME PRODUCTS

Fermented cereals can be a good source of antagonistic LAB strains, for example, *Lb. plantarum*, *Lb. fermentum*, and *W. cibaria* from Indian fermented foods like dosa, idli, and dhokla (Patel et al., 2012, 2013); *Lb. casei* and *Lb.* plantarum from different types of fermented maize products like *kenkey* and *ogi* (Olasupo et al., 1995); *Lb. plantarum* and *Lb. fermentum* from fermented maize dough *poto poto* (Omar et al., 2008); and

Lb. plantarum from fermented millet product *ben saalga* (Sánchez Valen-zuela et al., 2008). Certain antifungal compounds were also reported from fermented cereals by some research groups (Valerio et al., 2009; Dalié et al., 2010; Todorov, 2010). *Boza* is a low-alcoholic beverage obtained from the fermentation of barley, rice, maize, wheat, millet, or oat in Bulgaria. Few researchers reported bacteriocin producing inherent fermenting microflora from *boza* depicted in Table 6.3 (Ivanova et al., 2000; Kabadjova et al., 2000; Todorov and Dicks, 2004, 2005).

Many bacteriocins have been reported to produce from sourdough bacteria such as amylovorin L, pediocin, plantaricin ST31, BLIS C57, and bavaricin A (Messens and De Vuyst 2002; Corsetti and Settanni, 2004). In yeast-leavened breads, addition of cultured broths fermented with bacte-riocinogenic strains or sourdoughs led to prevent rope formation from *Bacillus* sp. (Menteş et al., 2007; Valerio et al., 2008). Nisin in combination with modified atmospheric packaging increased shelf life of ham pizza by suppressing the proliferation of spoilage LAB (Cabo et al., 2001). Fermen-tation of maize and millet flours is commonly practiced in African coun-tries. Though much is not published concerning fermentative LAB and their activities, outcome of few studies are indicative of promising strong inhibi-tory activities of these lactic bacteria against foodborne pathogen. Sánchez Valenzuela et al. (2008) narrated strong inhibitory activity *Lb. plantarum* 2.9 toward *E. coli* O157:H7, *S. enteric*, and *B. cereus* in malted millet flour. The strain was isolated from a fermented pearl millet product called *ben saalga*. Such strains might be employed as a starter culture to enhance the microbio-logical quality and safety of cereal-based fermented products.

Nisin in combination with bacteriocin producing *Lc. lactis* BFE 902 and *Enterococcus faecium* BFE 900-6a demonstrated antilisterial activity in homemade tofu, where these strains were employed as protective cultures (Schillinger et al., 2001). Grande et al. (2006b) added enterocin AS-48 (20–35 μg/g) to commercial infant rice-based gruel (which was dissolved in whole milk) and boiled rice and spiked the samples with *B. cereus*, both in vegetative and endospores forms, too. Growth of bacilli was totally suppressed as well as toxin production was also prevented during storage at 6–37 °C for 15 days. An improvement in the activity of enterocin was observed with addition of sodium lactate. Even combination of enterocin and heat treatments declined the *D* values for bacterial endospores. In one study, nisin-producing *Lc. lactis* subsp. *lactis* IFO12007 strain, originally isolated from *miso*, was employed as starter culture to ferment cooked rice and rice koji supplemented with soybean extract (Kato et al., 2001). The bacterial strain flourished well within the food matrixes (produced >10^9

TABLE 6.3 Bacteriocinogenic Strains Isolated from Cereal and Legume Products.

Bacteriocinogenic strains or bacteriocin	Product name/type	Antimicrobial activity and characteristics	References
Bacteriocin like substance produced by *Lc. lactis* subsp. *lactis* B14	*Boza*	*L. innocua* F, *Lb. plantarum* 73, *Lc. cremoris* 117, *E. coli*	Kabadjova et al. (2000), Ivanova et al. (2000)
Mesenetricin ST99 from *Leu. mesenteroides* subsp. *dextranicum* ST99	*Boza*	LAB, *B. subtilis, Enterococcus faecalis, L. innocua, L. monocytogenes, Pd. Pentosaceus*	Todorov and Dicks (2004)
Pediocin ST18 from *Pd. pentosaceus* ST18	*Boza*	*Pediococcus* sp., antilisterial activity	Todorov and Dicks (2005)
Lactobacillus plantarum	*Ben saalga*	*E. coli* U-9, *L. innocua, L. monocytogenes* CECT 4032, *S. aureus* CECT 192, *S. typhimurium, B. cereus* LWL1	Omar et al. (2006)
Lb. plantarum, Lb. casei	*Kenkey Ogi*	Foodborne pathogens	Olasupo et al. (1995)
Lb. fermentum and *Lb. plantarum*	*Poto poto*	*E. coli, En. aerogenes, S. typhi, S. aureus, B. cereus, L. monocytogenes, Enterococcus faecalis*	Omar et al. (2008)

cells/g) and represented adequate nisin production (1.28×10^5 AU/g) with significant inhibition of contaminating *B. subtilis* cells without affecting the growth of a mold, *Aspergillus oryzae* (desired fermentation flora) during the fermentation of koji.

Both raw and malted barley can be potential source for isolation of bacteriocinogenic strains. Vaughan et al. (2005) proposed that fermented worts containing bacteriocin could serve to control spoilage LAB in beer. Bacteriocin producing *Lb. plantarum*, *Pd. Pentosaceus*, and *O. oeni* had isolated from wine and wineyards (Lonvaud-Funel and Joyeux, 1993; Navarro et al., 2000; Rojo-Bezares et al., 2007). Todorov (2010) suggested that the isolated bacteriocinogenic strains of LAB from the identical fermented cereal product or beverage can be employed to prevent the growth of spoilage type as well as pathogenic bacteria, and it may also ensure the safety of foods for human consumption.

6.5.4 IN MEAT AND MEAT PRODUCTS

Meat products make a significant part of human diets and offer a potential source of proteins of high biological value. These food products get rapidly spoiled if proper preservative actions are not taken. The bacterial groups involved in the contamination of meat environment include *Enterobacteriaceae*, LAB (chiefly involves *Leuconostoc*, *Weissella*, and *Carnobacterium* genera) *Pseudomonas* and *Brochothrix*. The proliferation of bacteria leads to development of unpleasant odors and body-texture that makes the product undesirable for consumers. Some pathogenic strains of *E. coli*, *L. monocytogenes*, *Salmonella*, and *S. aureus* also get a chance to propagate in spoiled products. Ample of research and investigation have been carried out with the appliance of bacteriocins in raw meat, semiprocessed, and/or cooked meat products and fermented meat products. Bacteriocins have been tested for decontaminating carcass or to inhibit spoilage organisms in fresh meat while stored. Further, bacteriocin applied through various modes, that is, washing, dipping, or spraying in conjunction with other antimicrobials to potentiate the antimicrobial activity. The combinations of nisin with physical and chemical treatments offer suitable alternate to execute strong antimicrobial activity and ultimately prolong the shelf life of meat and meat products. Packaging of raw meat under various conditions like vacuum packaging, MAP, active packaging with O_2 scavengers, and CO_2 generating systems were tested to avoid crosscontamination. Studies

TABLE 6.4 Studies on Application of LAB Bacteriocins in Meat and Meat Products.

Bacteriocin in single or along with other treatments	Activity observed	References
Nisin and lactic or polylactic acids	Decreased initial microflora and prolonged the shelf life of vacuum packaged fresh meat	Ariyapitipun et al. (1999, 2000), Barboza de Martinez et al. (2002)
Nisin and lysozyme	In vacuum packaged pork, it strongly prevented the growth of *B. thermosphacta* and LAB	Nattress et al. (2001), Nattress and Baker (2003)
Nisin and EDTA	Together with MAP (65% CO_2, 30% N_2, and 5% O_2), influenced spoilage bacteria and reduced the amount of volatile compounds in fresh chicken meat up to 14 days during refrigeration storage	Economou et al. (2009)
Nisin (1000 IU/g) with essential oil (0.6%)	Declined *L. monocytogenes* count below the official limit prescribed by EU during 12 days of storage at 4 °C	Solomakos et al. (2008)
Nisin and oregano essential oil	Restricted the growth of *S. enteritidis* in sheep-minced meat during refrigeration storage	Govaris et al. (2010)
Pediocin PA-1/Ach	Exposure of raw Spanish meat surface as well as treatment to various conc. of bacteriocin decreased viable cell count of *L. monocytogenes* and *Cl. perfringens*	Nieto-Lozano et al. (2006)
Sakacin-P	Potentially inhibited *Listeria* population in vacuum packaged chicken cuts under refrigeration storage	Katla et al. (2002)
Nisin and organic acids	Viable cell count of *Salmonella* and *Staphylococci* spp. reduced in fresh pork sausages	Scannell et al. (1997)
Nisin with or lactacin 3147 sodium citrate or sodium lactate	Enhanced inhibition of *Listeria* and *C. perfringens* in fresh pork sausages	Scannell et al. (2000a)
Enterocins A and B (128 AU/g)	Inhibited slime production by *Lb. sakei* CTC746 strain, but not by *Leu. carnosum* CTC747 strain in vacuum packaged sliced cooked pork ham	Aymerich et al. (2002)
Bifidocin B and Lactococcin R	Incorporation to irradiated raw chicken breast restricted growth of *B. cereus* or *L. monocytogenes* for 3–4 weeks and 6–12 h at commercial refrigeration temperature and 22–25 °C, respectively	Yildirim et al. (2007)

TABLE 6.4 (Continued)

Bacteriocin in single or along with other treatments	Activity observed	References
Pentocin 31-1	Decreased the growth of *Pseudomonas* and *Listeria* as well as volatile nitrogenous compound in chill-stored nonvacuum tray-packaged pork meat	Zhang et al. (2010)
Nisin	Immobilized bacteriocin in alginate beads or in a palmitoylated alginate-based film prevented growth of *S. aureus* in sliced beef and ground beef meat through	Millette et al. (2007)
Nisin	Extended the lag phase of *L. monocytogenes* spiked in minced buffalo meat	Pawar et al. (2010)
Nisin (100 ppm) + butylated hydroxyanisole (BHA) 100 ppm	Significant inhibition of total viable count, anaerobic counts, as well as staphylococcal and streptococcal count in comparison to control and individual samples. The product was acceptable up to 5 days of storage at 35 ± 2 °C with 70–80% RH	Sureshkumar et al. (2010)
Pediocin, sodium diacetate and sodium lactate	When commercial beef franks were dipped in three antimicrobial solutions, i.e., pediocin (6000 AU), 3% sodium diacetate, and 6% sodium lactate combined, and a combination of the three, reduction of *L. monocytogenes* populations was ranged between 1 and 1.5 log units and 1.5 and 2.5 log units at 4 °C after 2 and 3 weeks of storage	Uhart et al. (2004)
Nisin in antimicrobial packaging together with HCl and EDTA	Combination of low temperature and active packaging system synergistically reduced the count of *Carnobacteria*, *B. thermosphacta*, LAB, and *Enterobacteriacae* in meat cuts during early stages of storage	Ercolini et al. (2010)
Enterocins, 2400 AU/g) and HHP (400 MPa, 10 min)	Avoided overgrowth of surviving *Listeria* upon a simulated cold-chain break event when the samples were stored at 1 °C, but not at 6 °C	Marcos et al. (2008a)
Immobilized enterocins in combination with HHP (400 MPa, 10 min)	Application of HHP potentiates the antilisterial activity of immobilized enterocin against *L. monocytogenes* in sliced ham and held the cell count below 1.5 log CFU/g at the end of storage for 30 days at 6 °C	Jofré et al. (2007)

TABLE 6.4 *(Continued)*

Bacteriocin in single or along with other treatments	Activity observed	References
Nisin	Effectively preserved bologna-type sausages against spoilage causing LAB	Davies and Delves-Broughton (1999)
Nisin	Growth of *L. monocytogenes* was inhibited in sucuk (type of Turkish fermented sausage)	Hampikyan and Ugur (2007)
Nisin with grape seed extract	Inhibited *L. monocytogenes* in turkey frankfurters during storage at 4 and 10 °C	Sivarooban et al. (2007)
Enterocins A and B (648 AU/g)	Diminished the viable counts of *L. innocua* below 50 CFU/g in Spanish sausage	Aymerich et al. (2000)
Enterocin AS-48 (450 AU/g)	Viable listeria were absent after 6 and 9 days of storage at 20 °C in a meat sausage model system	Ananou et al. (2005a)
Enterocin AS-48 (450 AU/g)	Viable count of *S. aureus* decreased below detectable levels at the end of storage in similar model	Ananou et al. (2005b)

illustrating application of LAB bacteriocins in meat and meat products are compiled in Table 6.4.

6.5.4.1 RAW MEAT PRODUCTS

Though nisin is largely used in meat products, some disadvantages associated with its application such as weak solubility, interaction with meat components like undenatured proteins, phospholipids, and glutathione are still under consideration (Stergiou et al., 2006). Prior to processing and packaging, surface decontamination of raw meats demonstrated positive outcomes. Total aerobic plate counts were reduced by applying combination of nisin and EDTA to MAP or vacuum packaging. It has also increased keeping quality of raw poultry meat for at least 4 days under aerobic condition and 9 days under vacuum packaged (Cosby et al., 1999). In several investigations, immobilization of nisin in various substrates like beads, liposomes, coating or films, etc. minimized the interaction between bacteriocin and food components and enzyme inactivation of bacteriocin, too. Such immobilized nisin or nisin plus citric acid/Tween 80/ EDTA, embedded in polyvinyl chloride, low-density polyethylene (LDPE), nylon, agar coatings, or calcium alginate gel, restricted the proliferation of *B. thermosphacta*, *L. monocytogenes*, *Salmonella typhimurium*, and *S. aureus* in refrigerated raw meat (Aymerich et al., 2008; Chen and Hoover, 2003; Galvez et al., 2007, 2008). Further, authors concluded that immobilization reduced the concentration or amount of bacteriocin required for inhibition of target microbial cell. Natrajan and Sheldon (2000a,b) reported inhibition of antibiotic resistant *S. tphimurium* on poultry drumstick skin through nisin in combinations with EDTA, Tween 80, and citric acid embedded in calcium-alginate, polyvinyl chloride, agar, LDPE, or nylon.

Other bacteriocins, like pediocin (PA-1/Ach), sakacins, bifidocins, lactocins, lactococcins, enterocins, carnobacteriocins, etc., in raw and poultry meats, in single or in combination with other hurdles also reported to delay growth of spoilage organisms efficiently. The population of *L. monocytogenes* reduced below detectable level in minced meat stored at 8 °C by addition of partially purified plantaricin was obtained from *Lb. plantarum* UG1 (Enan et al., 2002). A novel poly lactic acid/sawdust biodegradable film was developed by Woraprayote et al. (2013) with antilisterial effect by incorporating pediocin PA-1/Ach. This film prevented *L. monocytogenes* growth on raw sliced pork during storage at low temperature. Aymerich et al. (2008) suggested use of HHP, low dose irradiation, or exposure to UV light as an additional hurdle to reduce the spoilage bacteria in raw meat. All these treatments either alone or in combination may act in synergistic manner.

6.5.4.2 READY TO EAT, SEMIPROCESSED, AND COOKED MEAT PRODUCTS

The pH and other treatments like cooking, slicing, peeling, and packaging of semiprocessed or cooked meat favor the growth of postprocessing contaminants. Due to heat resistance, bacteriocins can apply to meat slurries before heat treatment. The other modes of bacteriocin application include surface application and coated film. The major group involved in the spoilage of vacuum-packaged meat is LAB which causes souring, ropiness/slime production, and gassiness. Three different bacteriocins namely nisin, enterocin, and sakacin K were tested against ropiness producing bacteria. Results revealed that only nisin was successful in controlling the growth of *Leu. carsnosum*, whereas *Lb. sakei* was inhibited by enterocin (Aymerich et al., 2002). The mixture of nisin with EDTA and lysozyme showed significant inhibition of LAB, *L. monocytogenes*, *E. coli* O157:H7, and *Salmonella* in ham and/or bologna sausages (Gill and Holley, 2000a,b).

Presurface appliance of nisin, nisin-lysozyme, Pediocin, or their combinations with sodium lactate or sodium diacetate in combination with post-packaging thermal treatments on products like frankfurters or turkey bologna have shown an efficient combination for *L. monocytogenes* control (Chen et al., 2004a; Mangalassary et al., 2008). Bacteriocin coating in packaging materials not only enhances the shelf life of the products but also acts as a barrier to restrict the external contamination. Nisin in association with other antimicrobial agents found most significant in the category of edible coating on vacuum packaged products (Franklin et al., 2004; Lungu and Johnson, 2005). Nguyen et al. (2008) studied the effect of 2500 IU/ml of nisin incorporated in edible bacterial cellulose film against total aerobic plate count and *L. monocytogenes*. Authors found lower counts up to 14 days under refrigeration storage compared to control samples. In a separate study, Santiago-Silva et al. (2009) analyzed the effect of Pediocin PA-1/Ach containing cellulose films against *L. innocua* and *Salmonella* sp. on sliced ham packaged under vacuum and stored at 12 °C, temperature that occurs normally in supermarkets. The system was more efficient in restricting *L. innocua* (2 log cycles reduction compared to control) than *Salmonella* (reduction of growth by 0.5 log cycle only relative to control), but it did not reduce the viable cell count of the inoculated bacteria. Similarly, enterocin 496K1 reduced *L. monocytogenes* count in frankfurters during initial 24 h of storage at 4 °C and 22 °C, but effect did not remain same afterward (Iseppi et al., 2008).

Sometimes, low bacteriocin concentration found more effective for inhibiting the growth of pathogenic bacteria, whereas higher concentration may exhibit a protective effect on microbial cell. Based on such results from various experiments, Maks et al. (2010) stated that the interactions between antimicrobial additives in formulations can vary at different concentrations and/or temperatures. Outcomes of several studies indicate the influence of storage temperature on the delicate balance between growth inhibition of surviving bacterial cells and repair of sublethal injury followed by further cell proliferation (Chen et al., 2004b; Marcos et al., 2008b).

The appliance of bacteriocin in combination with other nonthermal treatments HHP, PEF, and HIPEF followed with proper packaging can improve the keeping quality of ready to eat meat products even with poor refrigeration. Sausages exposed to Nisaplin dip followed by pulsed light (9.4 J/cm^2) reduced *L. monocytogenes* population by 4 log cycles (Uesugi and Moraru, 2009). Another approach is to give postpackaging irradiation. Bacteriocins activated film sensitizes target pathogens to radiation. Irradiation at 1.2 kGy or more in a combination of pediocin resulted in 50% decrease in *L. monocytogenes* in frankfurters (Chen et al., 2004b).

Enterocins in combination with HHP (400 MPa for 10 min) avoided overgrowth of *Lb. sakei* CTC746 strain during storage in cooked ham and improved the results than single treatment (Garriga et al., 2002). The interleavers (made up of polypropylene/polyamide layers) containing different bacteriocins with or without HHP treatment in cooked ham inoculated with *Salmonella* sp. showed that only nisin in combination with HHP resulted into complete inhibition of *Salmonella* during refrigeration storage from samples (Jofré et al., 2008).

6.5.4.3 FERMENTED MEAT AND MEAT PRODUCTS

Fermented meat products are quite safe for consumption compared to fresh meat due to low pH, which is unfavorable for majority of spoilage causing and pathogenic microbes. Incorporation of nisin would act as an additional hurdle in mild acidic fermented sausages where higher pH and moisture are present. Nisin, enterocin, pediocin, and leucocin studied in detail to prevent the growth of pathogens like *L. monocytogenes*, *Salmonella*, and *S. aureus*. The solubility of nisin increases at low pH that also enhances the efficiency of bacteriocin. The efficiency of nisin increases in combination with other antimicrobial agents, like organic acids or EDTA-lysozyme in fermented meats (Scannell et al., 1997; Gill and Holley 2000a). Enterocin 416K1 (10 AU/g, a concentrated supernatant) decreased the count of *L. monocytogenes* by 2.5 log CFU/g during 3 days of the drying period in Italian sausages, but during ripening could not suppress the growth of pathogen (Sabia et al., 2003).

6.5.5 IN FISH AND SEAFOOD PRODUCTS

Fish and seafood are considered as significant source of vital nutrients like omega-3 fatty acids, proteins, taurine, and various micronutrients. Microorganisms are associated with major illnesses and spoilage in seafood industry, and according to one of the reports, microbial spoilage leads to loss of nearly 25% of all the seafood produced (Baird-Parker, 2000). Drying and salting are the common practices in seafood industry to eliminate microorganisms, whereas other methods involve use of chlorinated water, disinfectants, quaternary ammonium compounds, iodophores, aldehydes, and H_2O_2. Due to consumer awareness for chemical free foods, in current times, biological preservatives have gained much importance. Various species of Lactobacilli, *Carnobacterium*, and Enterococci have been employed to control the proliferation of pathogenic and spoilage microbiota. The activity of employed bacteriocin could be affected by inherent factors like temperature, pH, and salt concentration and environmental parameters during the processing of sea foods (Galvez et al., 2007). Most of the researches and studies in context to seafood products have emphasized on prevention of *L. monocytogenes*; however, much focus should be given to other pathogens such as *Clostridium botulinum*, *Vibrio cholera*, *S. aureus*, and other enteric bacteria as well as spoilage bacteria such as *Shewanella putrefaciens*, *Pseudomonas* sp., and *Aeromonas* sp. The combination of bacteriocin together with nonthermal techniques like HHP, PEF, or HIPEF is not successful for seafood products due to the much delicate nature, and therefore, none of the research represents application of such treatments still date. Major studies reporting application of LAB derived bacteriocins in fish and seafood products are compiled in Table 6.5.

Fish and other seafood products, in comparison to nisin, other bacteriocinogenic strains and their bacteriocins, revealed significant inhibition of both spoilage and pathogenic types. Different bacteriocins such as carnobacteriocin B2, carnocin U149, piscicocin V1a, V1b, piscicocin CS526,

TABLE 6.5 Studies on Application of LAB-derived Bacteriocins in Fish and Seafood Products.

Bacteriocin in single or along with other treatments	Activity observed	References
Nisin (200 IU/g)	Significantly extended the shelf life of gilthead seabream fillets packed under modified atmosphere up to 48 days as compared control (10 days) at 0 °C. Nisin together with other antimicrobials increased inactivation or delayed the growth of spoilage organisms in sardines and in fish muscle extract	Tsironi and Taoukis (2010)
Nisin (1 and 15 mg/g)	Diminished *B. cereus* level by 2.5 and 25 fold, with 1 and 15 mg/g concentration, respectively, at 16 °C after 48 h in salmon. However, effect of added nisin on enterotoxin production was not reported	Labbé and Rahmati (2012)
Nisaplin	Incorporation at 60 and 600 µg/g decreased a number of eight different strains of *L. monocytogenes* in Karashi-mentaiko during 4 °C storage	Hara et al. (2009)
Nisin (400 or 1250 IU/g) or ALTA™ 2341 (1%)	Surface treatment decreased *L. monocytogenes* on smoked salmon sliced, and when packaged in 100% CO_2, *L. monocytogenes* count decreased below detectable concentration (i.e., 2 logs) during storage at 4 °C/21 days with both antimicrobials	Szabo and Cahill (1999)
Nisin and Microgard™	Increased product shelf life by reducing the total bacterial counts and delayed *L. monocytogenes* growth in fresh-chilled salmon	Zuckerman and Ben Avraham (2002)
Nisin and heat treatment	Successfully prevented *L. monocytogenes* contamination in cold-pack lobster cans. Nisin (25 mg/kg of can content) along with mild heating (60 °C internal temperature/5 min) reduced *L. monocytogenes* by 3–5 log relative to nisin and/or heat treatment alone which achieved 1- to 3-log cycle reduction only	Budu-Amoako et al. (1999)
Nisin and moderate heat treatment (60 °C/ 3 min)	Synergistically decreased total mesophiles and *L. monocytogenes* count. No *L. monocytogenes* cells were recovered from caviar treated with nisin (750 IU/ml) and heat at 4 °C after 28 days	Al-Holy et al. (2005)
Nisin and sodium lactate (120–180 IU/g and 1.8%)	Restricted growth of *L. monocytogenes* and mesophilic aerobes in cold smoked products in both the case of applications, i.e., before or after smoking	Nykanen et al. (2000)

TABLE 6.5 (Continued)

Bacteriocin in single or along with other treatments	Activity observed	References
Nisaplin, lysozyme, e-polylysine, and chitosan	Nisaplin inhibited growth of *L. monocytogenes* during storage in minced tuna fish (500 ppm) and in salmon roe (250 ppm)	Takahashi et al. (2011)
Lc. lactis PSY2 derived bacteriocin	Bacteriocin retained total viable count within maximum limit up to 21 days compared to control in reef cod fillets during 4 °C storage. The maximum activity observed against *S. aureus* and *Pseudomonadaceae*	Sarika et al. (2012)
Sakacin P (3.5 µg/g or 12 ng/g)	Inhibited *L. monocytogenes* growth in vacuum packaged CSS during 3 weeks of storage at 10 °C	Aasen et al. (2003)
Divergicin M35(50 µg/g) or *C. divergens* M35	Inhibitiom of *L. monocytogenes* growth with bacteriocin and concentrated supernatant at beginning and up to 21 days of storage, while LAB count remain unchanged	Tahiri et al. (2009a,b)
Enterocin AS-48 (250 ppm)	Significantly decreased the viable count of *S. aureus* and the levels of the biogenic amines like histamine and tyramine and producer LAB strains	Ananou et al. (2014)
Enterocin 1071A and enterocin b1071B	Suppressed the proliferation of aerobic mesophilic bacteria in fish spreads during storage at low temperature	Dicks et al. (2006)
Nisin (500 IU/cm^2) in chitosan coated plastic films	Retarded *L. monocytogenes* growth on CSS by about 1 log unit during storage at room temperature for 10 days	Ye et al. (2008)
Nisin + sodium lactate (2.3 mg/cm^2) in chitosan coated plastic films	Combination of nisin with sodium lactate inhibited *L. monocytogenes* cells in vacuum-packaged CSS during storage at 4 °C for up to 6 weeks	Ye et al. (2008)
Bacteriocin from *Lactobacillus curvattus* CWBI-B28	Packaging of CSS in bacteriocin-coated plastic film inactivated *L. monocytogenes* during storage at refrigeration temperature	Ghalfi et al. (2006)

TABLE 6.5 *(Continued)*

Bacteriocin in single or along with other treatments	Activity observed	References
Sakacin P and/or *Lb. sake* cultures (sakacin P producer)	Bacteriocin demonstrated inhibitory effect toward *L. monocytogenes* in CSS. Initial inhibition was attributed to Sakacin P on proliferation of *L. monocytogenes*, whereas the direct culture of *Lb. sake* showed bacteriostatic action for 4 weeks at 10 °C	Katla et al. (2001)
C. piscicola A9b	Showed antilisterial activity in salmon juice	Nilsson et al. (2004)
C. piscicola CS526	Showed antilisterial activity CSS	Yamazaki et al. (2003)
E. mundtii strain	Prevented *L. monocytogenes* growth on CSS during 4 weeks	Bigwood et al. (2012)
Lc. lactis subsp. *lactis* strain CWBI B1410	Diminished the counts of enteric bacteria in the fermented fish (guedj) with rapid acidification	Diop et al. (2009)
Bacteriocin (5%) from *Lb. fermentum* UN01	Low number of colonies in bacteriocin-treated sample in comparison to control and improved the shelf life of fish	Udhayashree et al. (2012)

divercin V41, and divergicin M35 are reported from Carnobacteria isolated form marine environment (Bhugaloo-Vial et al., 1996; Metivier et al., 1998; Suzuki et al., 2005; Tahiri et al., 2004). Species of *Carnobacterium* commonly found in fish products have less acidifying potential and ability to survive and synthesize bacteriocin at high salt concentration, low temperature, and low carbohydrate content without altering the sensory properties of food (Buchanan and Bagi, 1997). Some species also showed potential for probiotic candidature (Leisner et al., 2007). Similarly, enterococci have also gained much interest for bio-control of *L. monocytogenes* in the processed seafoods. Bacteriocins like enterocin P and enterocin B from *Enterococcus faecium* (Arlindo et al., 2006; Pinto et al., 2009) and other unknown bacteriocins from *Enterococcus mundtii* (Bigwood et al., 2012; Campos et al. 2006; Valenzuela et al., 2010) have shown successful applications.

Combination of nisin and NaCl followed by CO_2-packaging of cold-smoked salmon (CSS) showed 1 log to 2 log cycle reduction in the population of *L. monocytogenes*. The subsequent lag phase was 8 and 20 days in CSS with two different concentration of nisin, that is, 500 and 1000 IU/g, respectively (Nilsson et al., 1997). In another approach, effect of nisin, sodium lactate, or their combination was evaluated against *L. monocytogenes* and mesophilic bacteria on vacuum-packaged, cold-smoked rainbow trout (fish) (Niskanen and Nurmi, 2000). Individually, both nisin and sodium lactate retarded *L. monocytogenes* growth; however, their combination was even more effective, which reduced the count from 3.3 to 1.8 log CFU/g during 16 days of storage at 8 °C. Nisin together with salts and organic acids showed significant inhibition of microorganisms during storage in comparison to single treatment. For example, beheaded and peeled fresh shrimps dipped in nisin (500 IU/ml), EDTA (0.02 M), sodium benzoate (3%, w/v), sodium diacetate (3%, w/v), or potassium sorbate (3%, w/v) either singly or in their mixture were vacuum packaged and stored at 4 °C for 7 days (Wan-norhana et al., 2012). The combination of nisin–EDTA–potassium sorbate and nisin–EDTA–sodium diacetate effectively lowered count of *L. monocytogenes*, aerobic, and psychrotrophic bacteria throughout storage.

Tome et al. (2008) applied five different bacteriocin producing strains, *Lb. curvatus* ET30, *Lb. curvatus* ET06, *Lb. delbrueckii* ET32, *Enterococcus faecium* ET05, and *Pd. acidiliactici* ET34 against *L. innocua* 2030c on salmon fillets before and after cold-smoking followed by storage under vacuum pack. All bacteriocinogenic strains showed antilisterial activity, but strain ET105 showed strong inhibition of target cells among all in vacuum-packaged CSS. In another investigation, surface application of a bacteriocin preparation obtained from *Lc. lactis* PSY2 (originally isolated from marine perch) reduced proliferation of target bacteria during storage at 4 °C on fillets of reef cod. The strain PSY2 effectively inhibited the growth of *Staphylococcus* sp. and *Pseudomonadaceae* in fillets (Sarika et al., 2012). Ready-to-eat food products, like cold-smoked foods, which are not further cooked prior to consumption, pose health risk to the sensitive consumers due to contamination and subsequent growth of *L. monocytogenes*. Incorporation of bacteriocins either through mixing into food matrixes, spraying, injection, or immobilization demonstrated promising results and avoided the growth of target pathogens in various studies. In one of such studies, purified sakacin P incorporated at the rate 12 ng/g or 3.5 µg/g resulted into partial-to-inclusive suppression of *L. monocytogenes* growth in vacuum-packaged CSS (Aasen et al. 2003). During storage, authors observed a decrease in bacteriocin titers ascribed to degradation of bacteriocin by proteases within salmon tissue.

Seasonal factors together with food processing environment influence the efficiency of bacteriocins. For instance, Vaz-Velho et al. (2005) immersed salmon-trout fillets for 30 s in diluted *C. divergens* V41 supernatant before the cold-smoking process and took two trials, first in summer during which the smoking process the temperature reached to 33 °C, whereas the second in winter during which low temperature was prevalent. A strong listericidal effect was observed in the second trial, and no cells of *L. innocua* were found after smoking or at the end of storage period during the first trial; the viable count of *L. innocua* reduced by 3-log cycles after 1 week followed by subsequent regrowth of the bacterium.

Immobilization of bacteriocins on packaging material is another approach to enhance and extend their action for the preservation of seafoods. In one approach on vacuum-packed CSS, nisin-coated plastic film diminished cocktail of *L. monocytogenes* strains by 3.9 log CFU/cm^2 both at 4 °C as well as 10 °C after 56 days and 49 days during storage, respectively (Neetoo et al., 2008). The method successfully restricted proliferation of other spoilage bacteria such as aerobes, anaerobes, and LAB as well in a concentration-dependent style. In another study, combinations of nisin and lysozyme (from hen egg white and oysters) were employed within calcium alginate coating on the surface of CSS (Datta et al. 2008). In comparison to the control samples, *L. monocytogenes* and *Salmonella anatum* growth were restricted and ranged from 2.2 to 2.8 log CFU/g with nisin–lysozyme–calcium alginate coatings after 35 days of storage at 4 °C. The combination of nisin (500 IU/ml) and radio-frequency (RF; 27 MHz) heating led complete destruction of *L. innocua* cells from salmon (*Oncorhynchus keta*) as well as sturgeon (*Acipenser transmontanus*) caviar (Al-Holy et al., 2004).

Recently, bacteriocinogenic enterococcal strain was isolated from *Odontesthes platensis* and evaluated for antimicrobial spectrum regarding possible relevance for fish preservation. Bacteriocin in combination with chitosan and sodium lactate created synergistic action on the inhibition of *L. innocua*, *Shewanella putrefaciens*, and other psychrophilic microflora isolated from fish (Schelegueda et al., 2015).

Studies on biopreservation of fermented fish and seafoods through bacteriocins are very less due to production of biogenic amines by bacteriocinogenic LAB. It is believed that salted fermented foods are rich in amino acids, and several LABs can generate relatively significant amounts of biogenic amines within the food matrix (Mah et al. 2003). In this context, Mah and Hwang (2009) applied *Staphylococcus xylosus* in form of protective culture to reduce the generation of biogenic amines by starter LAB in salted and fermented anchovy. Matamoros et al. (2009) investigated several nonbiogenic amine producer strains of *Carnobacterium alterfunditum*, *Lc. piscium*, *Lb. fuchuensis*, and *Leu. gelidum* as possible protective cultures during fish preservation.

6.5.6   IN EGG PRODUCTS

The commercially available liquid whole egg is thermal processed with regard to enhance the shelf life as well as inactivation of foodborne pathogens. Traditionally, it was pasteurized, but now, it is ultrapasteurized (more than 60 °C for less than 3.5 min) together with aseptic filling and packaging. Such preparations have minimum 10 weeks shelf life at 4 °C. Egg products do impose risk of pathogens like *salmonella* and *L. monocytogenes*, and therefore, few studies do indicate use of bacteriocins to preserve liquid egg preparations. As revealed from Table 6.6, the application of bacteriocin in conjunction with other bacteriocin (pediocin), lysozyme, PEF, and HHP significantly reduced the spoilage and pathogenic microorganisms in liquid egg products. Further, immobilization of nisin with polymers also enhanced liquid egg products shelf life (Jin, 2010; Jin et al., 2013).

6.6   REGULATORY STANDARD AND OTHER FACTORS CHALLENGED TO USE OF BACTERIOCINS

At present, nisin is the single bacteriocin approved to be used as food preservative worldwide. Though nisin is authentically used in more than 80

TABLE 6.6 Studies on Bacteriocin Applications in Egg Products.

Bacteriocin in single or along with other treatments	Activity observed	References
Nisin (200 IU/ml)	Keeping quality of traditionally pasteurized liquid whole egg was extended by 9 days to 11 days at 6 °C as compared to control samples	Delves-Broughton et al. (1992)
Nisin (1000 IU/ml)	Diminished *L. monocytogenes* count by 1.6 to over 3.3 log CFU/ml in pH-adjusted ultrapasteurized liquid whole egg. It also prevented or delayed the proliferation of pathogen at 4 °C and 10 °C for 8 to 12 weeks	Schuman and Sheldon (2003)
Nisin (10 mg/l)	Significantly decreased the decimal reduction times (*D*-values) of *L. monocytogenes*	Knight et al. (1999)
Nisin	During pasteurization, the heat susceptibility of *S. enteritidis* PT4 in liquid whole egg and egg white was increased	Boziaris et al. (1998)
Nisin and lysozyme	Delayed the average onset of growth of *B. cereus* until 10 h at 25 °C or approximately 30 h in samples stored at 16 °C	Antolinos et al. (2011)
Nisin and HHP	Enhanced the inhibition of viable *E. coli* and *L. innocua* cells in liquid egg samples	Ponce et al. (1998)
Nisin and PEF	Synergistically inactivated *L. innocua* in liquid egg depending on the concentration of bacteriocin, intensity of electric field and pulse counts	Calderón-Miranda et al. (1999a,b)
Nisin (250 mg) in polylactic acid (PLA) coating	Rapidly decreased the *Listeria* count in liquid egg white below detectable levels after 1 day, and the bacterium remained undetectable throughout the whole storage periods of 48 days at 10 °C and 70 days at 4 °C	Jin (2010)
Nisin (250 mg) plus allyl isothiocyanate in PLA coating	Reduced the population of a three-strain *S. enterica* cocktail inoculated in liquid egg white to an undetectable level after storage of 21 days	Jin et al. (2013)

countries, the occurrence as well as level of it in different foods differs in national legislation. While in the United States, in 1988, Food and Drug Agency approved its application in pasteurized processed cheese spreads, and further, the primary approval was followed by other licensed relevance like Food Safety Inspection Service and Inspection Service 2002. There exist few diverse regulations regarding the applications of bacteriocin in foodstuffs, viz. based on the category of food, the mechanism of action (e.g. crude preparation of bacteriocin or direct incorporation of protective bacterial cultures), and the regulations of each different country. These regulations include direct food laws along with other related aspects of packaging, labeling, export, and genetically modified organisms (Gálvez et al., 2011). At present, major countries follow Codex Alimentarius standards (WHO/FAO) for application of nisin in different food products for harmony. In India, Food Safety and Standards Authority of India (FSSAI) specified use of nisin as food additive. It is categorized as a class 2 preservative (INS 234) for dairy products (like canned rasgulla, paneer, and processed cheese); however, it does not specify its use and dose for meat and other food products in India.

It is noteworthy that different countries have different regulations to make use of bacteriocin in foods. In European Union, nisin is considered either food ingredient or food additive based on its form (such as lyophilised powder, liquid concentrated preparation, or live bacteriocin producing cultures), and accordingly, its maximum permissible amount is defined for specific food product. The regulation also varies according to the kind of bacteriocin application in food product such as either applied directly in the food or applied during the active food packaging. On the other hand, in the Unites States, it is specified as shelf-life extenders that can be applied in a variety of food products. Bacteriocinogenic strains fall within the category of microbial cultures if viewed from a regulatory point. In the United States, a novel strain of microorganism intended for use in food should be classified either as a GRAS substance or as an additive (Wessels et al. 2004), whereas in the European Union, the microbiological cultures having long time span of secure use are recognized as conventional food ingredients and consequently covered under general European food law (European Parliament and Council, 2002).

The different forms of Nisaplin™ and Chrisin™ are the example of commercially available nisin in the market as lyophilized preparation after microbial fermentation. Christian Hansen (Denmark) and Danisco (Denmark) are the example of companies which markets a series of bacteriocin producing strains including *Lb. plantarum*, *Lb. casei*, *Lb. para-casei*, *Lb. sakei*, and *Propionibacterium freudenreichii* subsp. *shermanii* for definite applications in dairy and meat products and genetic modification not been subjected in such strains.

Though Codex Alimentarius standards specified maximum limit for nisin for different food products, this value differs in different countries for the same product due to their own legislative bodies. For instance, in Australia and New Zealand, maximum limit for processed meat is 12.5 mg/kg, whereas in China, it is 500 mg/kg. At the same time, there is no value defined for meat products or seafood products in India. The probable reason could be different environmental conditions, food product types (food components and even inherent fermenting microflora in case of fermented products), and their physicochemical parameters (pH, temperature, and moisture) which ultimately affect the activity and stability of bacteriocins. There exists a tremendous difference in the climatic conditions, kind of substrates, level of hygiene, and sanitary practices followed in different countries. Thus, it is difficult to implement a common guideline or standards for the use of bacteriocin or bacteriocinogenic strains for all countries. However, it can be harmonized up to certain extent.

6.7 CONCLUSION

Consumer awareness toward the chemical preservative free foods has created enormous interest of researchers and enforced the dairy and food industry to employ biological preservation methods. LAB and their metabolites are generally recognized as safe and making use of bacteriocinogenic LAB strains and/or bacteriocins represent viable alternate over chemical preservatives. In several food products, bacteriocins alone showed outstanding effects, and while in others, their efficiency found to enhanced through combination with organic acids, salts, heat, and/or nonthermal treatments (HHP, PEF, and HIPEF) without any detrimental effect on overall body, texture, and sensory properties of the food products. Several inherent factors together with environmental conditions influence the activity of bacteriocins or survival of bacteriocinogenic strains, and it is important to continue to expand our knowledge as well as understanding in this direction to improve their performance for future applications in food model systems followed by adequate application as flourishing biopreservatives. In addition, outcome of a number of researches confirmed productions of new promising bacteriocins with prospective antagonistic attributes, and therefore, more efforts should make to define, characterize, and/or classify such novel bacteriocins in coming years. Nisin is the sole legally permitted bacteriocin till date in numerous countries; however, other bacteriocins have also established enormous potential for commercial applications that should be considered by legal authorities and governing bodies.

KEYWORDS

- **antimicrobial activity**
- **bacteriocins**
- **biopreservation**
- **enterocin**
- **lactic acid bacteria**
- **nisin, non-thermal treatments**
- **pediocin**

REFERENCES

Aasen, I. M.; Markussen, S.; Moretro, T.; Katla, T.; Axelsson, L.; Naterstad, K. Interactions of the Bacteriocins Sakacin P and Nisin With Food Constituents. *Int. J. Food Microbiol.* **2003,** *87,* 35–43.

Abriouel, H.; Lucas, R.; Ben Omar, N.; Valdivia, E.; Galvez, A. Potential Applications of the Cyclic Peptide Enterocin AS-48 in the Preservation of Vegetable Foods and Beverages. *Prob. Antimicrob. Protect.* **2010,** *2,* 77–89.

Alasalvar, C. *Seafood Quality, Safety and Health Applications.* John Wiley and Sons: USA, 2010; pp 203. ISBN 978-1-4051-8070-2.

Al-Holy, M.; Lin, M.; Rasco, B. Destruction of *Listeria monocytogenes* in Sturgeon (*Acipenser transmontanus*) Caviar by a Combination of Nisin with Chemical Antimicrobials or Moderate Heat. *J. Food Protect.* **2005,** *68,* 512–520.

Al-Holy, M.; Ruiter, J.; Lin, M.; Kang, D. -H.; Rasco, B. Inactivation of *Listeria innocua* in Nisin-Treated Salmon (*Oncorhynchus keta*) and Sturgeon (*Acipenser transmontanus*) Caviar Heated by Radio Frequency. *J. Food Protect.* **2004,** *67,* 1848–1854.

Alpas, H.; Bozoglu, F. The Combined Effect of High Hydrostatic Pressure, Heat and Bacteriocins on Inactivation of Foodborne Pathogens in Milk and Orange Juice. *World J. Microbiol. Biotechnol.* **2000,** *16,* 387–392.

Ananou, S.; Garriga, M.; Hugas, M.; Maqueda, M.; Martínez-Bueno, M.; Gálvez, A.; Valdivia, E. Control of *Listeria monocytogenes* in Model Sausages by Enterocin AS-48. *Int. J. Food Microbiol.* **2005a,** *103,* 179–190.

Ananou, S.; Maqueda, M.; Martínez-Bueno, M.; Gálvez, A.; Valdivia, E. Control of *Staphylococcus aureus* in Sausages by Enterocin AS-48. *Meat Sci.* **2005b,** *71,* 549–576.

Ananou, S.; Zentar, H.; Martínez-Bueno, M.; Galvex, A.; Maqueda, M.; Valdivia, E. The Impact of Enterocin AS-48 on the Shelf Life and Safety of Sardines (*Sardina pilchardus*) Under Different Storage Conditions. *Food Microbiol.* **2014,** *44,* 185–195.

Antolinos, V.; Muñoz, M.; Ros-Chumillas, M.; Aznar, A.; Periago, P. M.; Fernández, P. S. Combined Effect of Lysozyme and Nisin at Different Incubation Temperature and Mild Heat Treatment on the Probability of Time to Growth of *Bacillus cereus*. *Food Microbiol.* **2011,** *28,* 305–310.

Ariyapitipun, T.; Mustapha, A.; Clarke, A.D. Survival of *Listeria monocytogenes* Scott A on Vacuum-Packaged Raw Beef Treated with Polylactic Acid, Lactic Acid, and Nisin. *J. Food Protect.* **2000,** *63,* 131–136.

Ariyapitipun, T.; Mustapha, A.; Clarke, A. D. Microbial Shelf Life Determination of Vacuum Packaged Fresh Beef Treated with Polylactic Acid, Lactic Acid, and Nisin Solutions. *J. Food Protect.* **1999,** *62,* 913–920.

Arlindo, S.; Calo, P.; Franco, C.; Prado, M.; Cepeda, A.; Barros-Velázquez, J. Single Nucleotide Polymorphism Analysis of the Enterocin P Structural Gene of *Enterococcus faecium* Strains Isolated from Non Fermented Animal Foods. *Mol. Nutr. Food Res.* **2006,** *50,* 1229–1238.

Arques, J. L.; Fernandez, J.; Gaya, P.; Nunez, M.; Rodriguez, E.; Medina, M. Antimicrobial Activity of Reuterin in Combination with Nisin Against Food-Borne Pathogens. *Int. J. Food Microbiol.* **2004,** 95, 225–229.

Arqués, J. L.; Rodríguez, E.; Gaya, P.; Medina, M.; Nuñez, M. Effect of Combinations of High-Pressure Treatment and Bacteriocin-Producing Lactic acid Bacteria on the Survival of *Listeria monocytogenes* in Raw Milk Cheese. *Int. Dairy J.* **2005a,** *15,* 893–900.

Arqués, J. L.; Rodríguez, E.; Gaya, P.; Medina, M.; Guamis, B.; Nuñez, M. Inactivation of *Staphylococcus aureus* in Raw Milk Cheese by Combinations of High-Pressure Treatments and Bacteriocin-Producing Lactic Acid Bacteria. *J. Appl. Microbiol.* **2005b,** *98,* 254–260.

Avons, L.; Van Uyten, E.; De Vuyst, L. Cell Growth and Bacteriocin Production of Probiotic *Lactobacillus* Strains in Different Media. *Int. Dairy J.* **2004,** *14,* 947–955.

Aymerich, M. T.; Garriga, M.; Costa, S.; Monfort, J. M.; Hugas, M. Prevention of Ropiness in Cooked Pork by Bacteriocinogenic Cultures. *Int. Dairy J.* **2002,** *12,* 239–246.

Aymerich, T.; Garriga, M.; Ylla, J. Application of Enterocins as Biopreservatives Against *Listeria innocua* in Meat Products. *J. Food Protect.* **2000,** *63,* 721–726.

Aymerich, T.; Picouet, P. A.; Monfort, J. M. Decontamination Technologies for Meat Products. *Meat Sci.* **2008,** *78,* 114–129.

Baird-Parker, T. The Production of Microbiologically Safe and Stable Foods. In *The Microbiological Safety and Quality of Food.* Lund, B. M., Baird-Parker, T., Gaithersburg, G. G. W., Eds.; Springer: USA, 2000; pp 3–18.

Bali, V.; Panesar, P. S.; Bera, M. B. Effect of Bacteriocin Extracted from *Enterococcus faecium bs* 13 on Shelf Life of Paneer and Khoya. *Int. J. Food Nutr. Sci.* **2013,** *2*(1), 5–11.

Barboza de Martinez, Y.; Ferrer, K.; Salas, E. M. Combined Effects of Lactic Acid and Nisin Solution in Reducing Levels of Microbiological Contamination in Red Meat Carcasses. *J. Food Protect.* **2002,** *65,* 1780–1783.

Bari, M. L.; Ukuku, D. O.; Kawasaki, T.; Inatsu, Y.; Isshiki, K.; Kawamoto, S. Combined Efficacy of Nisin and Pediocin with Sodium Lactate, Citric Acid, Phytic Acid, and Potassium Sorbate and EDTA in Reducing the *Listeria monocytogenes* Population of Inoculated Fresh-Cut Produce. *J. Food Protect.* **2005,** *68,* 1381–1387.

Bartowsky, E. J. Bacterial Spoilage of Wine and Approaches to Minimize It. *Lett. Appl. Microbiol.* **2009**, *48,* 149–156.

Basanta, A.; Sanchez, J.; Gomez-Sala, B.; Herranz, C.; Hernanadez, P. E.; Cintas, L. M. Antimicrobial Activity of *Enterococcus faecium* L50, a Strain Producing Enterocins L50 (L50A and L50B), P and Q, Against Beer-Spoilage Lactic Acid Bacteria in broth, Wort (Hopped and Unhopped), and Alcoholic and Non-Alcoholic Lager Beers. *Int. J. Food Microbiol.* **2008**, *125,* 293–307.

Bauer, R.; Nel, H. A.; Dicks, L. M. T. Pediocin PD-1 as a Method to Control Growth of *Oenococcus oeni* in Wine. *Am. J. Enol. Viticult.* **2003**, *54,* 86–91.

Bennik, M. H. J.; Smid, E. J.; Gorris, L. G. M. Vegetable-Associated *Pediococcus parvulus* Produces Pediocin PA-1. *Appl. Environ. Microbiol.* **1997**, *63*(5): 2074–2076.

Bennik, M. H. J.; Van Overbeek, W.; Smid, E. J.; Gorris, L. G. M. Biopreservation in Modified Atmosphere Stored Mungbean Sprouts: The Use of Vegetable-Associated Bacteriocinogenic Lactic Acid Bacteria to Control the Growth of *Listeria monocytogenes*. *Lett. Appl. Microbiol.* **1999**, *28,* 226–232.

Benz, R.; Jung, G.; Sahl, H.-G. Mechanism of Channel Forming Lantibiotics in Black Lipid Membranes. In *Nisin and Novel Lantibiotics.* ESCOM, Scientific Publishers BV: Leiden, The Netherlands, 1991; pp 359–372.

Bhugaloo-Vial, P.; Dousset, X.; Metivier, A.; Sorokine, O.; Anglade, P.; Boyaval, P.; Marion, D. Purification and Amino Acid Sequences of Piscicocins V1a and V1b, Two Class IIa Bacteriocins Secreted by *Carnobacterium piscicola* V1 that Display Significantly Different Levels of Specific Inhibitory Activity. *Appl. Environ. Microbiol.* **1996**, *62*(12), 4410–4416.

Bigwood, T.; Hudson, J. A.; Cooney, J.; McIntyre, L.; Billington, C.; Heinemann, J. A.; Wall, F. Inhibition of *Listeria monocytogenes* by *Enterococcus mundtii* Isolated from Soil. *Food Microbiol.* **2012**, *32,* 354–360.

Black, E. P.; Kelly, A. L.; Fitzgerald, G. F. The Combined Effect of High Pressure and Nisin on Inactivation of Microorganisms in Milk. *Innov. Food Sci. Emerg. Technol.* **2005**, *6,* 286–292.

Boots, J.-W.; Floris, R. Lactoperoxidase: From Catalytic Mechanism to Practical Applications. *Int. Dairy J.* **2006**, *16,* 1272–1276.

Boussouel, N.; Mathieu, F.; Benoit, V.; Linder, M.; Revol-Junelles, A. M.; Milliere, J. B. Response Surface Methodology, an Approach to Predict the Effects of a Lactoperoxidase System, Nisin, Alone or in Combination, on Listeria Monocytogenes in Skim Milk. *J. Appl. Microbiol.* **1999**, *86,* 642–652.

Boziaris, I. S.; Humpheson, L.; Adams, M. R. Effect of Nisin on Heat Injury and Inactivation of *Salmonella enteritidis* PT4. *Int. J. Food Microbiol.* **1998**, *43,* 7–13.

Breidt, F.; Crowley, K. A.; Fleming, H. P. Controlling Cabbage Fermentations with Nisin and Nisin-Resistant *Leuconostoc mesenteroides*. *Food Microbiol.* **1995**, *12,* 109–116.

Brotz, H.; Bierbaum, G.; Leopold, K.; Reynolds, P. E.; Sahl, H. G. The Lantibiotic Mersacidin Inhibits Peptidoglycan Synthesis by Targeting Lipid II. *Antimicrobial Agents Chemother.* 1998, *42,* 154–160.

Buchanan, R. L.; Bagi, L. K. Microbial Competition: Effect of Culture Conditions on the Suppression of *Listeria monocytogenes* Scott A by *Carnobacterium piscicola*. *J. Food Protect.* **1997**, *60*(3), 254–261.

Budu-Amoako, E.; Ablett, R. F.; Harris, J.; Delves-Broughton, J. Combined Effect of Nisin and Moderate Heat on Destruction of *Listeria monocytogenes* in Cold-Pack Lobster Meat. *J. Food Protect.* **1999**, *62,* 46–50.

Cabo, M. L.; Pastoriza, L.; Sampedro, G.; Gonzalez, M.; Murado, M. A. Joint Effect of Nisin, CO2, and EDTA on the Survival of *Pseudomonas aeruginosa* and *Enterococcus faecium* in a Food Model System. *J. Food Protect.* **2001,** *64,* 1943–1948.

Cabo, M. L.; Torres, B.; Herrera, J. J.; Bernardez, M.; Paztoriza, L. Application of Nisin and Pediocin against Resistance and Germination of *Bacillus* spores in Sous Vide Products. *J. Food Protect.* **2009,** *72,* 515–623.

Calderón-Miranda, M. L.; Barbosa-Cánovas, G. V.; Swanson, B. G. Inactivation of *Listeria innocua* in Liquid Whole Egg by Pulsed Electric Fields and Nisin. *Int. J. Food Microbiol.* **1999,** *51 (1),* 7–17.

Campos, C. A.; Rodriguez, O.; Calo-Mata, P.; Prado, M.; Barros-Velázquez, J. Preliminary Characterization of Bacteriocins *from Lactococcus lactis, Enterococcus faecium* and *Enterococcus mundtii* Strains Isolated from Turbot (*Psetta maxima*). *Food Res. Int.* **2006,** *39,* 356–364.

Cao-Hoang, L.; Chaine, A.; Grégoire, L.; Wache, Y. Potential of Nisin-Incorporated Sodium Caseinate Films to Control *Listeria* in Artificially Contaminated Cheese. *Food Microbiol.* **2010,** *27,* 940–944.

Capellas, M.; Mor-Mur, M.; Gervilla, R.; Yuste, J.; Guamis, B. Effect of High Pressure Combined with Mild Heat or Nisin on Inoculated Bacteria and Mesophiles of Goats' Milk Fresh Cheese. *Food Microbiol.* **2000,** *17,* 633–641.

Chang, J. Y.; Chang, H. C. Growth Inhibition of Foodborne Pathogens by Kimchi Prepared with Bacteriocin-Producing Starter Culture. *J. Food Sci.* **2011,** *76,* M72–M78.

Chawla, R.; Singh, A.; Patil, G. R. Shelf Life Enhancement of Functional Doda Burfi (Indian Milk Cake) with Biopreservatives Application. *Int. J. Res. Sci. Technol.* **2015,** *5*(2), 26–40.

Chen, C. M.; Sebranek, J. G.; Dickson, J. S.; Mendonca, A. F. Combining Pediocin (ALTA 2341) with Post Packaging Thermal Pasteurization for Control of *Listeria monocytogenes* on Frankfurters. *J. Food Protect.* **2004a,** *67,* 1855–1865.

Chen, C. M.; Sebranek, J. G.; Dickson, J. S.; Mendonca, A. F. Combining Pediocin with Post Packaging Irradiation for Control of *Listeria monocytogenes* on Frankfurters. *J. Food Protect.* **2004b,** *67,* 1866–1875.

Chen, H.; Hoover, D. G. Bacteriocins and Their Food Applications. *Comprehen. Rev. Food Sci. Food Saf.* **2003,** *2,* 82–100.

Chen, Y.; Wu, H.; Yanagida, F. Isolation and Characteristics of Lactic Acid Bacteria Isolated from Ripe Mulberries in Taiwan. *Braz. J. Microbiol.* **2010,** *41,* 916–921.

Choi, M. H.; Park, Y. H. Selective Control of Lactobacilli in Kimchi with Nisin. *Lett. Appl. Microbiol.* **2000,** *30,* 173–177.

Claeys, W. L.; Cardoen, S.; Daube, G.; De Block, J.; Dewettinck, K.; Dierick, K.; De Zutter, L.; Huyghebaert, A.; Imberechts, H.; Thiange, P.; Vandenplas, Y.; Herman, L. Raw or Heated Cow Milk Consumption: Review of Risks and Benefits. *Food Control.* **2013,** *31,* 251–262.

Cleveland, J.; Montville, T. J.; Nes, I. F.; Chikindas, M. L. Bacteriocins: Safe, Natural Anti-microbials for Food Preservation. *Int. J. Food Microbiol.* **2001,** *71,* 1–20.

Corsetti, A.; Settanni, L.; Van Sinderen, D. Characterization of Bacteriocin-Like Inhibitory Substances (BLIS) from Sourdough Lactic Acid Bacteria and Evaluation of Their In Vitro and In Situ Activity. *J. Appl. Microbiol.* **2004,** *96,* 521–534.

Cosby, D. E.; Harrison, M. A.; Toledo, R. T. Vacuum or Modified Atmosphere Packaging and EDTA-Nisin Treatment to Increase Poultry Product Shelf-Life. *J. Appl. Poultry Res.* **1999,** *8,* 185–190.

Cotter, P. D.; Hill, C.; Ross, R. P. Food Microbiology: Bacteriocins: Developing Innate Immunity for Food. *Nat. Rev. Microbiol.* **2005,** *3,* 777–788.

Daeschel, M. A.; McGuire, J.; Al-Makhlafi, H. Antimicrobial Activity of Nisin Adsorbed to Hydrophilic and Hydrophobic Silicon Surfaces. *J. Food Protect.* **1992,** *55,* 731–735.

Dal Bello, B.; Cocolin, L.; Zeppa, G.; Field, D.; Cotter, P. D.; Hill, C. Technological Characterization of Bacteriocin Producing *Lactococcus lactis* Strains Employed to Control *Listeria monocytogenes* in Cottage Cheese. *Int. J. Food Microbiol.* **2011,** 153, 58–65.

Dalie, D. K. D.; Deschamps, A. M.; Richard-Forget, F. Lactic Acid Bacteria–Potential for Control of Mould Growth and Mycotoxins: A Review. *Food Control.* **2010,** *21,* 370–380.

Datta, S.; Janes, M. E.; Xue, Q. G.; Losso, J.; La Peyre, J. F. Control of *Listeria monocytogenes* and *Salmonella anatum* on the Surface of Smoked Salmon Coated with Calcium Alginate Coating Containing Oyster Lysozyme and Nisin. *J. Food Sci.* **2008,** *73,* M67–M71.

Davies, E. A.; Delves-Broughton, J. Nisin. In Encyclopedia of Food Microbiology; Robinson R, Batt C, Patel P, Eds.; Academic Press: London, 1999; pp 191–198.

Davies, E. A.; Bevis, H. E.; Delves-Broughton, J. The Use of the Bacteriocin, Nisin, as a Preservative in Ricotta-Type Cheeses to Control the Food-Borne Pathogen *Listeria monocytogenes. Lett. Appl. Microbiol.* **1997,** *24,* 343–346.

de Arauz, L. J.; Jozala, A. F.; Mazzola, P. G.; Penna, T. C. V. Nisin Biotechnological Production and Application: A Review. *Trends Food Sci. Technol.* **2009,** *20,* 146–154.

De Buyser, M. L.; Dufour, B.; Maire, M.; Lafarge, V. Implication of Milk and Milk Products in Foodborne Diseases in France and in Different Industrialized Countries. *Int. J. Food Microbiol.* **2001,** 67, 1–17.

Deegan, L. H.; Cotter, P. D.; Hill, C.; Ross, P. Bacteriocins: Biological Tools for Bio-Preservation and Shelf-Life Extension. *Int. Dairy J.* **2006,** *16,* 1058–1071.

Delgado, A.; Brito, D.; *Peres, C.;* Arroyo, F. N.; Garrido-Fernández, A. Bacteriocin Production by *Lactobacillus pentosus* B96 can be Expressed as a Function of Temperature and NaCl Concentration. *Food Microbiol.* **2005,** *22,* 521–528.

Delves-Broughton, J.; Williams, G.C.; Wilkinson, S. The Use of Bacteriocin, Nisin, as a Preservative in Pasteurized Liquid Whole Egg. *Lett. Appl. Microbiol.* **1992,** *15,* 133–136.

Delves-Broughton, J.; Blackburn, P.; Evans, R. J.; Hugenholtz, *J.* Applications of the Bacteriocin, Nisin. *Antonie van Leeuwenhoek,* **1996,** *69,* 193–202.

Dicks, L. M. T.; Heunis, T. D. J.; van Staden, D. A.; Brand, A.; Sutyak Noll, K.; Chikindas, M. L. Chapter 19: Medical and Personal Care Applications of Bacteriocins Produced by Lactic Acid Bacteria. In *Prokaryotic Antimicrobial Peptides: From Genes to Applications*; Drider, D., Rebuffat, S., Eds.; Springer: New York, 2011; pp 391–421.

Dicks, L. M. T.; Todorov, S. D.; van der Merwe, M. P. Preservation of Fish Spread with Enterocins 1071A and 1071B, Two Antimicrobial Peptides Produced by *Enterococcus faecalis* BFE 1071. *J. Food Saf.* 2006, *26,* 173–183.

Diels, A. M. J.; Taeye, J. D.; Michiels, C. W. Sensitisation of *Escherichia coli* to Antibacterial Peptides and Enzymes by High-Pressure Homogenisation. *Int. J. Food Microbiol.* **2005,** *105,* 165–175.

Diop, M.B.; Dubois-Dauphin, R.; Destain, J.; Tine, E.; Thonart, P. Use of a Nisin-Producing Starter Culture of *Lactococcus lactis* subsp, *lactis* to Improve Traditional Fish Fermentation in Senegal. *J. Food Protect.* **2009,** *72,* 1930–1934.

Dortu, C.; Thonart, P. Les bactériocines des bactéries lactiques: Caractéristiques et intérêts pour labioconservation des produits alimentaires. *Biotechnol. Agronomy Soc. Environ.* **2009,** *13*(1), 143–154.

Economou, T.; Pournis, N.; Ntzimani, A.; Savvaidis, I. N. Nisin-EDTA Treatments and Modifi ed Atmosphere Packaging to Increase Fresh Chicken Meat Shelf-Life. *Food Chem.* **2009,** *114,* 1470–1476.

El-Ziney, M. G.; Debevere, J. M. The Effect of Reuterin on Listeria Monocytogenes and *Escherichia coli* O157:H7 in Milk and Cottage Cheese. *J. Food Protect.* **1998,** 61, 1275–1280.

Enan, G.; Alalyan, S.; Debevere, J. Inhibition of *Listeria monocytogenes* LMG10470 by Plantaricin UG1 In Vitro and In Beef meat. *Nahrung,* **2002,** *46,* 411–414.

Ercolini, D.; Ferrocino, I.; La Storia, A.; Mauriello, G.; Gigli, S.; Masi, P.; Villani, F. Development of Spoilage Microbiota in Beef Stored in Nisin Activated Packaging. *Food Microbiol.* **2010,** *27,* 137–143.

European Parliament and Council. Regulation (EC) No 178/2002 of 28 January 2002 Laying Down the General Principles and Requirements of Food Law, Establishing the European Food Safety Authority and Laying Down Procedures in Matters of Food safety. O J **2002,** L31:1–37.

Franklin, N. B.; Cooksey, K. D.; Getty, K. J. Inhibition of *Listeria monocytogenes* on the Surface of Individually Packaged Hot Dogs with a Packaging Film Coating Containing Nisin. *J. Food Protect.* **2004,** *67,* 480–485.

Franz, C. M. A. P.; Schillinger, U.; Holzapfel, W. H. Production and Characterization of Enterocin 900, A Bacteriocin Produced by *Enterococcus faecium* BFE 900 from Black Olives. *Int. J. Food Microbiol.* **1996,** *29,* 255–270.

Gabrielsen, C.; Brede, D. A.; Nes, I. F.; Diep, D. B. Circular Bacteriocins: Biosynthesis and Mode of Action. *Appl. Environ. Microbiol.* **2014,** *80,* 6854–6862.

Galvez, A.; Abriouel, H.; Lucas, R. L.; Omar, N. B. Bacteriocin-Based Strategies for Food Biopreservation. *Int. J. Food Microbiol.* **2007,** *120,* 51–70.

Galvez, A.; Abriouel, H.; Omar, N. B.; Lucas, R. Chapter 18: Food Applications and Regulation. In *Prokaryotic Antimicrobial Peptides: From Genes to Applications*; Drider, D., Rebuffat, S., Eds.; Springer: New York, 2011; pp 353–390.

Galvez, A.; Lucas, R. L.; Abriouel, H.; Valdivia, E.; Omar, N. B. Application of Bacteriocins in the Control of Foodborne Pathogenic and Spoilage Bacteria. *Crit. Rev. Biotechnol.* **2008,** *28,* 125–152.

Gao, Y.; Jia, S.; *Gao, Q.*; Tan, Z. L. A Novel Bacteriocin With a Broad Inhibitory Spectrum Produced by *Lactobacillus sake* C2, Isolated from Traditional Chinese Fermented Cabbage. *Food Control.* **2010,** 21, 76–81.

Garcia-Graells, C.; Masschalck, B.; Michiels, C. W. Inactivation of *Escherichia coli* in Milk by High-Hydrostatic-Pressure Treatment in Combination with Antimicrobial Peptides. *J. Food Protect.* **1999,** *62,* 1248–1254.

Garriga, M.; Aymerich, M. T.; Costa, S.; Monfort, J. M.; Hugas, M. Bactericidal Synergism through Bacteriocins and High Pressure in a Meat Model System During Storage. *Food Microbiol.* **2002,** *19,* 509–518.

Ghalfi, H.; Allaoui, A.; Destain, J.; Benkerroum, N.; Thonart, P. Bacteriocin Activity by *Lactobacillus curvatus* CWBI-B28 to Inactivate *Listeria monocytogenes* in Cold-Smoked Salmon During 4 °C Storage. *J. Food Protect.* **2006,** *69,* 1066–1071.

Gilbreth, S. E.; Somkuti, G. A. Thermophilin 110: A Bacteriocin of *Streptococcus thermophilus* ST110. *Curr. Microbiol.* **2005,** *51,* 175–182.

Gill, A. O.; Holley, R. A. Inhibition of Bacterial Growth on Ham and Bologna by Lysozyme, Nisin and EDTA. *Food Res. Int.* **2000a,** *33,* 83–90.

Gill, A. O.; Holley, R. A. Surface Application of Lysozyme, Nisin, and EDTA to Inhibit Spoilage and Pathogenic Bacteria on Ham and Bologna. *J. Food Protect.* **2000b**, *63*, 1338–1346.

Govaris, A.; Solomakos, N.; Pexara, A.; Chatzopoulou, S. The Antimicrobial Effect of Oregano Essential Oil, Nisin and Their Combination Against *Salmonella enteritidis* in Minced Sheep Meat during Refrigerated Storage. *Int. J. Food Microbiol.* **2010**, 137, 175–180.

Grande, M. J.; Abriouel, H.; Lucas, R.; Valdivia, E.; Ben Omar, N.; Martínez-Cañamero, M.; Gálvez, A. Efficacy of Enterocin AS-48 Against Bacilli in Ready to- Eat Vegetable Soups and Purees. *J. Food Protect.* **2007a**, *70*, 2339–2345.

Grande, M. J.; Lucas, R.; Abriouel, H.; Omar, N. B.; Maqueda, M.; MartiNez-Bueno, M.; Martínez-Cañamero, M.; Valdivia, E.; Gálvez A. Control of *Alicyclobacillus acidoterrestris* in Fruit Juices by Enterocin AS-48. *Int. J. Food Microbiol.* **2005**, *104*, 289–297.

Grande, M. J.; Lucas, R.; Abriouel, H.; Valdivia, E.; Ben Omar, N.; Maqueda, M.; Martínez-Bueno, M.; Martínez-Cañamero, M.; Gálvez, A. Inhibition of *Bacillus licheniformis* LMG **19409** from Ropy Cider by Enterocin AS-48. *J. Appl. Microbiol.* **2006b**, *101*, 422–428.

Grande, M. J.; Lucas, R.; Abriouel, H.; Valdivia, E.; Ben Omar, N.; Maqueda, M.; Martínez-Bueno, M.; Martínez-Cañamero, M.; Gálvez, A. L. Inhibition of Toxicogenic *Bacillus cereus* in Rice Based Foods by Enterocin AS-48. *Int. J. Food Microbiol.* **2006a**, *106*, 185–194.

Grande, M. J.; Lucas, R. L.; Abriouel, H.; Valdivia, E.; Ben Omar, N.; Maqueda, M.; Martinez-Canamero, M.; Galvez, A. Treatment of Vegetable Sauces with Enterocin AS-48 Alone or in Combination With Phenolic Compounds to Inhibit Proliferation of *Staphylococcus aureus*. *J. Food Protect.* **2007b**, *70*, 405–411.

Gupta, R. K.; Prasad, N.; Prasad, D. N. Use of Nisin in Dairy Industry, *Indian Dairyman.* **1989**, *41*(7), 229–233.

Hampikyan, H.; Ugur, M. The Effect of Nisin on *L. monocytogenes* in Turkish Fermented Sausages (sucuks). *Meat Sci.* **2007**, *76*, 327–332.

Hara, H.; Ohashi, Y.; Sakurai, T.; Yagi, K.; Fujisawa, T.; Igimi, S. Effect of Nisin (Nisaplin) on the Growth of *Listeria monocytogenes* in Karashi-Mentaiko (Red-Pepper Seasoned Cod Roe). *Shokuhin Eiseigaku Zasshi.* **2009**, *50*, 173–177.

Hassan, M.; Kjos, M.; Nes, I. F.; Diep, D. B.; Lotfipour, F. Natural Antimicrobial Peptides from Bacteria: Characteristics and Potential Applications to Fight Against Antibiotic Resistance. *J. Appl. Microbiol.* **2012**, *113*, 723–736.

He, L.; Chen, W. Synergetic Activity of Nisin with Cell-Free Supernatant of *Bacillus licheniformis* ZJU12 against Food-Borne Bacteria. *Food Res. Int.* **2006**, *39*, 905–909.

Heng, N. C. K.; Wescombe, P. A.; Burton, J. P.; Jack, R. W.; Tagg, J. R. The Diversity of Bacteriocins in Gram-Positive Bacteria. In *Bacteriocins;* Riley, M. A., Chavan, M. A., Eds.; Springer: Berlin Heidelberg, 2007; pp 45–92.

Holo, H.; Faye, T.; Brede, D. A.; Nilsen, T.; Ødegård, I, Langsrud, T.; Brendehaug, J.; Ne, I. F. Bacteriocins of Propionic acid Bacteria. *Le Lait.* **2002**, *82*, 59–68.

Holo, H.; Faye, T.; Brede, D. A.; Nilsen, T.; Odigard, I.; Langsrud. T.; Brendehaug, J.; Nes, I. F. Bacteriocins of Propionic acid Bacteria. *Le Lait.* **2002**, *82*, 59–68.

Huang, Y.; Luo, Y.; Zhai, Z.; Zhang, H.; Yang, C.; Tian, H.; Li, Z.; Hao, Y. Characterization and Application of an Anti-*Listeria* Bacteriocin Produced by *Pediococcus pentosaceus* 05-10 Isolated from Sichuan Pickle, A Traditionally Fermented Vegetable Product from China. *Food Control.* **2009**, *20*, 1030–1035.

Ibrahim, E. M. A.; Elbarbary, H. A. Effect of Bacteriocin Extracted from *Lactobacillus acidophilus* on the Shelf-Life of Pasteurized Milk. *J. Am. Sci.* **2012**, *8*(2), 620–626.

Iseppi, R.; Pilati, F.; Marini, M.; Toselli, M.; de Niederhhausern, S.; Guerrieri E.; Messi, P.; Sabia, C.; Manicardi, G.; Anacarso, I.; Bondi, M. Anti-Listerial Activity of a Polymeric Film Coated with Hybrid Coatings Doped with Enterocin 416K1 for Use as Bioactive Food Packaging. *Int. J. Food Microbiol.* **2008**, *123*, 281–287.

Ivanova, I.; Kabadjova, P.; Pantev, A.; Danova, S.; Dousset, X. Detection, Purification and Partial Characterization of a Novel Bacteriocin Substance Produced by *Lactococcus lactis* subsp. *lactis* B14 Isolated From Boza-Bulgrian Traditional Cereal Beverage. *Biocatal–Vestnik Moskov univ Kimia.* **2000**, *41*, 47–53.

Jimenez-Diaz, R.; Rios-Sanchez, R. M.; Desmazeaud, M.; Holo, H.; Nes, I. F.; Sletten. K. H.; Warner, P. J. Plantaricins S and T, Two New Bacteriocins Produced by *Lactobacillus plantarum* LPCO10 Isolated from a Green Olive Fermentation. *Appl. Environ. Microbiol.* **1993**, *59*, 1916–1924.

Jin, T. Inactivation of *Listeria monocytogenes* in Skim Milk and Liquid Egg White by Antimicrobial Bottle Coating with Polylactic Acid and Nisin. *J. Food Sci.* **2010**, *75*, M83–M88.

Jin, T. Z.; Gurtler, J. B.; Li, S. Q. Development of Antimicrobial Coatings for Improving the Microbiological Safety and Quality of Shell Eggs. *J. Food Protect.* **2013**, *76*, 779–785.

Jofré, A.; Aymerich, T.; Garriga, M. Assessment of the Effectiveness of Antimicrobial Packaging Combined with High Pressure to Control *Salmonella* sp. in Cooked Ham. *Food Control.* **2008**, *19*, 634–638.

Jofré, A.; Garriga, M.; Aymerich, T. Inhibition of *Listeria monocytogenes* in Cooked Ham Through Active Packaging with Natural Antimicrobials and High-Pressure Processing. *J. Food Protect.* **2007**, *70*, 2498–2502.

Kabadjova, P.; Gotcheva, I.; Ivanova, I.; Dousset, X. Investigation of Bacteriocin Activity of Lactic acid Bacteria Isolated from Boza. *Biotechnol. Biotechnol. Equip.* **2000**, *14*, 56–59.

Katla, T.; Moretro, T.; Aasen, I. M.; Holck, A.; Axelsson, L.; Naterstad, K. Inhibition of *Listeria monocytogenes* in Cold Smoked Salmon by Addition of Sakacin P and/or Live *Lactobacillus sakei* Cultures. *Food Microbiol.* **2001**, *18*, 431–439.

Katla, T.; Møretrø, T.; Sveen, I.; Aasen, I. M.; Axelsson, L.; Rørvik, L. M. Inhibition of *Listeria monocytogenes* in Chicken Cold Cuts by Addition of Sakacin P and Sakacin P-producing *Lactobacillus sakei. J. Appl. Microbiol.* **2002**, *93*, 191–196.

Kato, T.; Inuzuka, L.; Kondo, M.; **Matsuda, T.** Growth of Nisin-Producing Lactococci in Cooked Rice Supplemented with Soybean Extract and its Application to Inhibition of *Bacillus subtilis* in Rice Miso. *Biosci. Biotechnol. Biochem.* **2001**, *65*, 330–337.

Klaenhammer, T. R. Genetics of Bacteriocins Produced by Lactic acid Bacteria. *FEMS Microbiol. Rev.* **1993**, *12*, 39–85.

Knight, K. P.; Bartlet, F. M.; McKellar, R. C.; Harris, L. J. Nisin Reduces the Thermal Resistance of *Listeria monocytogenes* Scott A in Liquid Whole Egg. *J. Food Protect.* **1999**, *62*, 999–1003.

Komitopoulou, E.; Boziaris, I. S.; Davies, E. A.; Delves-Broughton, J.; Adams, M. R. *Alicyclobacillus acidoterrestris* in Fruit Juices and Its Control by Nisin. *Int. J. Food Sci. Technol.* **1999**, *34*, 81–85.

Kozakova, D.; Holubova, J.; Plockova, M.; Chumchalova, J.; Curda, L. Impedance Measurement of Growth of Lactic Acid Bacteria in the Presence of Nisin and Lysozyme. *Eur. Food Res. Technol.* **2005**, *221*, 774–778.

Kumar, N.; Prasad, D. N. Preservative Action of Nisin in Lassi under Different Storage Temperatures. *Indian J. Anim. Sci.* **1996**, *66*(5), 525–528.

Kumar, R.; Sarkar, S.; Misra, A. K. Effect of Nisin on the Quality of Dahi. *J. Dairying Food Home Sci.* **1998**, *17*, 13–16.

Labbe, R.; Rahmati, T. Growth of Enterotoxigenic *Bacillus cereus* on Salmon (*Oncorhynchus nerka*). *J. Food Protect.* **2012**, *75*, 1153–1156.

Leisner, J. J.; Laursen, B. G.; Prévost, H.; Drider, D.; Dalgaard, P. *Carnobacterium*: Positive and Negative Effects in the Environment and in Foods. *FEMS Microbiol. Rev.* **2007**, *31*, 592–613.

Leisner, J. J.; Laursen, B. G.; Prevost, H.; Drider, D.; Dalgaard, P. *Carnobacterium*: Positive and Negative Effects in the Environment and in Foods. *FEMS Microbiol. Rev.* **2007**, *31*(5), 592–613.

Leverentz, B.; Conway, W. S.; Camp, M. J.; Janisiewicz, W. J.; Abuladze, T.; Yang, M.; Saftner, R.; Sulakvelidze, A. Biocontrol of *Listeria monocytogenes* on Freshcut Produce by Treatment with Lytic Bacteriophages and A Bacteriocin. *Appl. Environ. Microbiol.* **2003**, *69*, 4519–4526.

Liang, Z.; Cheng, Z.; Mittal, G. S. Inactivation of Spoilage Microorganisms in Apple Cider Using a Continuous Flow Pulsed Electric Field System. *LWT Food Sci. Technol.* **2006**, *39*, 351–357.

Liang, Z.; Mittal, G. S.; Griffiths, M. W. Inactivation of *Salmonella* Typhimurium in Orange Juice Containing Antimicrobial Agents by Pulsed Electric Field. *J. Food Protect.* **2002**, *65*, 1081–1087.

Linnett, P. E.; Strominger, J. L. Additional Antibiotic Inhibitors of Peptidoglycan Synthesis. *Antimicrobial Agents Chemother.* **1973**, *4*, 231–236.

Lonvaud-Funel A.; Joyeux, A. Antagonism between Lactic acid Bacteria of Wines: Inhibition of *Leuconostoc oenos* by *Lactobacillus plantarum* and *Pediococcus pentosaceus. Food Microbiol.* **1993**, *10*, 411–419.

López-Pedemonte, T. J.; Roig-Sagues, A. X.; Trujillo, A. J. Inactivation of Spores of *Bacillus cereus* in Cheese by High Hydrostatic Pressure with the Addition of Nisin of Lysozyme. *J. Dairy Sci.* **2003**, *86*, 3075–3081.

Lucas, R. L.; Grande, M. J.; Abriouel, H.; Galvez, A. Application of the Broad-Spectrum Bacteriocin Enterocin AS-48 to Inhibit *Bacillus coagulans* in low-pH Canned Fruit and Vegetable Foods. *Food Chem. Toxicol.* **2006**, *44*, 1774–1781.

Lungu, B.; Johnson, M. G. Fate of *Listeria monocytogenes* Inoculated Onto the Surface of Model Turkey Frankfurter Pieces Treated with Zein Coatings Containing Nisin, Sodium Diacetate, and Sodium lactate At 4 °C. *J. Food Protect.* **2005**, *68*, 855–859.

Lynch, M. F.; Tauxe, R. V.; Hedberg, C. W. The Growing Burden of Foodborne Outbreaks due to Contaminated Fresh Produce: Risks and Opportunities. *Epidemiol. Infect.* **2009**, *137*, 307–315.

Mah, J. H.; Hwang, H. J. Inhibition of Biogenic Amine Formation in a Salted and Fermented Anchovy by *Staphylococcus xylosus* as a Protective Culture. *Food Control.* **2009**, *20*, 796–801.

Mah, J. H.; Ahn, J. B.; Park, J. H.; Huang, H. J. Characterization of Biogenic Amine-Producing Microorganisms Isolated from Myeolchi-Jeot, Korean Salted and Fermented Anchovy. *J. Microbiol. Biotechnol.* **2003**, *13*, 692–699.

Maks, N.; Zhu, L.; Juneja, V. K.; Ravishankar, S. Sodium Lactate, Sodium Diacetate and Pediocin: Effects and Interactions on the Thermal Inactivation of *Listeria monocytogenes* on Bologna. *Food Microbiol.* **2010**, *27*, 64–69.

Malheiros, P. S.; Sant'Anna, V.; Barbosa, M. S.; Brandeli, A.; Melo Franco, B. D. G. Effect of Liposome-Encapsulated Nisin and Bacteriocin-Like Substance P34 on *Listeria monocytogenes* Growth in Minas Frescal Cheese. *Int. J. Food Microbiol.* **2012**, *156*(3), 272–277.

Mangalassary, S.; Han, I.; Rieck, J.; Dawson, P. L. Effect of Combining Nisin and/or Lyso-zyme With in Package Pasteurization for Control of *Listeria monocytogenes* in Ready-to-Eat Turkey Bologna During Refrigerated Storage. *Food Microbiol.* **2008**, *25*, 866–870.

Mansour, M.; Milliere, J. B. An Inhibitory Synergistic Effect of a Nisin–Monolaurin Combi-nation on *Bacillus* sp. Vegetative Cells in Milk. *Food Microbiol.* **2001**, *18*, 87–94.

Mansour, M.; Amri, D.; Bouttefroy, A.; Linder, M.; Milliere, J. B. Inhibition of *Bacillus licheniformis* Spore Growth in Milk by Nisin, Monolaurin, and pH Combinations. *J. Appl. Microbiol.* **1999**, *86*, 311–324.

Marcos, B.; Aymerich, T.; Monfort, J. M.; Garriga, M. High-Pressure Processing and Anti-microbial Biodegradable Packaging to Control *Listeria monocytogenes* During Storage of Cooked Ham. *Food Microbiol.* **2008a**, *25*, 177–182.

Marcos, B.; Jofré, A.; Aymerich, T.; Monfort, J. M.; Garriga, M. Combined Effect of Natural Antimicrobials and High Pressure Processing to Prevent *Listeria monocytogenes* Growth After a Cold Chain Break During Storage of Cooked Ham. *Food Control.* **2008b**, *19*, 76–81.

Martínez Viedma, P.; Abriouel, H.; Ben Omar, N. Anti-Staphylococcal Effect of Enterocin AS-48 in Bakery Ingredients of Vegetable Origin, Alone and in Combination with Selected Antimicrobials. *J. Food Sci.* **2009a**, 74, M384–M389.

Martínez Viedma, P.; Abriouel, H.; Ben Omar, N.; Lucas, R. L.; Galvez, A. Effect of Enterocin EJ97 Against *Geobacillus stearothermophilus* Vegetative Cells and Endospores in Canned Foods and Beverages. *Eur. Food Res. Technol.* **2010a**, *230*, 513–519.

Martínez Viedma, P.; Ercolini, D.; Ferrocino, I.; Abriouel, H.; Ben Omar, N.; Lucas, R. L.; Gálvez, A. Effect of Polythene Film Activated With Enterocin EJ97 in Combination with EDTA Against *Bacillus coagulans*. *LWT—Food Sci. Technol.* **2010b**, *43*, 514–518.

Martínez-Viedma, P.; Abriouel, H.; Ben Omar, N.; Lucas, R. L.; Valdivia, E.; Gálvez, A. Assay of Enterocin AS-48 for Inhibition of Foodborne Pathogens in Desserts. *J. Food Protect.* **2009b**, *72*, 1654–1659.

Martínez-Viedma, P.; Abriouel, H.; Ben Omar, N.; Valdivia, E.; Lucas, R. L.; Gálvez, A. Inactivation of Exopolysaccharide and 3-Hydroxypropionaldehyde-Producing Lactic Acid Bacteria in Apple Juice and Apple Cider by Enterocin AS-48. *Food Chem. Toxicol.* **2008a**, *46*, 1143–1151.

Martínez-Viedma, P.; Abriouel, H.; Sobrino, A.; Ben Omar, N.; Lucas Lopez, R.; Valdivia, E.; Martin Belloso, O.; Galvez, A. Effect of Enterocin AS-48 in Combination with High-Intensity Pulsed-Electric Field Treatment Against the Spoilage Bacterium *Lactobacillus diolivorans* in Apple Juice. *Food Microbiol.* **2009c**, *26*, 491–496.

Matamoros, S.; Pilet, M. F.; Gigout, F.; Prevost, H.; Leroi, F. Selection and Evaluation of Seafood-Borne Psychrotrophic Lactic Acid Bacteria as Inhibitors of Pathogenic and Spoilage Bacteria. *Food Microbiol.* **2009**, *26*, 638–644.

Menteş, O.; Ercan, R.; Akçelik, M. Inhibitor Activities of Two *Lactobacillus* strains, Isolated from Sourdough, Against Rope-Forming *Bacillus* strains. *Food Control.* **2007**, *18*, 359–363.

Messens, W.; De Vuyst, L. Inhibitory Substances Produced by Lactobacilli Isolated from Sourdoughs—A Review. *Int. J. Food Microbiol.* **2002**, *72*, 31–43.

Metivier, A.; Pilet, M. F.; Dousset, X.; Sorokine, O.; Anglade, P.; Zagorec, M.; Piard, J. C.; Marion, D.; Cenatiempo, Y.; Fremaux, C. Divercin V41, A New Bacteriocin with Two Disulphide Bonds Produced by *Carnobacterium divergens* V41: Primary Structure and Genomic Organization. *Microbiology.* **1998**, *14*, 2837–2844.

Millette, M.; Le Tien, C.; Smoragiewicz, W.; Lacroix, M. Inhibition of *Staphylococcus aureus* on Beef by Nisin-Containing Modified Alginate Films and Beads. *Food Control.* **2007**, *18*, 878–884.

Molinos, A.; Abriouel, H.; Ben Omar, N.; Valdivia, E.; Lucas, R. L.; Maqueda, M.; Cañamero, M. M.; Gálvez, A. Effect of Immersion Solutions Containing Enterocin AS-48 on *Listeria monocytogenes* in Vegetable Foods. *Appl. Environ. Microbiol.* **2005**, *71*, 7781–7787.

Molinos, A.; Abriouel, H.; Lucas, L. R.; Valdivia, E.; Ben Omar, N.; Gálvez, A. Combined Physico-Chemical Treatments Based on Enterocin AS-48 for Inactivation of Gram-Negative Bacteria in Soybean Sprouts. *Food Chem. Toxicol.* **2008**, *46*, 2912–2921.

Molinos, A.; Abriouel, H.; Lucas, L. R.; Ben Omar, N.; Valdivia, E.; Gálvez, A. Enhanced Bactericidal Activity of Enterocin AS-48 in Combination with Essential Oils, Natural Bioactive Compounds and Chemical Preservatives Against Listeria Monocytogenes in Ready-to-Eat Salad. *Food Chem. Toxicol.* **2009a**, *47*, 2216–2223.

Molinos, A.; Lucas, L. R.; Abriouel, H.; Ben Omar, N.; Valdivia, E.; Gálvez, A. Inhibition of *Salmonella enterica* Cells in Delitype Salad by Enterocin AS-48 in Combination with Other Antimicrobials. *Probiotics Antimicrobial Prot.* **2009b**, *1*, 85–90.

Moll, G. N.; Clark, J.; Chan, W. C.; Bycroft, B. W.; Roberts, G. C. K.; Konings, W. N.; Driessen, A. J. Role of Transmembrane pH Gradient and Membrane Binding in Nisin Pore Formation. *J. Bacteriol.* **1997**, *179*, 135–140.

Morgan, S. M.; Garvin, M.; Ross, R. P.; Hill, C. Evaluation of a Spray-Dried Lacticin 3147 Powder For the Control of Listeria Monocytogenes and *Bacillus cereus* in a Range of Food Systems. *Lett. Appl. Microbiol.* **2001**, *33*, 387–391.

Natrajan, N.; Sheldon, B. W. Inhibition of *Salmonella* on Poultry Skin Using Protein- and Polysaccharide-Based Films Containing a Nisin Formulation. *J. Food Protect.* **2000b**, *63*, 1268–1272.

Natrajan, N.; Sheldon, B. W. Efficacy of Nisin-Coated Polymer Films to Inactivate *Salmonella* typhimurium on Fresh Broiler Skin. *J. Food Protect.* **2000a**, *63*, 1189–1196.

Nattress, F. M.; Baker, L. P. Effects of Treatment with Lysozyme and Nisin on the Microflora and Sensory Properties of Commercial Pork. *Int. J. Food Microbiol.* **2003**, *85*, 259–267.

Nattress, F. M.; Yost, C. K.; Baker, L. P. Evaluation of the Ability of Lysozyme and Nisin to Control Meat Spoilage Bacteria. *Int. J. Food Microbiol.* **2001**, *70*, 111–119.

Navarro, L.; Zarazaga, M.; Sáenz, J.; Ruiz-Larrea, F.; Torres, C. Bacteriocin Production by Lactic Acid Bacteria Isolated from Rioja Red Wines. *J. Appl. Microbiol.* **2000**, 88, 44–51.

Neetoo, H.; Ye, M.; Chen. H. Potential Antimicrobials to Control *Listeria monocytogenes* in Vacuum-Packaged Cold-Smoked Salmon pâté and Fillets. *Int. J. Food Microbiol.* **2008b**, *123*, 220–227.

Neetoo, H.; Ye, M.; Chen, H.; Joerger, R. D.; Hicks, D. T.; Hoover, D. G. Use of Nisin-Coated Plastic Films to Control *Listeria monocytogenes* on Vacuum-Packaged Cold-Smoked Salmon. *Int. J. Food Microbiol.* **2008a**, *122*, 8–15.

Nes, I. F.; Diep, D. B.; Havarstein, L. S.; Brurberg, M. B.; Eijsink, V.; Holo, H. Biosynthesis of Bacteriocins in Lactic Acid Bacteria. *Antonie van Leeuwenhoek.* **1996**, *70*, 113–128.

Nguyen, P.; Mittal, G. S. Inactivation of Naturally Occurring Microorganisms in Tomato Juice Using Pulsed Electric Field (PEF) with and without Antimicrobials. *Chem. Eng. Process.* **2007**, *46*, 360–365.

Nguyen, V. T.; Gidley, M. J.; Dykes, G. A. Potential of a Nisin-Containing Bacterial Cellulose Film to Inhibit *Listeria monocytogenes* on Processed Meats. *Food Microbiol.* **2008**, *25*, 471–478.

Nieto-Lozano, J. C.; Reguera-Useros, J. I.; Peláez-Martínez, M. C.; Torre, A. H. Effect of a Bacteriocin Produced by *Pediococcus acidilactici* against *Listeria monocytogenes* and *Clostridium perfringens* on Spanish Raw Meat. *Meat Sci.* **2006**, *72*, 57–61.

Nilsson, L.; Christiansen, N. Y. Y.; Jorgensen, J. N.; Jørgensen, B. L.; Grótinum, D.; Gram, L. The Contribution of Bacteriocin to Inhibition of *Listeria monocytogenes* by *Carnobacterium piscicola* Strains in Cold-Smoked Salmon Systems. *J. Appl. Microbiol.* **2004**, *96*, 133–143.

Nilsson, L.; Huss, H. H.; Gram, L. Inhibition of *Listeria monocytogenes* on Cold-Smoked Salmon by Nisin and Carbon Dioxide Atmosphere. *Int. J. Food Microbiol.* **1997**, *38*, 217–227.

Niskanen, A.; Nurmi, E. Effect of Starter Culture on Staphylococcal Enterotoxin and Thermonuclease Production in Dry Sausage. *Appl. Environ. Microbiol.* **2000**, *31*, 11–20.

Nykanen, A.; Weckman, K.; Lapvetelainen, A. Synergistic Inhibition of *Listeria monocytogenes* on Cold-Smoked Rainbow Trout by Nisin and Sodium Lactate. *Int. J. Food Microbiol.* **2000**, *61*, 63–72.

O'Sullivan, L.; O'Connor, E. B.; Ross, R.P.; Hill, C. Evaluation of Live-Culture-Producing Lacticin 3147 as a Treatment for the Control of *Listeria monocytogenes* on the Surface of Smear-Ripened Cheese. *J. Appl. Microbiol.* **2006**, *100*, 135–143.

Ogden, K.; Tubb, R. S. Inhibition of Beer-Spoilage Lactic Acid Bacteria by Nisin. *J. Inst. Brewing.* **1985**, *91*, 390–392.

Olasupo, N. A.; Olukoya, D. K.; Odunfa, S. A. Studies on Bacteriocinogenic *Lactobacillus* Isolates from Selected Nigerian Fermented Foods. *J. f Basic Microbiol.* **1995**, *35*(5), 319–324.

Omar, N.B.; Abriouel, H.; Keleke, S.; Venuezela, A.S.; Martinez-Canamero, M.; Lopez, R.L.; Ortega, E.; Galvez, A. Bacteriocin-Producing *Lactobacillus* strains Isolated From *poto poto*, A Congolese Fermented Maize Product, and Genetic Fingerprinting of Their Plantaricin Operons. *Int. J. Food Microbiol.* **2008**, *127*, 18–25.

Omar, N.B.; Abriouel, H.; Lucas, R.; Martinez-Canamero, M.; Guyot, J.; Galvez, A. Isolation of Bacteriocinogenic *Lactobacillus plantarum* Strains from Ben Saalga, A Traditional Fermented Gruel from Burkina Faso. *Int. J. Food Microbiol.* **2006**, *112*, 44–50.

Parada, J.; Caron, C.; Medeiros, A.; Soccol, C. Bacteriocins from Lactic Acid Bacteria: Purification, Proprieties and Use As Biopreservatives. *Braz. Arch. Biol. Technol.* **2007**, *50*(3), 521–542.

Patel, A.; Lindström, C.; Patel, A.; Prajapati, J. B.; Holst, O. Screening and Isolation of Exopolysaccharide Producing Lactic acid Bacteria from Vegetables and Indigenous Fermented Foods of Gujarat, India. *Int. J. Fermented Foods*, **2012**, *1*(1), 77–86.

Patel, A.; Shah, N.; Ambalam, P.; Prajapati, J. B.; Holst, O.; Ljungh, A. Antimicrobial Profile of Lactic Acid Bacteria Isolated from Vegetables and Indigenous Fermented Foods of India Against Clinical Pathogens Using Microdilution Method. *Biomed. Environ. Sci.* **2013**, *26*(9), 759–764.

Pathanibul, P.; Taylor, T. M.; Davidson, P. M.; Harte, F. Inactivation of *Escherichia coli* and *Listeria innocua* in Apple and Carrot Juices Using High Pressure Homogenization and Nisin. *Int. J. Food Microbiol.* **2009**, *129*, 316–320.

Pawar, D. D.; Malik, S. V. S.; Bhilegaonkar, K. N.; Barbuddhe, S. B. Effect of Nisin and Its Combination with Sodium Chloride on the Survival of *Listeria monocytogenes* Added to Raw Buffalo Meat Mince. *Meat Sci.* **2000**, *56*, 215–219.

Pena, W. E.; de Massaguer, P. R.; Teixeira, L. Q. Microbial Modeling of Thermal Resistance of *Alicyclobacillus acidoterrestris* CRA7152 Spores in Concentrated Orange Juice with Nisin Addition. *Braz. J. Microbiol.* **2009**, *40*, 601–611.

Penna, T.C.V.; Moraes, D. A. The Influence of Nisin on the Thermal Resistance of *Bacillus cereus*. *J. Food Protect.* **2002**, *65*, 415–418.

Perez, R.H.; Zendo, T.; Sonomoto, K. Novel Bacteriocins from Lactic Acid Bacteria: Various Structures and Applications. *Microbial Cell Factor.* **2014**, *13*, S3.

Pinto, A.L.; Fernandes, M.; Pinto, C.; Albano, H.; Castilho, F.; Teixeira, P.; Gibbs, P.A. Characterization of Anti-*Listeria* Bacteriocins Isolated from Shellfish: Potential Antimicrobials to Control Non-Fermented Seafood. *Int. J. Food Microbiol.* **2009**, *129*, 50–58.

Plockova, M.; Stepanek, M.; Demnerova, K.; Curda, L.; Svirakova, E. Effect of Nisin for Improvement in Shelf Life and Quality of Processed Cheese. *Adv. Food Sci.* **1996**, *18*, 78–83.

Pol, I. E.; Mastwijk, H.C.; Slump, R.A.; Popa, M.E.; Smid, E.J. Influence of Food Matrix on Inactivation of *Bacillus cereus* by Combinations of Nisin, Pulsed Electric Field Treatment, and Carvacrol. *J. Food Protect.* **2001**, *64*, 1012–1018.

Ponce, E.; Pla, R.; Sendra, E.; Guamis, B.; Mor-Mur, M. Combined Effect of Nisin and High Hydrostatic Pressure on Destruction of *Listeria innocua* and *Escherichia coli* in Liquid Whole Egg. *Int. J. Food Microbiol.* **1998**, *43*, 15–19.

Radha, K. Nisin as a Biopreservative for Pasteurized Milk. *Indian J. Vet. Anim. Sci. Res.* **2014**, *43*(6), 436–444.

Radler, F. Possible Use of Nisin in Wine-Making. II Experiments to Control Lactic Acid Bacteria in the Production of Wine. *Am. J. Enol. Viticult.* **1990**, *41*, 7–11.

Randazzo, C.L.; Pitino, I.; Scifo, G.O.; Caggia, C. Biopreservation of Minimally Processed Iceberg Lettuces using a Bacteriocin Produced by *Lactococcus lactis* Wild Strain. *Food Control.* **2009**, *20*, 756–763.

Rao, K. V. S. S. A Process for the Manufacture of Paneer Like Product, Ph.D Thesis submitted to N.D.R.I.; Karnal (Haryana), India, 1990.

Rao, K. V. S. S.; Mathur, B. N. Thermal Death Kinetics of *Bacillus stearothermophilus* Spores in a Nisin Supplemented Acidified Concentrated Buffalo Milk System. *Milchwissenschaft.* **1996**, *51*, 186–191.

Reina, L.D.; Breidt, F.; Fleming, H. P.; Kathariou, S. Isolation and Selection of Lactic Acid Bacteria as Biocontrol Agents for Non Acidified, Refrigerated Pickles. *J. Food Sci.* **2005**, *70*, 7–11.

Reisinger, H.; Seidel, H.; Tschesche, P.; Hammes, W. P. The Effect of Nisin on Murein Synthesis. *Arch. Microbiol.* **1980**, *127*, 187–193.

Reviriego, C.; Fernández, A.; Horn, N.; Rodríguez, E.; Marín, M.L.; Fernández, L.; Rodríguez, J. M. Production of Pediocin PA-1, and Coproduction of Nisin A and Pediocin PA-1, by Wild *Lactococcus lactis* Strains of Dairy Origin. *Int. Dairy J.* **2005**, *15*, 45–49.

Roberts, R. F.; Zottola, E. A. Shelf Life of Pasteurized Process Cheese Spreads Made from Cheddar Cheese Manufactured With a Nisin Producing Starter Culture. *J. Dairy Sci.* **1993**, *76*, 1829–1836.

Roberts, R.F.; Zottola, E.A.; McKay, L. L. Use of Nisin-Producing Starter Cultures Suitable for Cheddar Cheese Manufacture. *J. Dairy Sci.* **1992**, *75*, 2353–2363.

Rodriguez, J. M.; Martinez, M. I.; Kok, J. Pediocin PA-1, a Wide-Spectrum Bacteriocin from Lactic Acid Bacteria. *Crit. Rev. Food Sci. Nutr.* **2002**, *42*, 91–121.

Rojo-Bezares, B.; Saenz, Y.; Zarazaga, M.; Torres, C.; Ruiz-Larrea, F. Antimicrobial Activity of Nisin Against *Oenococcus oeni* and Other Wine Bacteria. *Int. J. Food Microbiol.* **2007**, *116*, 32–36.

Sabia, C.; de Niederhausern, S.; Messi, P.; Manicardi, G.; Bondi, M. Bacteriocin-Producing *Enterococcus casselifl avus* IM 416K1, A Natural Antagonist for Control of *Listeria monocytogenes* in Italian Sausages ("cacciatore"). *Int. J. Food Microbiol.* **2003**, *87*, 173–179.

Sanchez Valenzuela, A.; Díaz Ruiz, G.; Ben Omar, N.; Abriouel, H.; Rosario Lucas, R.L.; Cañamero, M.M.; Ortega, E.; Galvez, A. Inhibition of Food Poisoning and Pathogenic Bacteria by *Lactobacillus plantarum* Strain 2.9 Isolated from Ben Saalga, Both in a Culture Medium and in Food. *Food Control.* **2008**, *19*, 842–848.

Santiago-Silva, P.; Nilda, F.F.; Nobrega, J.E.; Marcus, A.W. Junior, Barbosa, K.B.F.; Volp, A.C.P.; Evelyn, R.M.A.; Zerdas, Wurlitzer, N. J. Antimicrobial Efficiency of Film Incorporated With Pediocin (ALTA ® 2351) on Preservation of Sliced Ham. *Food Control.* **2009**, *20*, 85–89.

Sarika, A.R.; Lipton, A.P.; Aishwarya, M.S.; Dhivya, R.S. Isolation of a Bacteriocin-Producing *Lactococcus lactis* and Application of Its Bacteriocin to Manage Spoilage Bacteria in High-Value Marine Fish Under Different Storage Temperatures. *Appl. Biochem. Biotechnol.* **2012**, *167*, 1280–1289.

Scannell, A.G.; Ross, R.P.; Hill, C.; Arendt, E.K. An Effective Lacticin Biopreservative in Fresh Pork Sausage. *J. Food Protect.* **2000a**, *63*, 370–375.

Scannell, A. G. M.; Hill, C.; Buckley, D. J.; Arendt E. K. Determination of the Influence of Organic Acids and Nisin on Shelf-Life and Microbiological Safety Aspects of Fresh Pork Sausage. *J. Appl. Microbiol.* **1997**, *83*, 407–412.

Schelegueda, L.I.; Vallejo, M.; Gliemmo, M.F.; Marguet, E.R.; Campos, C.A. Synergistic Antimicrobial Action and Potential Application for Fish Preservation of a Bacteriocin Produced by *Enterococcus mundtii* Isolated from *Odontesthes platensis. LWT—Food Sci. Technol.* **2015**, *64*(2), 794–801.

Schillinger, U.; Becker, B.; Vignolo, G. et al. Efficacy of Nisin in Combination with Protective Cultures against *Listeria monocytogenes* Scott A in Tofu. *Int. J. Food Microbiol.* **2001**, *71*, 159–168.

Schillinger, U.; Geisen, R.; Holzapfel, W. H. Potential of Antagonistic Microorganisms and Bacteriocins for the Biological Preservation of Foods. *Trends Food Sci. Technol.* **1996**. *7*, 158–164.

Schuller, F.; Benz, R.; Sahl, H. G. The Peptide Antibiotic Subtilin Acts by Formation of Voltage-Dependent Multi-State Pores in Bacterial and Artificial Membranes. *Eur. J. Biochem.* **1989**, *182*, 181–186.

Schuman, J. D.; Sheldon, B. W. Inhibition of *Listeria monocytogenes* in pH-Adjusted Pasteurized Liquid Whole Egg. *J. Food Protect.* **2003**, *66*, 999–1006.

Settanni, L.; Corsetti, A. Application of Bacteriocins in Vegetable Food Biopreservation. *Int. J. Food Microbiol.* **2008**, *121*, 123–138.

Shin, M. S.; Han, S. K.; Ryu, J. S.; Kim, K. S.; Lee, W. K. Isolation and Partial Characterization of a Bacteriocin Produced by Pediococcus Pentosaceus K23-2 Isolated from Kimchi. *J. Appl. Microbiol.* **2008**, *105*(2), 331–339.

Silveira, A. C.; Conesa, A.; Aguayo, E.; Artes, F. Alternative Sanitizers to Chlorine for Use on Freshcut "Galia" (*Cucumis melo* var. *catalupensis)* Melon. *J. Food Sci.* **2008**, *73*, 405–411.

Singh, L.; Mohan, M. S.; Sankaran, R. "Nisin as an Aid for Thermal Preservation of Indian Dishes-*upma* and *kheer*", *J. Food Sci. Technol.* **1987**, *24*(6), 277–280.

Smith, K.; Mittal, G. S.; Griffiths, M. W. Pasteurization of Milk using Pulsed Electrical Field and Antimicrobials. *J. Food Sci.* **2002**, *6*, 2304–2308.

Sivarooban, T.; Hettiarachchy, N. S.; Johnson, M. G. Inhibition of *Listeria monocytogenes* Using Nisin with Grape Seed Extract on Turkey Frankfurters Stored at 4 and 10 °C. *J. Food Protect.* **2007**, *70*, 1017–1020.

Snyder, A. B.; Worobo, R. W. Chemical and Genetic Characterization of Bacteriocins: Antimicrobial Peptides for Food Safety. *J. Sci. Food Agric.* **2013**, *94*, 28–44.

Sobrino-Lopez, A.; Martin-Belloso, O. Use of Nisin and Other Bacteriocins for Preservation of Dairy Products. *Int. Dairy J.* **2008**, *18*, 329–343.

Sobrino-Lopez, A.; Raybaudi-Massilia, R.; Martin-Belloso, O. Enhancing Inactivation of *Staphylococcus aureus* in Skim Milk by Combining High Intensity Pulsed Electric Fields and Nisin. *J. Food Protect.* **2006**, *69*, 345–353.

Solomakos, N.; Govaris, A.; Koidis, P.; Botsoglou, N. The Antimicrobial Effect of Thyme Essential Oil, Nisin, and Their Combination Against *Listeria monocytogenes* in Minced Beef During Refrigerated Storage. *Food Microbiol.* **2008**, *25*, 120–127.

Somkuti, G. A.; Steinberg, D. H. Pediocin Production in Milk by *Pediococcus acidilactici* in Coculture with *Streptococcus thermophilus* and *Lactobacillus delbrueckii* subsp. *bulgaricus*. *J. Indian Microbiol. Biotechnol.* **2010**, *37*, 65–69.

Stamer, J. R.; Stoyla, B. O.; Dunckel, B. A. Growth Rates and Fermentation Patterns of Lactic Acid Bacteria Associated With the Sauerkraut Fermentation. *J. Milk Food Technol.* **1971**, *34*, 521–525.

Stergiou, V. A.; Thomas, L. V.; Adams, M. R. Interactions of Nisin with Glutathione in a Model Protein System and Meat. *J. Food Protect.* **2006**, *69*, 951–956.

Sureshkumar, S.; Kalaikannan, A.; Dushyanthan, K.; Venkataramanujam, V. Effect of Nisin and Butylated Hydroxy Anisole on Storage Stability of Buffalo Meat Sausage. *J. Food Sci. Technol.* **2010**, *47*(3), 358–363.

Suzuki, M.; Yamamoto, T.; Kawai, Y.; Inoue, N.; Yamazaki, K. Mode of Action of Piscicocin CS526 Produced by *Carnobacterium piscicola* CS526. *J. Appl. Microbiol.* **2005**, *98*(5), 1146–1151.

Szabo, E. A.; Cahill, M. E. Nisin and ALTA 2341 Inhibit the Growth of *Listeria monocytogenes* on Smoked Salmon Packaged Under Vacuum or 100 % CO_2. *Lett. Appl. Microbiol.* **1999**, *28*, 373–377.

Tahiri, I.; Desbiens, M.; Benech, R.; Kheadr, E.; Lacroix, C.; Thibault, S.; Ouellet, D.; Fliss, I. Purification, Characterization and Amino Acid Sequencing of Divergicin M35: A Novel Class IIa Bacteriocin Produced by *Carnobacterium divergens* M35. *Int. J. Food Microbiol.* **2004**, *97*(2), 123–136.

Tahiri, I.; Desbiens, M.; Kheadr, E.; Kheadr, E.; Fliss, I. Growth of *Carnobacterium divergens* M35 and Production of Divergicin M35 in Snow Crab By-Product, A Natural-Grade Medium. *LWT—Food Sci. Technol.* **2009a**, *42*, 624–632.

Tahiri, I.; Desbiens, M.; Kheadr, E.; Lacroix, C.; Fliss, I. Comparison of Different Application Strategies of Divergicin M35 for Inactivation of *Listeria monocytogenes* in Cold-Smoked Wild Salmon. *Food Microbiol.* **2009b**, *26*, 783–793.

Takahashi, H.; Kuramoto, S.; Miya, S.; Koiso, S.; Kuda, H.; Bon, K. T. Use of Commercially Available Antimicrobial Compounds for Prevention of *Listeria monocytogenes* Growth in Ready-to-Eat Minced Tuna and Salmon Roe During Shelf Life. *J. Food Protect.* **2011**, *74*, 994–998.

Thakral, S.; Prasad, M. M.; Ghodeker, D. R. Effect of Incorporation of Potassium Sorbate and Nisin for Improvement in Shelf Life of Paneer. In Brief Communications of the XXIII International Dairy Congress, Montreal, October 8-12, I, 1990, 150.

Thomas, L. V.; Clarkson, M. R.; Delves-Broughton, J. Nisin. In *Natural Food Antimicrobial Systems*; Naidu. A. S. Ed.; CRC: Boca Raton, FL, 2000; pp 463–524.

Thomas, L. V.; Ingram, R. E.; Bevis, H. E.; Davies, A.; Milne, C. F.; Delves-Broughton, J. Effective use of Nisin to Control *Bacillus* and *Clostridium* Spoilage of a Pasteurized Mashed Potato Product. *J. Food Protect.* **2002**, *65*, 1580–1585.

Todorov, S. D.; Dicks, L. M. T. Characterization of Mesenetricin ST99, A Bacteriocin Produced by *Leuconostoc mesenteroides* subsp. *dextranicum* ST99 Isolated from Boza. *J. Ind. Microbiol. Biotechnol.* **2004**, *31*, 323–329.

Todorov, S. D.; Dicks, L. M. T. Pediocin ST18, An Anti-Listerial Bacteriocin Produced by *Pediococcus pentosaceus* ST18 Isolated from Boza, A Traditional Cereal Beverage from Bulgaria. *Process Biochem.* **2005**, *40*, 365–370.

Todorov, S. D. Diversity of Bacteriocinogenic Lactic Acid Bacteria Isolated from Boza, A Cereal-Based Fermented Beverage from Bulgaria. *Food Control.* **2010**, *21*, 1011–1021.

Tome, E.; Gibbs, P. A.; Teixeira, P. C. Growth Control of *Listeria innocua* 2030c On Vacuum Packaged Cold-Smoked Salmon by Lactic Acid Bacteria. *Int. J. Food Microbiol.* **2008**, *121*, 285–294.

Tsironi, T. N.; Taoukis, P. S. Modeling Microbial Spoilage and Quality of Gilthead Seabream Fillets: Combined Effect of Osmotic Pre-Treatment, Modified Atmosphere Packaging, and Nisin on Shelf Life. *J. Food Sci.* **2010**, *75*, M243–M251.

Udhayashree, N.; Senbagam, D.; Senthilkumar, B.; Nithya, K.; Gurusamy, R. Production of Bacteriocin and Their Application in Food Products. *Asian Pacific J. Trop. Biomed.* **2012**, S406–S410.

Uesugi, A. R.; Moraru, C. I. Reduction of *Listeria* on Ready-to-Eat Sausages After Exposure to a Combination of Pulsed Light and Nisin. *J. Food Protect.* **2009**, *72*, 347–353.

Uhart, M.; Ravishankar, S.; Maks, N. D. Control of *Listeria monocytogenes* with Combined Antimicrobials on Beef Franks Stored at 4 °C. *J. Food Protect.* **2004**, *67*, 2296–2301.

Uhlman, L.; Schillinger, U.; Rupnow, J. R.; Holzapfel, W. H. Identification and Characterization of Two Bacteriocin-Producing Strains of *Lactococcus lactis* Isolated from Vegetables. *Int. J. Food Microbiol.* **1992**, *16*, 141–151.

Ukuku, D. O.; Fett, W. F. Effect of Nisin in Combination with EDTA, Sodium Lactate, and Potassium Sorbate for Reducing *Salmonella* on Whole and Fresh-Cut Cantaloupes. *J. Food Protect.* **2004**, *67*, 2143–2150.

Ukuku, D. O.; Ban, M. L.; Kawamoto, S.; Isshiki, K. Use of Hydrogen Peroxide in Combination with Nisin, Sodium Lactate and Citric Acid for Reducing Transfer of Bacterial Pathogens from Whole Melon Surfaces to Fresh-Cut Pieces. *Int. J. Food Microbiol.* **2005**, *104*, 225–233.

Ukuku, D. O.; Zhang, H.; Huang, L. Growth Parameters of *Escherichia coli* O157:H7, *Salmonella* spp.; *Listeria monocytogenes*, and Aerobic Mesophilic Bacteria of Apple Cider Amended with Nisin-EDTA. *Foodborne Pathogens Dis.* **2009**, *6*, 487–494.

Valerio, F.; De Bellis, P.; Lonigro, S. L.; Visconti, A.; Lavermicocca, P. Use of *Lactobacillus plantarum* Fermentation Products in Bread-Making to Prevent *Bacillus subtilis* Ropy Spoilage. *Int. J. Food Microbiol.* **2008**, *122*, 328–332.

Valerio, F.; Favilla, M.; De Bellis, P.; Sisto, A.; de Candia, S.; Lavermicocca, P. Antifungal Activity of Strains of Lactic Acid Bacteria Isolated from a Semolina Ecosystem Against *Penicillium roqueforti*, *Aspergillus niger* and *Endomyces fibuliger* Contaminating Bakery Products. *Syst. Appl. Microbiol.* **2009**, *32*, 438–448.

Vaughan, A.; O'Sullivan, T.; van Sinderen, D. Enhancing the Microbiological Stability of Malt and Beer–A Review. *J. Inst. Brew.* **2005**, *111*, 355–371.

Vaz-Velho, M.; Todorov, S.; Ribeiro, J.; Gibbs, P. Growth Control of *Listeria innocua* 2030c During Processing and Storage of Cold-Smoked Salmon-Trout by *Carnobacterium divergens* V41 Culture and Supernatant. *Food Control.* **2005**, *16*, 541–549.

Walker, M.; Phillips, C. A. The Effect of Preservatives on *Alicyclobacillus acidoterrestris* and *Propionibacterium cyclohexanicum* in Fruit Juice. *Food Control.* **2008**, *19*, 974–981.

Wan Norhana, M. N.; Poole, S. E.; Deeth, H. C. Effects of Nisin, EDTA and Salts of Organic Acids on *Listeria monocytogenes*, *Salmonella* and Native Microflora on Fresh Vacuum Packaged Shrimps Stored at 4 °C. *Food Microbiol.* **2012**, *31*, 43–50.

Wessels, S.; Axelsson, L.; Bech Hansen, E.; de Vuyst, L.; Laulund, S.; Lähteenmäki, L.; Lindgren, S.; Mollet, B.; Salminen, S.; von Wright, A. The Lactic Acid Bacteria, the Food Chain, and Their Regulation. *Trends Food Sci. Technol.* **2004**, *15*, 498–505.

Wirjantoro, T. I.; Lewis, M. J. Effect of Nisin and High Temperature Pasteurization on Shelf Life of WHOLE milk. *J. Soc. Dairy Technol.* **1996**, *49*, 99–102.

Woraprayote, W.; Kingcha, Y.; Amonphanpokin, P.; Kruenate, J.; Zendo, T.; Sonomoto, K.; Benjakul, S.; Visessanguan, W. Anti-Listeria Activity of Poly(Lactic Acid)/Sawdust Particle Biocomposite Film Impregnated with Pediocin PA-1/AcH and Its Use in Raw Sliced Pork. *Int. J. Food Microbiol.* **2013**, 2167, 229–235.

Wu, Y.; Mittal, G. S.; Griffiths, M. W. Effect of Pulsed Electric Field on the Inactivation of Microorganisms in Grape Juices With and Without Antimicrobials. *Biosyst. Eng.* **2005**, 90, 1–7.

Yamazaki, K.; Mukarami, M.; Kawai, Y.; Inoue, N.; Matsuda, T. Use of Nisin for Inhibition of *Alicyclobacillus acidoterrestris* in Acidic Drinks. *Food Microbiol.* **2000**, *17*, 315–320.

Yamazaki, K.; Suzuki, M.; Kawai, Y.; Inoue, N.; Montville, T. J. Inhibition of *Listeria monocytogenes* in Cold Smoked Salmon by *Carnobacterium piscicola* CS526 Isolated from Frozen Surimi. *J. Food Protect.* **2003**, *66*, 1420–1425.

Yang, E.; Fan, L.; Jiang, Y.; Doucette, C.; Fillmore, S. Antimicrobial Activity of Bacteriocin-Producing Lactic Acid Bacteria Isolated from Cheeses and Yogurts. AMB Exp. **2012**, *2*, 48.

Ye, M.; Neetoo, H.; Chen, H. Effectiveness of Chitosan-Coated Plastic Films Incorporating Antimicrobials in Inhibition of *Listeria monocytogenes* on Cold-Smoked Salmon. *Int. J. Food Microbiol.* **2008**, *127*, 235–240.

Yildirim, Z.; Yildirim, M.; Johnson, M. G. Effects of Bifidocinb and Lactococcinr on the Growth of *Listeria monocytogenes* and *Bacillus cereus* on Sterile Chicken Breast. *J. Food Saf.* **2007**, *27*, 373–385.

Yuste, J.; Fung, D. Y. Inactivation of *Salmonella typhimurium* and *Escherichia coli* O157:H7 in Apple Juice by a Combination of Nisin and Cinnamon. *J. Food Protect.* **2004**, *67*, 371–377.

Zhang, J.; Liu, G.; Li, P.; Qu, Y. Pentocin 31-1, A Novel Meat-Borne Bacteriocin and Its Application as Biopreservative in Chill-Stored Tray-Packaged Pork Meat. *Food Control.* **2010**, *21*, 198–202.

Zhao, L.; Wang, Y.; Wang, S.; Li, H.; Huang, W.; X. Liao. Inactivation of Naturally Occurring Microbiota in Cucumber Juice by Pressure Treatment. *Int. J. Food Microbiol.* **2014**, *174*, 12–18.

Zuckerman, H.; Ben Avraham, R. Control of Growth of *L. monocytogenes* in Fresh Salmon using Microgard and Nisin. *Lebensmittel-Wissenschaft and-Technology.* **2002**, *35*, 543–548.

PART II
Soil Microbiology

CHAPTER 7

BIOFERTILIZERS AND PGPR FOR EVERGREEN AGRICULTURE

HARSHA N. SHELAT*, R. V. VYAS, and Y. K. JHALA

Department of Agricultural Microbiology & Bio-fertilizers Projects, B. A. College of Agriculture, Anand Agricultural University, Anand 388110, Gujarat, India

Corresponding author. E-mail: hnshelat@aau.in; hnshelat@gmail.com

CONTENTS

ABSTRACT

To answer to the food demand, all farmers try to maximize the crops they can cultivate on their land. For decades, farmers have used fertilizers to increase their crop yields. With the green revolution, more environment friendly but still effective biofertilizers have been introduced, and they are widely used all over the world, especially in areas where the usage of synthetic fertilizers has ruined the natural nutrient levels of the soil. Biofertilizers are a way to increase crop production by naturally optimizing the nutrient levels of the soil. A biofertilizer is a substance which contains living microorganisms which, when applied to seed, plant surfaces, or soil, colonizes the rhizosphere or the interior of the plant, and promotes growth by increasing the supply or availability of primary nutrients to the host plant. The microorganisms in biofertilizers restore the soil's natural nutrient cycle and build soil organic matter. Through the use of biofertilizers, healthy plants can be grown, while enhancing the sustainability and the health of the soil. As they play several roles, a preferred scientific term for such beneficial bacteria is "plant-growth promoting rhizobacteria." Therefore, they are extremely advantageous in enriching soil fertility and fulfilling plant nutrient requirements by supplying the organic nutrients through microorganism and their by-products.

7.1 INTRODUCTION

Microorganisms are omnipresent and are extremely important. A tip of a needle can hold more than 1 lakh bacteria. Similarly, one spoonful weighing a gram of soil harbors more than 10 crore bacteria and about 10 lakh fungal populations. The soil is a pool for millions of microorganisms of which more than 85% are beneficial to plants. Thus, the soil is a resilient ecosystem. On an average, good quality soil posses 93% mineral and 7% bioorganic substances. The bioorganics are mainly 85% humus, 10% roots, and 5% edaphon. Humus is derived of the synthetic and decomposing activities of the microflora, existing in the dynamic state. Similarly, edaphon consists of microbes, fungi, bacteria, earthworms, microfauna, and macrofauna of which 40% fungi + algae, 40% bacteria + actinomycetes, 12% earthworms, 5% Macrofauna, 3% micro + mesofauna. Such soil microorganisms provide valuable life to soil systems promoting plant growth. These microorganisms work silently to maintain the ecological balance by lively participation in carbon, nitrogen, sulfur, and phosphorous cycles in nature. Soil microbes play a fundamental role both in the evolution of agriculturally useful soil conditions and in stimulating plant growth.

7.2 THE CONCEPT OF BIOFERTILIZER

India is a country with ancient history of organic farming. There are reports in Ramayana, Mahabharata, Kautilya Arthsashtra, Bruhad Samhita, Rig-Veda, Atharvaveda, Vrukshayurveda, Quran, and many more of organic cultivation in India, describing importance of natural resources, cow excreta, recycling of residues, and microbial degradation of materials for crop production. Ancient human being started agriculture near river bank using natural resources. Millennia before farmers used rudimentary means of inoculation such as the transfer of soil from paddocks growing well-nodulated legumes to others that were legume free. Later in 1920, Australian farmers were encouraged to inoculate Lucerne seed with a mixture of glue and sieved air-dried soil taken from paddocks containing well-nodulated plant of targeted legume. The biofertilizer concept goes back to 300 B.C. when our ancestors realized the importance of legume crops bearing nodules. Theophrastus 370–285 B.C. said, "Beans Best Reinvigorates soil, beans are not burdensome crops to the ground."

The term biofertilizer refers to preparation containing living microorganisms, which help in enhancing the soil fertility by fixing atmospheric nitrogen, solubilization/mineralization of phosphorous, potash, zinc, or decomposing organic wastes or by producing plant growth substances. Biofertilizers or microbial inoculants are preparations containing sufficient number in viable or latent state and when inoculated to soil or seed augment nutrient availability to plant by enhancing the growth and proliferation of microbes. Biofertilizer is an organic product containing specific microorganism in concentrated form 10^7 to 10^8 g^{-1} or l^{-1} which is derived from plant parts or from rhizosphere, that is, root zone. Biofertilizer may be referred to as inoculants after the name of the microorganisms they contain, for example, *Azotobacter*, *Azospirillum*, *Rhizobium*, *Acetobacter*, and so on.

7.3 WHY USE BIOFERTILIZERS?

Main challenge for agriculture development is to increase agriculture production and productivity of land. The green revolution brought impressive gains in food production with the help of chemical fertilizer, but with insufficient concern for sustainability. Use of chemical fertilizer not only impoverished soils, destroyed ecological balances, and lead to environment damage but also adversely affected human health (Townsend et al., 2003) (Fig. 7.1).

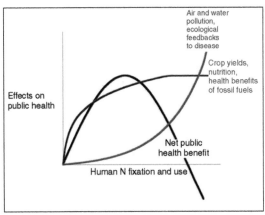

FIGURE 7.1   Effect of chemical fertilizer on human health. (Reprinted from Townsend, A. R.; Howarth, R. W.; Bazzaz, F. A.; Booth, M. S.; Cleveland, C. C. Human Health Effects of a Changing Global Nitrogen Cycle. *Front. Ecol. Environ.* **2003**, 1(5), 240–246. © 2003   With permission from Wiley.)

First description of nitrogen cycle unlocked the mechanism of production of nitrogenous fertilizer. Thereafter, first fertilizer factory in India started at Ranipet, Tamilnadu during 1906, that is, history of chemical fertilizer and its use is about a century old. After independence, use of chemical fertilizer was extensively promoted by Government providing subsidy that proved fruitful to achieve first green revolution; however, indiscriminate use of the fertilizers by farmers have now resulted in threats of soil, water, and air pollution with adverse effect on living entity (Fig. 7.2).

The point to be considered over here is that chemical fertilizers feed the world for the short-term but what about the Future? Indian agriculture has moved from traditional to intensive farming with indiscriminate use of chemical fertilizers and pesticides, which in turn has made our soils largely nonproductive and contaminated the ground water. Indian soils are also becoming poor in organic matter, and there is a growing concern for continuing regular application of organic manure and recycling crop residues to sustain productivity and high responses to NPK (nitrogen, phosphorus, and potassium) fertilizer. Dependence on chemical fertilizers for future agriculture growth would mean further loss in soil quality, possibilities of water contamination, and unsustainable burden on the fiscal system. Moreover, chemical fertilizers always falling short by 6–7 million tons are staggering, considering country's future demand for food. Biofertilizers offer a new technology to Indian agriculture holding a promise to balance many of the shortcomings of the conventional chemical-based technology. Incorporation of biofertilizers or microbial inoculants in combination with organic manure can reduce the detrimental effects of the current agricultural practices in an eco-friendly manner. Biofertilizers are the best supplement of chemical fertilizers. They are alternative means of accessing plant nutrient without polluting environment with sustainability. Biofertilizers are not only effective for that particular crop in which they are used, but also increased the production of next crops. Biofertilizers are cheap inputs, free from the environmentally adverse implications that chemical fertilizers have. Biofertilizers are cheap and have various benefits. It is very effective for sustainable agriculture development so the Government of India has been trying to promote an improved practice involving use of biofertilizers along with fertilizers (Vyas et al., 2008).

Dark red circles indicate oil spills and gray-shaded areas indicate sea pollution and land pollution from chemical fertilizers. Marisa Buxbaum worldprocessor.com

FIGURE 7.2   Environmental pollution due to chemical fertilizers.

Biofertilizers are best described as microorganisms, which add, conserve, and mobilize the crop nutrients in the soil. Use of biofertilizers such as *Rhizobium*, *Azospirillum*, *Azotobacter*, blue green algae (BGA), *Azolla*, mycorrhizae, phosphate solubilizers, and potash mobilizer can facilitate management of crop nutrients leading to long-term sustainability in crop production. The most striking relationship that these have with plants is symbiosis, in which the partners derive benefits from each other. Their addition also increases the fertilizer use efficiency of the crop plants. With ever-increasing population, it is necessary to increase crop productivity per unit area without causing further deterioration in soil health and maintaining micronutrient balance (Vora et al., 2008). Later half of the last century has witnessed remarkable development in the field of biofertilizers. The efforts have created a *cafeteria* of biofertilizers and revealed variety of attributes of these microbes implicating them in nitrogen fixation, phosphate, potash, and zinc solubilization or mobilization (Fig. 7.3), addition of organic matter, secretion of growth factors, improving the physical and chemical properties of the soil, better utilization of the chemical fertilizers, bioconcentration of nutrients in the rhizosphere, and acceleration of the process of composting.

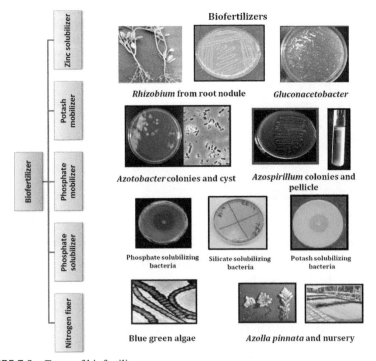

FIGURE 7.3 Types of biofertilizers.

7.4   TYPES OF BIOFERTILIZERS

7.4.1   NITROGEN (N₂) FIXERS

Nitrogen is essential element for all forms of life and a basic requisite for synthesizing nucleic acids, proteins, and other organic nitrogenous compounds. In atmosphere, there is about 78% of molecular nitrogen ($N \equiv N$) which cannot be directly utilized as N fertilizer by plants; it must first be converted to ammonia before it can be absorbed by plants to produce proteins, nucleic acids, and other biomolecules.

Globally, 255 million metric tonnes (MMT) nitrogen get fixed through physical, chemical, and biological processes (Fig. 7.4) of which chemical fixation accounts for about 20% (50 MMT) of total nitrogen fixation, whereas various physical processes such as lightening, volcano eruption, and so on contributes about 30 MMT of total nitrogen fixation. The large portion of total nitrogen fixation, that is, about 175 MMT is contributed by biological processes mediated by microorganisms (Rubio et al., 2005).

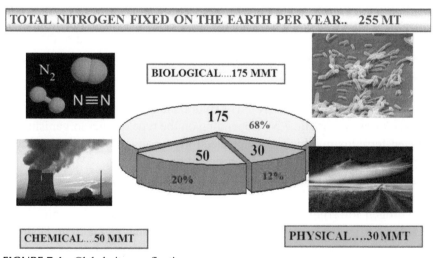

FIGURE 7.4   Global nitrogen fixation.

7.4.1.1   BIOLOGICAL N₂ FIXATION

The biological nitrogen fixation (BNF) system in bacteria represents an inexpensive and eco-friendly alternative to reduce usage of nitrogenous fertilizer and increases the potential nitrogen supply by fixing nitrogen directly with little or no loss. Nitrogen-fixing organisms are classified as (a) symbiotic including members of the family rhizobiaceae forming symbiosis with leguminous plants (e.g., *Rhizobia*) (Ahemad and Khan, 2012; Zahran, 2001) and nonleguminous trees (e.g., *Frankia*) and (b) nonsymbiotic (free living, associative, and endophytes) such as cyanobacteria (*Anabaena, Nostoc*), *Azospirillum, Azotobacter, Gluconoacetobacter diazotrophicus, Azoarcus*, and so on (Bhattacharyya and Jha, 2012).

7.4.1.2   MECHANISM OF BIOLOGICAL N_2 FIXATION

BNF can be summarized by the following equation wherein from 1 mole of nitrogen gas, 2 moles of ammonia are produced utilizing 16 moles of adenosine triphosphate (ATP):

$$N_2 + 8H^+ + 8e^- + 16 \text{ ATP} \longrightarrow 2NH_3 + H_2 + 16ADP + 16 \text{ Pi}$$

The reaction is carried out only by prokaryotes (the bacteria and related organisms), using an enzyme complex, that is, nitrogenase. This enzyme consists of two proteins–an iron (Fe) (60 kDa) and a molybdenum–iron (MoFe) (230 kDa). The reactions take place when N_2 is bound to the nitrogenase enzyme complex. Electrons donated by ferredoxin first reduce the Fe protein. Then, the reduced Fe protein binds ATP and reduces the MoFe protein, which donates electrons to N_2, producing HN=NH (Diazene) which is subsequently reduced to $H_2N–NH_2$ (Hydrazine), and this in turn is reduced to $2NH_3$ (ammonia)(Fig. 7.5). Depending on the type of microorganism, the reduced ferredoxin, which supplies electrons for this process, is generated by photosynthesis, respiration, or fermentation.

An unwanted reaction during nitrogen fixation is the reduction of H^+ to H_2 by nitrogenase. Here, ATP half of the ATP is wasted on production of hydrogen lowering the overall effectiveness of nitrogen fixation. Some diazotrophic strains contain hydrogenase enzyme that can take up H_2 from the atmosphere and convert it into H^+. Thus, presence of a hydrogen uptake system in a symbiotic diazotroph enhances its ability to increase plant growth by recycling the hydrogen gas that formed inside the nodule by the action of nitrogenase. Although it is advantageous to the plant to obtain its nitrogen from a symbiotic diazotroph that has a hydrogen uptake system, this trait is not common in naturally occurring rhizobial strains. Biological nitrogen fixers are known as diazotrophs which can grow without external sources of fixed nitrogen (Fig. 7.6). All diazotrophs contain Mo–Fe or

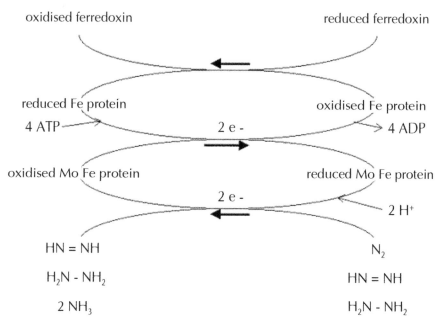

FIGURE 7.5 Oxidation and reduction reactions during BNF.

vanadium nitrogenase systems. The most studied organisms are *Klebsiella pneumoniae* and *Azotobacter vinelandii*, which are used because of their fast growth. There are about 38 genera of bacteria, 20 genera of cyanobacteria, and 87 species of archaea which are classified as diazotrophs (Stolp, 2008). Diazotrophs are phylogenetically diverse and include organisms with vastly different physiological properties. The capacity to perform BNF has been detected in various phototrophic microorganisms like aerobic Cyanobacteria (Young, 1992; Vaishampayan et al., 2001); anaerobic purple-sulfur phototrophs, *Chromatium*, and green-sulfur phototrophs; *Chlorobium* (Young, 1992); chemolithotrophic microorganism, for example, *Alcaligenes, Thiobacillus, Methanosarcina* (Madigan et al., 2000; Young, 1992), or *Azospirillum* (Malik and Schlegel, 1981); and in a great number of heterotrophic bacterial strains (anaerobes such as *Clostridium*, microaerophiles such as *Herbaspirillum* and *Azospirillum*) (Hill, 1992; Paul and Clark, 1996). Diazotrophs are also found in a wide variety of habitats including soil and water. Nitrogen-fixing microorganisms which are most commonly used as biofertilizers can be classified as shown in Figure 7.6.

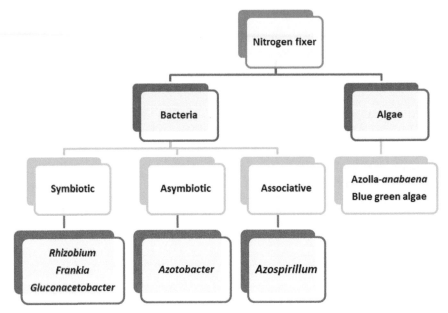

FIGURE 7.6   Classification of nitrogen fixing microbes.

7.4.1.3   RHIZOBIUM

Rhizobium is the familiar species of nitrogen-fixing bacteria that can infect the roots of leguminous plants, forming nodules where nitrogen fixation takes place. Rhizobia nitrogenase enzyme system supplies a constant source of reduced nitrogen to the host plant and the plant in turn provides nutrients and energy for the activities of the bacterium. About 90% of legumes found to be nodulated by *Rhizobium* group of bacteria. In the soil, the rhizobia are free living and motile, possess regular rod shape, cannot fix nitrogen, as against the bacteria found in root nodules which are irregular cells called bacteroids which are often club and Y-shaped.

All the rhizobial strains do not nodulate every legume crops. On the contrary, particular rhizobia form a symbiosis with specific legumes or groups of legumes. To get the maximum benefits of BNF, it is a must to provide farmers with the correct efficient rhizobia for their legume crop. Scientists have studied this matching system for many important food, forage, and tree legumes and have categorized different rhizobia and their legume partners into cross inoculation groups as mentioned in Table 7.1.

TABLE 7.1   Cross Inoculation Groups of *Rhizobium*.

Name of Rhizobia	Legume group	Legume cross inoculation group
R.leguminosarum bv. viceae	Pea	Peas (*Pisum* spp.); vetches (*Vicia* spp.); lentils (*Lens culinaris*); faba bean (*Vicia faba*)
R. leguminosarum bv. Phaseoli	Bean	beans (*Phaseolus vulgaris*); scarit runner bean (*P. coccineus*)
R. leguminosarum bv. trifolii	Clover	Clovers (*Trifolium* spp.)
R. meliloti	Alfalfa	Alfalfa (*Medicago* spp.); sweet clovers (*Melilotus* spp.); fenugreek (*Trigonella* spp.)
Rhizobium spp.	Chickpea	Chickpea (*Cicer arietinum*)
R. lupine	Lupin	lupine, *Lupinus* spp. serradella, *Ornithopus* spp.
Bradyrhizobium spp.	Cowpea	Pigeon pea (*Cajanus cajan*); peanut (*Arachis hypogaea*); cowpea, mungbean, black gram, rice bean (*Vigna* spp.); lima bean (*Phaseolus lunatus*); *Acacia mearnsii*; *A. mangium*; *Albizia* spp.; *Enterlobium* spp., *Desmodium* spp., *Stylosanthes* spp., Kacang bogor (*Voandzeia subterranea*), *Centrosema* spp., winged bean (*Psophocarpus tetragonolobus*), hyacinth bean (*Lablab purpureus*), siratro (*Macroptilium atropurpureum*), guar bean (*Cyamopsis tetragonoloba*), calopo (*Calopogonium mucunoides*), puero (*Pueraria phaseoloides*)

Rhizobium inoculation provides benefit in two ways, direct and indirect. Direct contribution is through nitrogen fixation to the tune of 50–200 kg for the inoculated crops as presented in Table 7.2, whereas indirect contribution is *residual* nitrogen effect to the succeeding crop (Table 7.3) and that is why the crop rotation regime of legume–cereal is common practice to get maximum benefit of nitrogen fixing ability of leguminous crops.

TABLE 7.2   Direct Contribution of *Rhizobium* for N Nutrition in Various Crops.

Crop	N_2 (kg/ha)	Crop	N_2 (kg/ha)
Alfalfa	100–200	Soybean	60–80
Groundnut	50–60	Cowpea	80–85
Chickpea	85–110	Pea	52–57
Clover	100–150	Greengram/Blackgram	50–55
Pigeon pea	168–200	Lentil	90–100

TABLE 7.3   Residual Effect of *Rhizobium*.

Preceding leguminous crop	Succeeding cereal crop	N₂ provided to succeeding crop (kg/ha)
Chickpea/Cowpea	Maize	65/60
Pigeonpea	Maize	20–67
Urdbean/Mungbean	Sorghum	40/40
Mungbean/Cowpea/Pigeonpea	Wheat	16–68/40/38
Cowpea/Chickpea	Pearl millet	60/40

Source: Wani et al. (1995).

7.4.1.4   FRANKIA

Frankia is a genus of the bacterial group termed actinomycetes—Grampositive nitrogen-fixing filamentous bacteria that are well known for their production of air-borne spores. *Frankia* resembles the antibiotic-producing *Streptomyces* spp. as both grow by branching and tip extension. *Frankia* spp. produces three cell types: sporangiospores, hyphae, and diazo-vesicles (spherical, thick walled, and lipid-enveloped cellular structures). The spores are "sporangiospores" borne in "multilocular" sporangia that are formed either terminally, or in an intercalary position on the hyphae. Hyphae measure about 0.5–1.5 μm wide. The diazo-vesicles are responsible for the supplying of sufficient nitrogen to the host plant during symbiosis.

Frankia is having great ecological interest for several reasons due to its wide distribution, ability to fix nitrogen, differentiation into sporangium, and to nodulate many plants of about 24 genera. *Frankia* species are slow-growing in culture and require specialized media, suggesting that they are specialized symbionts. *Frankia* are found in the soil and have a symbiotic relationship with certain woody angiosperms, called actinorhizal plants. They form nitrogen-fixing root nodules (sometimes called actinorhizae) with several woody plants of different families, such as Alder (*Alnus* species), Sea buckthorn (Hippophae rhamnoides, which is common in sand-dune environments), and Casuarina (a Mediterranean tree genus). The other host plants of *Frankie* include Ceanothus, Trevoa, Talguenea, Chamaebatia, Cercocarpus, Purshia, Myrica, Hippophae, Dryas, and so on.

Frankia can establish a nitrogen-fixing symbiosis with host plants where nitrogen is the limiting factor, and plants can grow without added nitrogen. Thus, actinorhizal plants often prosper in soils that are low in combined nitrogen.

7.4.1.5   GLUCONACETOBACTER

The family *Acetobacteriaceae* includes genera, *Acetobacter, Gluconobacter, Gluconoacetobacter*, and *Acidomonas*. Based on 16S rRNA sequence analysis, the name *Acetobacter diazotrophicus* has been changed to *G. diazotrophicus* (Yamada et al., 1997). *G. diazotrophicus* is a Gram-negative, obligate aerobe, acid-tolerant. The cells are straight rods with rounded ends (0.7–0.9 mm by 1–2 mm), arranged singly or in pair or chain-like structures without endospores. It can grow on high sucrose concentration of 30% and very low pH (3.0) and has an ability to fix N_2 under microaerophilic conditions. The optimum carbon source is 10% sucrose. As sucrose cannot be respired by *G. diazotrophicus*, it grows by secreting an extracellular enzyme, levansucrase that can hydrolyze sucrose into fructose and glucose. *Gluconacetobacter* is proved to be important inoculants for sugarcane (Gillis et al., 1989; Muthukumarasamy et al., 2000). *G. diazotrophicus* has been recognized as an aerotolerant diazotroph in which oxygen is instrumental for the generation of large quantities of ATP required for N fixation. Besides sugarcane, *G. diazotrophicus* was found to colonize cameroon grass, coffee, ragi, tea, pineapple, mango, banana, and so on *G. diazotrophicus* isolated from various sources does not exhibit much variation in the genetic diversity (Salgado et al., 1997; Cabellaro-Mellado and Martinez, 1994). However, Suman et al. (2001) found that the diversity of the isolates of *G. diazotrophicus* by RAPD (random amplification of polymorphic DNA) analysis was more striking than that reported on the basis of morphological and biochemical characters. Certain genetically related groups of *G. diazotrophicus* or its ancestors have acquired the capability of colonizing plants by themselves or with the aid of the vectors such as insects or fungi (Tapia-Hernandez et al., 2002). *G. diazotrophicus* has been found to possess plasmids of 2–170 kb. *Gluconoacetobacter* could colonize sugarcane varieties in India where the chemical N fertilizers are completely replaced by organic manures (Ashbolt and Inkerman, 1990).

7.4.1.6   AZOTOBACTER

The family *Azotobacteriaceae* includes two genera, namely, *Azomonas* (non-cyst forming) with three species (*A. agilis, A. insignis*, and *A. macrocytogenes*) and *Azotobacter* (cyst forming) comprising of six species, namely, *A. chroococcum, A. vinelandii, A. beijerinckii, A. nigricans, A. armeniacus*, and *A. paspali. Azotobacter* is generally regarded as free living as well as endophytic aerobic nitrogen fixer. *A. paspali*, which was first described by Dobereiner and Day (1975), has been isolated from the rhizosphere of *Paspalum notatum*, a tetraploid subtropical grass, and is highly host specific.

The genus *Azotobacter* comprises large, Gram-negative, obligately aerobic rods found in neutral to alkaline soils, capable of fixing N_2 asymbiotically. *Azotobacter* is also of interest because it has the highest respiratory rate among any other living organisms. *Azotobacters* possess a special resting structure called cyst. *Azotobacter* cells are very large with diameter of 2–4 µm or more and sometimes as large as yeast. Pleomorphism is very common and various cell shapes and sizes have been observed. They are motile by peritrichous flagella. On carbohydrate-containing media, extensive capsules or slime layers are produced by free-living N_2 fixing *Azotobacter*. They can utilize wide variety of carbon compounds like carbohydrates, alcohols, and organic acids. The metabolism is strictly oxidative, and acids or other fermentation products are rarely produced. *Azotobacter* can fix nitrogen and hence can grow on N_2 free media; however, growth also occurs on simple combined nitrogen like ammonia, urea, and nitrate.

7.4.1.7 *AZOSPIRILLUM*

Azospirillum fix nitrogen under microaerophilic conditions and are frequently associated with root and rhizosphere of a large number of agriculturally important crops and cereals. This group of associative rhizobacteria or endophytes includes 10 species, viz. *A. lipoferum*, *A. brasilense*, *A. amazonense*, *A. halopraeferens*, *A. irakense*, *A. largimobile*, *A. doebereinerae*, *A. oryzae*, *A. melinis*, and *A. canadensis*, each one classified according to its particular biochemical and molecular characteristics. *Azospirillum* strains are root colonizer but not specific to any crop and hence can be successfully used in any plants.

Azospirillum are Gram-negative, vibroid, and 1–1.5 µm in diameter, possessing peritrichous flagella for swarming and a polar flagellum for swimming, also contain poly-3-hydroxybutyrate granules. *Azospirillum* are essentially aerobic but able to grow and fix nitrogen under microaerophilic condition. Many *Azospirillum* strains produce plant hormones both in liquid culture and natural situation. The major hormone produced is indole-3-acetic acid (IAA) (Tien et al., 1979). *Azospirillum* converts tryptophan to indole acetic acid and also hormones like gibberlins and cytokinins are excreted by *A. brasilense* (Tien et al., 1979). Root exudates of plants contain tryptophan, which probably acts as precursor for the synthesis of

IAA source. Although the benefit from BNF is disputed, association of gramineae with *A. brasilense* or *A. lipoferum* has been reported to result in a more robust root system, increasing absorption of water and minerals from the soil and faster plant growth (Baldani et al., 1987; Okon, 1985). Some most significant modes of action for *Azospirillum* are nitrogen fixation, secretion of phytohormones, production of undefined signal molecules that can interfere with plant metabolism, and the enhancement of mineral uptake by plants (Okon and Itzigsohn, 1995). Thus, they exert beneficial effect on the growth of plants (Carrillo-Castaneda et al., 2003) and increase the yield of many crops of agronomic importance. *Azospirillum* are known to be highly pleomorphic and to change their metabolic activities swiftly in the face of changing environmental conditions (Berg et al., 1980). Production of cyst may represent a mechanism by which azospirilla can persists in the rhizosphere during unfavorable conditions such as desiccation, temperature, and nutrient limitation and convert into enlarged cyst forms (Sadasivan and Neyra, 1985).

The activity of associative diazotroph such as *Azospirillum* could meet the nitrogen nutrition of cereal crops. After establishing in the rhizosphere, *Azospirilla*, usually, promote the growth of plants (Okon, 1985; Tilak and Subba Rao, 1987; Bashan and Holguin, 1997). Although they possess N_2-fixing capability of about 1–10 kg/ha increase in yield is mainly attributed to improved root development due to the production of growth promoting substances and consequently increased rates of water and mineral uptake (Fallik et al., 1994). Isolation and characterization of bacterial diversity from endorhizosphere of sugarcane (*Saccharum* spp.) and rye grass (*Lolium perenne*) suggested that *Azospirillum* isolates from sugarcane and rye grass exhibited maximum nitrogenase activity among *Azospirillum* as compared to other isolates namely *Bacillus*, *Escherichia coli*, and *Pseudomonas* (Gangwar and Kaur 2009). Gupta and Shelat (2010) reported plant growth promoting activity through soil application as well as by foliar spray of endophytic *Azospirillum* on maize.

7.4.1.8 BLUE GREEN ALGAE (BGA)

BGA, commonly known as cyanobacteria, can carry out photosynthesis and nitrogen fixation simultaneously which make them ubiquitous in distribution. A specialized thick walled cells known as heterocyst in BGA carry nitrogenase enzymes and responsible for nitrogen fixation. Dried cyanobacteria are generally used as biofertilizer as means of adding fertility to the agricultural soils. The term algalization is generally used to define the inoculation of soil with cyanobacteria. BGA strains used for biofertilizer are *Anabaena variabilis*, *Nostoc muscorum*, *Aulosira fertilissima*, *Tolypothrix tenuis*, *Cylindrospermum*, *Westolliopsis*, and so on BGA inoculation in field could benefit up to 20–30 kg N/ha under ideal conditions. In India, Tamil Nadu, Uttar Pradesh, Jammu and Kashmir, Andhra Pradesh, Karnataka, Maharashtra, and Haryana are the major states where algalization in rice is being followed in. Same is being used in China, Egypt, Philippines, and erstwhile U.S.S.R.

7.4.1.9   AZOLLA-ANABAENA ASSOCIATION

BGA generally *Anabaena* can remain in symbiotic association with free floating water fern *Azolla*. This association is considered as miniature nitrogen fertilizer factories. An *Azolla-Anabena* system is ideal for the cultivation of rice under tropical conditions because of its ability to fix atmospheric nitrogen and capacity to multiply at faster rates. The relatively quick decomposition of the biomass and rapid availability of its nitrogen to the standing crop makes it agronomically outstanding. At present, many progressive farmers and nongovernmental organizations are employing *Azolla* as an invaluable input in agriculture, and some of them are actively engaged in its popularization as biofertilizers, green manure, poultry feed, and fodder. The benefits of *Azolla* application in the rice field are following:

- Basal application of green manure at 10–12 t/ha increases soil nitrogen by 50–60 kg/ha and reduces 30–35 kg of nitrogenous fertilizer requirement of rice crop.
- *Azolla* may be used for the production of hydrogen fuel and biogas, control of weeds and mosquitoes, increase water use efficiency in rice cultivation by reducing evaporation of water, and the reduction of ammonia volatilization that accompanies the application of chemical nitrogen fertilizer.

The most commonly found forms of *Azolla* are *A. pinnata*, *A. filiculoides*, *A. rubra*, *A. microphylla*, *A. imbricata*, and *A. caroliniana*. *A. pinnata* is the most widely distributed species in India and throughout the world in both tropical and temperate regions.

7.4.2 PHOSPHATE SOLUBILIZING BACTERIA (PSB)

Phosphorous is second important nutrient required for plant growth. Phosphorus is an essential element for plant development and growth making up about 0.2% of plant dry weight. Plants acquire P from soil solution as phosphate anions. It is applied in soil as single super phosphate (SSP), diaamonium phosphate fertilizer which is produced from a raw source rock phosphate. Rock phosphate is imported from countries like Morocco, Africa, China, and so on. India has a large reserve about 2500 lakh ton of rock phosphate which can easily be used along with phosphate solubilizing bacteria (PSB) to cut down import cost.

PSB otherwise called phosphobateria play a major role in the solubilization and uptake of native and applied soil phosphorus requirement of crops (Table 7.4). Efficiency of applied phosphorus in form of SSP or diaamonium phosphate is very low due to its fixation either in the form of aluminum or Fe phosphates in acidic soils or in the form of calcium phosphate in neutral or alkaline soils. The introduction of efficient "P" solubilizers in the rhizosphere of crops and in soils increases the availability of phosphorus. Phosphobacteria bring about dissolution of bound forms of phosphates in soil and hence advocated as biofertilizer for all crops.

7.4.2.1 MECHANISM OF P SOLUBILIZATION

The major mechanism used by PSB is synthesis of low molecular weight organic acids such as gluconic and citric acid (Bnayahu, 1991; Rodriguez et al., 2004). These are the organic acids which possess hydroxyl and carboxyl groups bind with insoluble phosphate and release soluble phosphate (Kpomblekou and Tabatabai, 1994). Other mechanisms are the release of H^+ (Illmer and Schinner, 1992), the production of chelating substances (Sperber, 1958; Duff and Webley, 1959), and inorganic acids (Hopkins and Whiting, 1916). Also, exopolysaccharides (EPS) synthesized by PSB contribute ultimately in the solubilization of tricalcium phosphates by binding free P in the medium (Yi et al., 2008).

The mineralization of organic P occurs through hydrolysis of phosphoric esters enzymes like phosphomonoesterase, phosphodiesterase, and phosphotriesterase (Rodriguez and Fraga, 1999). P solubilization and mineralization can coexist in the same bacterial strain (Tao et al., 2008). Among the phosphate solubilizing rhizobacterial genera *Bacillus*, *Pseudomonas*,

Serratia, Burkholderia, Enterobacter, and *Staphylococcus* are the most common and cultivated on artificial media (Richardson and Hadobas, 1997; Hussin et al., 2007; Shedova et al., 2008). Under favorable conditions, PSB can solubilize 20–30% of insoluble phosphate resulting in 10–20% yield increase.

TABLE 7.4 Major Groups of Phosphate Solubilizing Microorganisms

S. N.	Group	Example
1.	Bacteria	*Bacillus megaterium* var. *phosphaticum, B. circulans, B.subtilis, B. polymyxa, B. coagulans, Pseudomonas striata, P. liquifacines, P. putida, Burkholderia, Enterobacter* spp.
2.	Fungi	*Aspergillus awamori, A. fumigatus, A. flavus, Penicillium digitatum, Paecilomyces lilacinum, Trichoderma* spp.
3.	Yeast	*Torulospora globosa*

7.4.3 PHOSPHATE SOLUBILIZER AND MOBILIZER—VESICULAR ARBUSCULAR MYCORRHIZAE (VAM)

Mycorrhizae are a group of fungi that include a number of types based on the different structures formed inside or outside the root commonly known as vesicular arbuscular mycorrhizae (VAM). Mycorrhizae are obligate and saprophytic in nature which are totally biotroph and, hence, requires a living host for its survival. VAM aids in transfer of nutrients from soil into root system through specialized structures known as vesicles and arbuscules. Its primary function is the absorption of resources from the soil. Their symbiotic associations with plant roots help in the greater absorption of phosphorous, water, and other important macro- and essential micro-elements making them available to the plants in an organic form. Besides, they impart resistance to plants against drought and soil borne fungal pathogens and nematodes. In addition, mycorrhizal plants show higher tolerance to high soil temperatures, various soil- and root-borne pathogens, and heavy metal toxicity. VAM can be used in cereals, pulses, oil seeds, and fruit crops, but it cannot be used in cruciferous plants. Mycorrhizae are described as improving the absorption of several nutrients like phosphorous, potassium, magnesium, copper, zinc, calcium, and iron (Quilambo, 2003). Mycorrhizal fungi can increase yield by 30–40%.

7.4.4 POTASH MOBILIZING BACTERIA (KMB)

Potash is the third essential nutrient for plant growth and is extremely important for the productive farming of major agricultural crops. This mineral is required in high quantities by plant cells and possesses essential physiological and biochemical functions involving cell osmotic regulation and enzyme activation. Soil microbes have been reported to play a key role in the natural K cycle, and therefore, potassium solubilizing microorganisms present in the soil could provide an alternative technology to make potassium available for uptake by plants (Groudev, 1987; Rogers, 1998).

7.4.4.1 MECHANISM OF K SOLUBILIZATION

Potassium mobilizing/solubilizing bacteria were found to dissolve potassium, silicon, and aluminum from insoluble K-bearing minerals such as micas, illite, and orthoclases, by excreting organic acids which either directly dissolved rock K or chelated silicon ions to bring K into the solution (Parmar and Sindhu, 2013). A wide range of bacteria, namely *Pseudomonas, Burkholderia, Enterobacter, Acidothiobacillus ferrooxidans, Bacillus mucilaginosus, Bacillus edaphicus, B. circulans,* and *Paenibacillus* sp. *Frateuria, Citrobacter* and so on have been reported to release potassium in accessible form from potassium-bearing minerals in soil (Sheng, 2005; Lian et al., 2002; Li et al., 2006; Liu et al., 2012). These bacteria increase potash dissolution rate by producing and excreting by-products that interact with the mineral surface. Microbial respiration, degradation of particulate and dissolved organic carbonic acid concentration at mineral surfaces in soil and in ground water, leads to increase in the rates of mineral weathering by a proton-promoted dissolution (Barker et al., 1998). Silicate bacteria were found to dissolve potassium, silicon, and aluminum from insoluble minerals (Aleksandrov et al., 1967). Most of the potassium in soil exists in the form of silicate minerals. Liu et al. (2006) proved that the polysaccharides strongly adsorbed the organic acids and attached to the surface of the mineral, resulting in an area of high concentration of organic acids near the mineral. Also, EPS adsorb silica and create equilibrium between the mineral and fluid phases leading to the reaction toward SiO_2 and thereby K^+ solubilization.

7.4.5 ZINC SOLUBILIZING BACTERIA (ZSB)

Among all the micronutrients, zinc is not mobile in plants, and hence for optimum plant growth, supply of available zinc is mandatory. The function of zinc is to help the plant produce chlorophyll. Zinc deficiency is reflected by chlorosis of base of leaf, reduction in root growth and activity. Zn deficiency is widespread throughout the world, especially in soil orders of Aridisols, Alfisols, Mollisols, and Vertisols (Srivastava and Gupta, 1996). Lower microbial activity in the soil decreases zinc release from soil organic matter. External addition of soluble Zn to alleviate deficiency results in the transformation of about 96–99% to various fractions of unavailable forms and about 1–4% is left as available fraction in the soil. The deficiency of Zn in soils is usually attributed to low solubility of Zn rather than low total content of Zn in most agricultural soils. Plants take up Zn as (Zn^{2+}) divalent cation. Numerous microorganisms, especially those associated with roots, have the ability to increase crop growth and productivity by solubilization of unavailable mineral nutrients (Cunningham and Kuiack, 1992). Zinc solubilizing potential of few microbial genera such as *Bacillus*, *Pseudomonas*, and *Aspergillus* are widely explored.

7.4.5.1 MECHANISM OF ZN SOLUBILIZATION

The release of organic acids that sequester cations and acidify the microenvironment near root is thought to be a major mechanism of Zn solubilization. A number of organic acids such as acetic, citric, lactic, propionic, glycolic, oxalic, gluconic acid, and so on have been considered due to its effect in pH lowering by microorganisms (Cunningham and Kuiack, 1992). Organic acid secreted by microflora increase soil Zn availability in two ways; they are probably exuded both with protons and as counter ions and, consequently, reduce rhizospheric pH. In addition, the anions can chelate Zn and increase Zn solubility (Jones and Darrah, 1994) which results in the conversion of available form (Zn^{2+}) to plants.

7.4.6 SILICATE SOLUBILIZING BACTERIA (SSB)

Silicon (Si), an element available on earth in large quantities, is second only to oxygen (Ehrlich, 1981). It is the eighth most plentiful elements in the universe. Si content in soils ranges from <1 to 45% by dry weight (Sommer et al., 2006), whereas silica (SiO_2) varies from less than 10% to almost 100%. Silicon dioxide (SiO_2) known as silica comprises 50–70% of the soil mass. Among the plants, silica concentration are found to be higher in monocotyledons than in dicotyledons, and its level increased from legumes < fruit crops < vegetables < grasses < grain crops (Thiagalingam et al., 1977). Grasses accumulate 2–20% foliar dry weight as hydrated polymer or silica gel, whereas rice straw accumulates 4–20%. The Si found in soil is in unavailable polymerized form which is to be made soluble by means of biological or chemical reactions in soil. Plants absorb Si exclusively as mono silicic acid (ortho-silicic acid) by the diffusion and also by the influence of transpiration-induced root absorption. Silicate solubilizing bacteria or silicate bacteria, namely *Bacillus, Enterobacter, Pseudomonas* spp. are of great interest in recent times because of their role in solubilization of silicate minerals rendering silica and potassium available for crop uptake thus reducing the potash fertilizer requirement (Sheng, 2005) and also due to their role in desilication of ores like bauxite.

7.4.6.1 MECHANISM OF SI SOLUBILIZATION

Several mechanisms of dissolution of silicates by bacteria have been suggested. Acidolysis, alkaline hydrolysis, ligand degradation, enzymolysis, capsule adsorption, extracellular polysaccharides, and redox have been shown to play a role in microbial dissolution of silicates. But acidolysis is the main and largely accepted mechanism of weathering silicate minerals (Jongmans et al., 1977). Although all types of organic acids have been shown to be involved in dissolution of silicate, gluconic acid was identified as the most effective agent (Sheng et al., 2008).

7.5 PLANT GROWTH PROMOTING RHIZOBACTERIA (PGPR)

Plant-growth promoting rhizobacteria (PGPR) include a wide variety of soil bacteria when grown in association with a host plant, resulting in stimulation of growth of their host. According to Kloepper and Schroth (1978), PGPR promote plant growth by alteration of the whole microbial community in rhizospheric niche through the production of various substances (Kloepper and Schroth, 1978). Generally, PGPR promote plant growth directly by either facilitating nitrogen, phosphorus, or essential minerals acquisition as well as modulating plant hormone levels or indirectly by decreasing the inhibitory effects of various pathogens (Glick, 2012). PGPR role by direct mechanisms of nutrient acquisition is discussed at length in Section 7.4.

7.5.1 MODULATION OF PLANT HORMONAL LEVEL

Plant growth and development is also regulated by phytohormones produced by rhizospheric bacteria. Phytohormones are organic compounds which in extremely low concentrations influence biochemical, physiological, and morphological processes in plants, and their synthesis is finely regulated (Fuentes-Ramírez and Caballero-Mellado, 2006). With the production of different phytohormones like IAA, gibberellic acid, and cytokinins, PGPR can increase root surface and length and promote plant development (Kloepper et al., 2007). IAA (auxin) is the most quantitatively important phytohormone produced by PGPR, and treatment with auxin-producing rhizobacteria increased the plant growth (Vessey, 2003). Production of other phytohormones by biofertilizing PGPR has been identified, but not nearly to the same extent as bacteria which produce IAA (Vessey, 2003). A few PGPR strains *Bacillus licheniformis* and *Bacillus pumilus*, *Pseudomonas fluorescens*, and *Azospirillum* were reported to produce cytokinins and gibberellins/gibberellic acid (Gutiérrez-Mañero et al., 2001; Vessey, 2003). Cytokinin encourages tissue expansion, cell division, and cell enlargement in plant. Many PGPR have the capability to produce 1-aminocyclopropane-1-carboxylate (ACC) deaminase, an enzyme which cleaves ACC, the immediate precursor of ethylene in the biosynthetic pathway for ethylene in plants (Glick et al., 1998) imparting stress tolerance.

7.5.2 BIOLOGICAL CONTROL OF PHYTOPATHOGENS

Most of PGPR contribute in plant growth promotion by acting as biocontrol agents giants bacteria, fungi, and virus (Glick, 2012). This phenomenon is called induced systemic resistance (ISR). Competition for nutrients, exclusion of habitat, ISR, and production of antifungal metabolites like siderophores, hydrogen cyanide (HCN), phenazines, pyrrolnitrin, 2,4-diacetylphloroglucinol, pyoluteorin, viscosinamide, and tensin are the chief approaches of biocontrol activity in PGPR (Bhattacharyya and Jha, 2012). Thus, microbial inoculants biofertilizers/PGPR are an alternative approach to overcome constraints of synthetic fertilizer and to revive soil's fertility resulting in the intensive farming.

7.6 METHOD OF APPLICATION OF BIOFERTILIZERS

Biofertilizers should to be properly applied to the seeds, seedlings, or the soil. The important steps are summarized herein. First step is to purchase the right type and good quality of biofertilizer from a reputed source and use it before the expiry date. It should be freshly prepared. Some important suggestions while buying biofertilizers are:

- Different crops require different biofertilizers. Make sure to have the right choice. The name of the crop should be mentioned on the packet. For pulses and other legumes, only *Rhizobium* of the right type which is crop specific should be used.
- Be sure that the inoculant is fresh. Read for the expiry date and use before expiry date after which the inoculant may not give good results because it may not contain required number of bacteria.
- For carrier-based products, please see that the product is having sufficient moisture. Product, which has dried, should not be used.
- Use biofertilizer products of a reputed manufacturer with fertilizer control order (FCO) certification.

7.6.1 SEED TREATMENT

The method of inoculation of seed with microbial inoculant affects their efficacy. In general, the insouciant is mixed in clean water or in 5% solution of an adhesive to make thick slurry. Many a times, biofertilizer manufacturers add gum Arabic in their product; in this case, there is no need to use any adhesive. This slurry is then mixed with the required quantity of seeds to be sawn in an acre of land. The seeds are dried in shade and are stored in cool dry place before use. Generally, about 25–30 g inoculant is required to coat 1 kg seeds. However, exact quantity will depend upon size of the seed. Generally, large-seeded crop, namely, rajma, groundnut, and so on require 1.5 kg inoculant/ha, whereas medium-sized seed, namely gram, soybean and small-sized seeds, namely mungbean, uradbean, cowpea, Lucerne, and so on will need 1.0 and 0.5 kg inoculant, respectively. Seeds should be sown as soon as possible after treating with inoculants to take full benefit of the inoculation.

The seeds can also be inoculated by sprinkle method. In this case, the adhesive solution is first mixed uniformly with the seeds and extra solution, if any should be decanted from the seeds. Contents of inoculant packet are then sprinkled over the seeds and mixed uniformly with the seeds.

The methods treating seed with inoculants do not hold well, if the agroclimatic conditions are not favorable. To protect the inoculated seed from soil stresses, it is necessary to coat the inoculated seeds with inert materials. This technique is known as pelleting. Pelleting provides prolonged survival to the rhizobia on seeds. The most commonly used coating materials are calcium carbonate for acidic soils and gypsum for saline neutral soil conditions.

- Prepare the slurry of 200 g of biofertilizer in 200–500 ml of water.
- Pour this slurry slowly on 1–25 kg seeds. Mix the seeds evenly with hands to get uniform coating on all seeds.
- For liquid formulation, mix 5 ml culture with 1 l water and treat approx. 1 kg seeds.
- Dry the treated seeds in shade and sow immediately.

7.6.2   SEEDLING TREATMENT

- Prepare a suspension of 1–2 kg of biofertilizer in 10–15 l of water.
- For liquid formulation, mix 5 ml culture in 1 l water.
- Dip the seedlings from 10 to 15 kg of seeds into the suspension for 20–30 min and transplant treated seedlings immediately.

7.6.3   SOIL TREATMENT

- Prepare the mixture of 2–3 kg of biofertilizer in 40–50 kg of soil/compost.
- For liquid formulation, mix 250 ml of biofertilizer with compost/soil.
- Broadcast the mixture in one acre (0.4 ha) either at sowing or 24 h earlier.
- The broadcast may be done in standing crops just before irrigation.
- 4–5 kg biofertilizer in 120–150 kg of soil/compost should be used for long duration crops.
- Do not mix chemical fertilizers, insecticides, or pesticides directly with biofertilizers.

7.7   DO'S AND DON'TS WHILE USING BIOFERTILIZER

Biofertilizer contains live microorganisms so that we have to keep some points in mind as mentioned below.

Do's	Don'ts
Keep biofertilizer packets under shade	Don't use poor quality biofertilizer
Mix biofertilizer with seeds as per full requirement	Don't use packets on which crop, date, batch no., name of manufacturer, and expiry date is not mentioned
Dry-treated seeds in shade	Don't mix biofertilizer with insecticides, fungicides, herbicides, or fertilizer
Sow treated seeds immediately after drying. The better time of sowing is either the early morning or afternoon when heat is tolerable	Don't mix the biofertilizer with seed under direct sunlight

KEYWORDS

- cultivation
- green revolution
- biofertilizers
- rhizosphere
- rhizobacteria
- PGPR

REFERENCES

Ahemad, M.; Khan, M. S. Effects of Pesticides on Plant Growth Promoting Traits of *Mesorhizobium* Strain MRC4. *J. Saudi Soc. Agric. Sci.* **2012**, *11*, 63–71.

Aleksandrov, V. G.; Blagodyr, R. N.; Iiiev, I. P. Liberation of Phosphoric Acid from Apatite by Silicate Bacteria. *Mikrobiology Zh (Kiev).* **1967**, *29*, 111–114.

Ashbolt, N. J.; Inkerman, P. A. Acetic Acid Bacterial Biota of the Pink Sugarcane Mealy Bug, *Saccharococcus sacchari,* and its Environs. *Appl. Environ. Microbiol.* **1990**, *56*(3), 707–712.

Baldani, V. L. D.; Baldani, J. I.; Dobereiner, J. Inoculation of Field-Grown Wheat (*Triticum aestivum*) with *Azospirillum* spp. *Biol. Fertil. Soils.* **1987**, *4*, 37–40.

Barker, W. W.; Welch, S. A.; Chu, S.; Banfield, J. F. (1998). Experimental Observations of the Effects of Bacteria on Aluminosilicate Weathering. *Am. Mineral.*, **1998**, *83*(11–12), 1551–1563.

Bashan, Y.; Holguin, G. *Azospirillum*–Plant Relations: Environmental and Physiological Advances. *Can. J. Microbiol.* **1997**, *43*, 103–121.

Berg, R. H.; Tyler, M. E.; Novick, N. J.; Vasil, V.; Vasil, I. K. Biology of *Azospirillum*-sugarcane Association Enhancement of Nitrogenase Activity. *Appl. Environ. Microbiol.* **1980**, *39*, 642–649.

Bhattacharyya, P. N.; Jha, D. K. Plant Growth-Promoting Rhizobacteria (PGPR): Emergence in Agriculture. *World J. Microbiol. Biotechnol.* **2012**, *28*, 1327–1350.

Bnayahu, B. Y. Root Excretions and Their Environmental Effects: Influence on Availability Of Phosphorus. In *Plant Roots: The Hidden Half*; Waisel, Y; Eshel, A.; Kafkafi, U., Eds.; Dekker: New York, 1991, pp. 529–557.

Cabellaro-Mellado, J.; Martinez-Romero, E. Limited Genetic Diversity in the Endophytic Sugarcane Bacterium *Acetobacter diazotrophicus*. *Appl. Environ. Microbiol.* **1994**, *60*(5), 1532–1537.

Carrillo-Castaneda, G.; Munoz, J. J.; Peralta-Videa, J. R.; Gomez, E.; Gardea-Torresdey, J. L. Plant Growth-Promoting Bacteria Promote Copper and Iron Translocation from Root to Shoot in Alfalfa Seedlings. *J. Plant Nutr.* **2003**, *26*, 1801–1814.

Cunningham, J. E.; Kuiack, C. Production of Citric and Oxalic Acid and Solubilization of Calcium Phosphate by *Penicillium billai*. *Appl. Environ. Microbiol.* **1992**, *58*, 1451–1458.

Dobereiner, J.; Day, J. M. Nitrogen Fixation in Rhizosphere of Grasses. In *Nitrogen Fixation by Free-Living Microorganisms;* Stewart, W. D. P., Ed.; Cambridge University Press: Cambridge, 1975, pp. 39–56.

Duff, R. B.; Webley, D. M. 2-Ketogluconic Acid As a Natural Chelator Produced by Soil Bacteria. *Chem. Ind.* **1959**, 1376–1377.

Ehrlich, H. L. *Geomicrobiology*; Marcel Dekker Inc: New York, 1981, p. 393.

Fallik, E.; Sarig, S.; Okon, Y. Morphology and Physiology of Plant Roots Associated with *Azospirillum*. In *Azospirillum–Plant Associations*; Okon, Y. Ed.; CRC Press: Boca Raton, 1994, pp 77–84.

Fuentes-Ramírez, L. E.; Caballero-Mellado, J. Bacterial Bio-fertilizers. In *PGPR: Biocontrol and Biofertilization*; Siddiqui, Z. A. Ed.; Springer: Netherlands, 2006, pp. 143–172.

Gangwar, M.; Kaur, G. Isolation and Characterization of Endophytic Bacteria from Endorhizosphere of Sugarcane and Ryegrass. *Int. J. Microbiol.* **2009**, *7*(1), 5–8.

Gillis, M.; Kersters, K.; Hoste, B.; Janssens, D.; Kroppenstedt, R. M.; Stephan, M. P.; Teixeira, K. R. S.; Dobereiner, J.; De Ley, J. *Acetobacter diazotrophicus* sp. nov.; A Nitrogen Fixing Acetic Acid Bacterium Associated with Sugarcane. *Int. J. Syst. Bacteriol.* **1989**, *39*, 361–364.

Glick, B. R. Plant Growth-Promoting Bacteria: Mechanisms and Applications., *Scientifica*, **2012**, *2012*, 1–15.

Glick, B. R.; Penrose, D. M.; Li, J.; A Model for the Lowering of Plant Ethylene Concentrations by Plant Growth Promoting Bacteria. *J. Theor. Biol.* **1998**, *190*, 63–68.

Groudev, S. N. Use of Heterotrophic Microorganisms in Mineral Biotechnology. *Acta Biotechnol.* **1987**, *7*, 299–306.

Gupta, D. G.; Shelat, H. N. Biomass Increase in Rabi Maize Through Foliar Application of IAA and GA3 Producing *Azospirillum* and *Acetobacter* Endophytes. *Green Farm.* 2010, *1*(1), 64–66.

Gutiérrez-Mañero, F. J.; Ramos-Solano, B.; Probanza, A.; Mehouachi, J. R.; Tadeo, F.; Talon, M.; The Plant Growth-promoting Rhizobacteria *Bacillus pumilus* and *Bacillus licheniformis* Produce High Amounts of Physiologically Active Gibberellins. *Physiol. Plant.* **2001**, *111*, 206–211.

Hill, S. Physiology of Nitrogen Fixation in Free-living Heterotrophs. In *Biological Nitrogen Fixation*. Chapman & Hall: New York, 1992, pp. 87–134. http://www.marketsandmarkets.com/. (accessed on Oct 10, 2014).

Hopkins, C. G.; Whiting, A. L. Soil Bacteria and Phosphates. III. *Agric. Exp. Bull.* **1916**, *190*, 395–406.

Hussin, A. S. M.; Farouk, A-E.; Greiner, R.; Salleh, H. M.; Ismail, A. F. Phytate Degrading Enzyme Production by Bacteria Isolated from Malaysian Soil. *World J. Microbiol. Biotechnol.* **2007**, *23*, 1653–1660.

Illmer, P.; Schinner, F. Solubilization of Inorganic Phosphates by Microorganisms Isolated from Forest Soil. *Soil Biol. Biochem.* **1992**, *24*, 389–395.

Jones, D. L.; Darrah, P. R. Role of Root Derived Organic Acids in the Mobilization of Nutrients from the Rhizosphere. *Plant and Soil.* **1994**, *166*, 247–257.

Jongmans, A. G.; van Breemen, N.; Lundstrom, U.; Finlay, R. D.; van Hees, P. A. W.; Giesler, R.; Melkerude, P. A.; Olson, M.; Srinivasan, M.; Unestam, T. Rock eating fungi: a true case of mineral plant nutrition. *Nature.* **1997**, *389*, 682–683.

Kloepper J. W.; Schroth M. N. Plant Growth-Promoting Rhizobacteria on Radishes. In *Proceedings of the 4th International Conf. on Plant Pathogenic Bacteria, Station de Pathologie Vegetable et Phytobacteriologie*, INRA, Angers: France, 1978, pp. 879–882.

Kloepper, J. W.; Gutiérrez-Estrada, A.; McInroy, J. A. Photoperiod Regulates Elicitation of Growth Promotion but not Induced Resistance by Plant Growth-Promoting Rhizobacteria. *Can. J. Microbiol.* **2007**, *53*, 159–167.

Kpomblekou, K.; Tabatabai, M. A. Effect of Organic Acids on Release of Phosphorus from Phosphate Rocks. *Soil Sci.* **1994**, *158*, 442–453.

Li, F. C.; Li, S.; Yang, Y. Z.; Cheng, L. J. Advances in the Study of Weathering Products of Primary Silicate Minerals, Exemplified by Mica and Feldspar. *Acta Petrol Miner.* **2006**, *25*, 440–448.

Lian, B.; Fu, P. Q.; Mo, D. M.; Liu, C. Q. A Comprehensive Review of the Mechanism of Potassium Release by Silicate Bacteria. *Acta Miner. Sin.* **2002**, *22*, 179–183.

Liu, D.; Lian, B.; Dong, H. Isolation of *Paenibacillus* sp. and Assessment of Its Potential for Enhancing Mineral Weathering. *Geomicrobiol. J.* **2012**, *9*, 413–421.

Liu, W.; Xu, X.; Wu, S.; Yang, Q.; Luo, Y.; Christie, P. Decomposition of Silicate Minerals by *Bacillus Mucilaginosus* in Liquid Culture. *Environ. Geochem. Health.* **2006**, *28*, 133–140.

Madigan, M. T.; Martinko, J. M.; Parker, J. *Brock Biology of Microorganisms.* (9th ed). Prentice Hall: Upper Saddle River, NJ, 2000.

Malik, K. A.; Schlegel, H. G. Chemolithoautotrophic Growth of Bacteria Able to Grow Under N_2-Fixing Conditions. *Feder. Eur. Soc. Microbiol. Lett.* **1981**, *11*, 63–67.

Muthukumarasamy, R.; Revathi, G.; Vadivelu, M. *Acetobacter diazoptrophicus*: Prospects and Potentialities–An Overview. In *Recent Advances in Bio-fertilizers Technology, Society for Promotion & Utilization Resources & Technology*; Yadav, A. K.; Motsara, M. R.; Ray Chaudhury, S. Eds.; Vedams eBooks: New Delhi, 2000, pp. 126–153.

Okon, Y. *Azospirillum* as a Potential Inoculant for Agriculture. *Trends Biotechnol.* **1985**, *3*(9), 223–228.

Okon, Y.; Itzigsohn, R. The Development of *Azospirillum* as a Commercial Inoculant for Improving Crop Yields. *Biotechnol. Adv.* **1995**, *13*, 415–424.

Parmar, P.; Sindhu, S. S. Potassium Solubilization by Rhizosphere Bacteria: Influence of Nutritional and Environmental Conditions. *J. Microbiol. Res.* **2013**, *3*(1), 25–31.

Paul, E. A.; Clark, F. E. Soil Microbiology and Biochemistry (2nd ed.). Academic Press: San Diego, CA, 1996.

Quilambo, O. A. The Vesicular-Arbuscular Mycorrhizal Symbiosis. *Afr. J. Biotechnol.* **2003**, *2*(12), 539–546.

Richardson, A. E.; Hadobas, P. A. Soil Isolates of *Pseudomonas* spp. that Utilize Inositol Phosphates. *Can. J. Microbiol.* **1997**, *43*, 509–516.

Rodriguez, H.; Fraga, R. Phosphate Solubilizing Bacteria and Their Role in Plant Growth Promotion. *Biotechnol. Adv.* **1999**, *17*, 319–339.

Rodriguez, H.; Gonzalez, T.; Goire, I.; Bashan, Y. Gluconic Acid Production and Phosphate Solubilization by the Plant Growth-Promoting Bacterium *Azospirillum* spp. *Naturwissenschaften.* **2004**, *91*, 552–555.

Rogers, J. R.; Bennett, P. C.; Choi, W. J. Feldspars as a Source of Nutrients for Microorganisms. *Am. Mineral.* **1998**, *83*, 1532–1540.

Rubio, P.; Ludden, W.; Luis, M. Maturation of Nitrogenase: A Biochemical Puzzle. *J. Bacteriol.* **2005**, *187*, 405–414.

Sadasivan, L.; Neyra, C. A. Flocculation in *Azospirillum brasilense* and *Azospirillum lipoferum:* Exopolysaccharides and Cyst formation. *J. Bacteriol.* **1985**, *163*, 716–723.

Salgado, J. T.; Fuentes-Ramirez, L. E.; Hernandez, T. A.; Mascarua, M. A.; Martinez-Romero, E.; Caballero-Mellado, J. *Coffea arabica* L. a New Host Plant for *Acetobacter diazotrophicus* and Isolation of Other Nitrogen Fixing Acetobacteria. *Appl. Environ. Microbiol.* **1997**, *63*(9), 3676–3683.

Shedova, E.; Lipasova, V.; Velikodvorskaya, G.; Ovadis, M.; Chernin, L.; Khmel, I. Phytase Activity and its Regulation in a Rhizospheric Strain of *Serratia plymuthica. Folia Microbiol.* **2008**, *53*, 110–114.

Sheng, X. F. Growth Promotion and Increased Potassium Uptake of Cotton and Rape by a Potassium Releasing Strain of *Bacillus edaphicus. Soil Biol. Biochem.* **2005**, *37*, 1918–1922.

Sheng, X. F.; Zhao, F, He, I. Y, Qiu, G and Chen, I. Isolation and Characterisation of Silicate Mineral Solubilising *Bacillus globisporus* Q12 from the Surface of Weathered Feldspar. *Can. J. Microbiol.* **2008**, 54,1064–1068.

Sommer, M. D.; Fuzyakov, M. D.; Breuer, J. Silicon Pools and Fluxes in Soils and Landscapes-A Review. *J. Plant Nutr. Soil Sci.* **2006**, *169*, 310–329.

Sperber, J. I. **1958**. The Incidence of Apatite-Solubilizing Organisms in the Rhizosphere and Soil. *Aust. J. Aric Res.* **2006**, *9*, 778–781.

Srivastava, P. C.; Gupta, U. C. *Trace Elements in Crop Production.* Science Publishers: New Hampshire, 1996.

Stolp, H. *Microbial Ecology: Organisms, Habitats, Activities.* Cambridge University Press: New York, 2008.

Suman, A.; Shasany, A. K.; Singh, M.; Shahi, H. N.; Gaur, A.; Khanuja, S. P. S. Molecular Assessment of Diversity in Endophytic Diazotrophs of Sub-Tropical Indian Sugarcane. *World J. Microbiol. Biotechnol.* **2001**, *17*(1), 39–45.

Tao, G.; Tian, S.; Cai, M.; Xie G. Phosphate Solubilizing and Mineralizing Abilities of Bacteria Isolated from Soils. *Pedosphere.* **2008**, *18*, 515–523.

Tapia-Hernandez, A.; Bustillos-Cristales, M. R.; Jimenezsalgado, T.; Cabellaro-Mellado, J.; Fuentes-Ramirez, L. E. Endophytic *nif*H Gene Diversity in African Sweet Potato. *Can. J. Microbiol.* **2002**, *44*, 162–167.

Thiagalingam, K.; Silva, J. A.; Fox, R. L. Effect of Calcium Silicate on Yield and Nutrient Uptake in Plant Growth on a Humic Ferriginous Latosol. In *Proceedings of Conference on Chemistry And Fertility of Tropical Soils*, Malaysian Society Of Soil Science: Kuallalumpur, Malaysia, 1977, pp. 149–155.

Tien, T. M.; Gaskins, M. H.; Hubbell, D. H. Plant Growth Substances Produced by *Azospirillum brasilense* and their Effect on the Growth of Pearl Millet (*Pennisetium americanum*). *Appl. Environ. Microbiol.* **1979**, *37*, 1016–1024.

Tilak, K. V. B. R.; Subba Rao, N. S. Association of *Azospirillum brasilense* with Pearl Millet (*Pennisetum americanum* (L) *Leeke*). *Biol. Fertil. Soils.* **1987**, *4*(1–2), 97–102.

Townsend, A. R.; Howarth, R. W.; Bazzaz, F. A.; Booth, M. S.; Cleveland, C. C. Human Health Effects of a Changing Global Nitrogen Cycle. *Front. Ecol. Environ.* **2003**, *1*(5), 240–246.

Vaishampayan, A.; Sinha, R. P.; Hader, D. P.; Dey, T.; Gupta, A. K.; Bhan, U.; Rao, A. L. Cyanobacterial Bio-fertilizers in Rice Agriculture. *Botanical Rev.* **2001**, *67*, 453–516.

Vessey, J. K. Plant Growth Promoting Rhizobacteria as Bio-fertilizers. *Plant Soil.* **2003**, *255*, 571–586.

Vora, M. S.; Shelat H. N.; Vyas R. V. *Handbook on Bio-fertilizers and Microbial Pesticides*; Satish Serial Publishing House: Delhi, 2008, pp. 251.

Vyas, R. V.; Shelat, H. N.; Vora, M. S. Bio-Fertilizers Techniques for Sustainable Production of Major Crops for Second Green Revolution in Gujarat–An Overview. *Green Farm.* **2008**, *1*, 68–72.

Wani, S. P.; Rupela, D. P.; Lee, K. K. Sustainable Agriculture in the Semi-Arid Tropics Through Biological Nitrogen Fixation in Grain Legumes. *Plant Soil.* **1995**, *174*, 29–49.

Yamada, Y.; Hoshino, K.; Ishikawa, T. The Phylogeny of Acetic Acid Bacteria Based on the Partial Sequences of 16S Ribosomal RNA: The Elevation of the Subgenus *Gluconoacetobacter* to Generic Level. *Biosci. Biotechnol. Biochem.* **1997**, *61*(8), 1244–1251.

Yi, Y.; Huang, W.; Ge, Y. Exopolysaccharide: A Novel Important Factor in the Microbial Dissolution Of Tricalcium Phosphate. *World J. Microbiol. Biotechnol.* **2008**, *24*, 1059–1065.

Young, J. P. W. Phylogenetic Classification of Nitrogen-fixing Organisms. In *Biological Nitrogen Fixation*. Chapman and Hall: New York, 1992, pp. 43–86.

Zahran, H. H. Rhizobia from Wild Legumes: Diversity, Taxonomy, Ecology, Nitrogen Fixation and Biotechnology. *J. Biotechnol.* **2001**, *91*, 143–153.

PART III
Environmental Microbiology

CHAPTER 8

VAM FUNGI: RECENT ADVANCES IN CROP DISEASES MANAGEMENT

RAKESH KUMAR*, D. V. PATHAK, and S. K. MEHTA

District Extension Specialist, CCS Haryana Agricultural University, Krishi Vigyan Kendra, Fatehabad 125120, Haryana, India

*Corresponding author. E-mail: drrakeshchugh@rediffmail.com

CONTENTS

ABSTRACT

The potential role of mycorrhizal fungi biocontrol agent for the control of fungal and nematode plant diseases has recently attracted considerable attention. Vesicular arbuscular–mycorrhizal infections generally inhibit or sometimes increase and occasionally have no effect on disease caused by fungal pathogens. The damage due to nematode diseases has also generally decreased in mycorrhizal plants. Establishment of vesicular arbuscular mycorrhizae (VAM) in plants generally confers resistance to nematode parasitism or adversely affects nematode reproduction. Although the improvement of plant nutrition, compensation for pathogen damage, and competition for photosynthates or infection sites have been claimed to play a protective role in VAM symbiosis. Information is scare, fragmentary, or even controversial, particularly concerning other mechanisms such as anatomical or morphological and biochemical VAM induced changes in the root system, microbial change in the rhizosphere populations of mycorrhizal plants, and local elicitation of plant defense mechanisms by VAM fungi. In the present chapter, interactions between VAM and phytopathogens (fungi and nematode) reported during past 2 decades are reviewed.

8.1 INTRODUCTION

Arbuscular mycorhizal (AM) fungi occur over a wide range of agro-climatic conditions and are geographically ubiquitous. They form symbiotic relationship with the roots of about 90% land plants in natural and agricultural ecosystems (Brundrett, 2002). The AM association has been observed in 200 families of plants representing 1000 genera and about 300,000 plant species (Bagyaraj, 1991). It is as normal for the roots of plants to be mycorrhizal as it is for the leaves to photosynthesize (Mosse, 1986). The AM fungi are included in the phylum Zygomycota, order Glomales (Redecker et al., 2000), but now they are placed in phylum "Glomeromycota" (Schussler et al., 2001). The Glomeromycota is divided into 4 orders, 8 families, 10 genera, and 150 species. The common genera are Aculospora, Gigaspora, Glomus, and scutellospora (Schussler, 2005). They are characterized by the presence of mycelium-branched haustoria like structure within the corticle cells, termed as arbucules and are the main site of nutrient transfer between the two symbiotic partners (Hock and Verma, 1995; Smith and Read, 1997). The AM fungi colonize plant roots and penetrate into surrounding soil, extending the root depletion zone and the root system. They supply water and mineral nutrients from oil to the root system. The AM fungi are associated with improved growth of host plant species due to increased nutrient uptake, production of growth promoting substances, tolerance to drought, salinity, and interaction with other beneficial microorganisms (Sreenivasa and Bagyaraj, 1989).

Agricultural operation that disturbs the natural ecosystem will have repercussions on the mycorrhizal system (Mosse, 1986). The inclusion of nonmycorrhizal crops within rotations decreases both the fungal colonization and yield of the subsequent crops (Douds et al., 1997; Arihawa and Karasawa, 2000). In addition to crop sequence, variety selection, cultivation, and fallowing have been shown to affect mycorrhizal activity (Ocampo et al., 1980; Hetrick et al., 1996; McGonigle and Miller, 2000). Biological control of plant pathogens is currently accepted as a key practice in sustainable agriculture because it is based on the management of natural resource, that is, certain rhizosphere organisms, common components of ecosystems, known to develop antagonistic activities against harmful organisms (bacteria, fungi, nematode, etc.).

The potential role of mycorrhizal fungi as biocontrol agent for the control of fungal and nematode plant diseases has recently received considerable attention. Literature on various aspects of AM research such as AM–soil-borne pathogen interactions (Azcon-Aguilar and Brea, 1996; Quarles, 1999; Chandra and Kehri, 2004; Grosch et al., 2005; Berg et al., 2007; Siddiqui and Akhtar, 2008; Tahat et al., 2010), interaction of VA–mycorrhizal fungi with nematode pathogens (Mohanty and Sahoo, 2003; Pinochet et al., 1996), interaction of AM with rhizosphere and soil microorganisms (Azcon-Aguilar and Barea, 1992; Fitter and Garbaye, 1994), vesicular arbuscular mycorrhizae (VAM)–agrochemicals (Vyas, 1995), and further of VAM commercialization (Wood and Cummings, 1992) and trends in mycorrhizal research (Sahay et al., 1998) has been reviewed. Useful data generated on AM in recent years have opened up the prospects for their practical applications.

Arbuscular mycorrhizal infections generally inhibit or sometimes increase and occasionally has no effect on diseases caused by fungal pathogens. The damage due to nematode diseases also has generally decreased in mycorrhizal plants. In several studies, mycorrhizal fungi have shown an antagonistic influence on the population of plant parasitic nematodes. Establishment of AM in plants generally confers resistance to nematode parasitism or adversely affects nematode reproduction. It is difficult, therefore, to generalize the effects of AM on plant diseases. Apart from disease control, mycorrhizal association is beneficial to plants in many folds. It increases water and nutrient absorption by the plants, increase tolerance to adverse environmental conditions and biomass, and productivity of plants. In recent years, major interest has centered on its relevance in biological control of soil-borne diseases caused by fungi and nematodes which are otherwise difficult to control by conventional methods. In the present review, interactions between AM and phytopathogens (fungal and nematodes) were reported after 1990s.

8.2 INTERACTION OF VAM WITH FUNGAL PATHOGENS

More than 10,000 species of fungi are known to cause diseases of plants and are common in soil, air, and on plant surfaces throughout the world (Agrios, 2005). The disease caused by fungal pathogens persists in soil matrix and in residues on the soil surface and is called soil-borne diseases. Soil is a reservoir of inoculum of these pathogens. Damage to root tissues is often hidden until the above ground parts of the plants are severely affected and are showing symptoms such as stunting, wilting, chlorosis, and death. Fungal diseases are difficult to control because pathogen survives in soil for long period without the presence of crop host and have a wide host range. The occurrence of AM fungi and plant pathogenic fungi in roots of different crops and their dependence for nutrition on the host generally results in the interactions of AM fungi, plant pathogenic fungi, and host plants. Therefore, AM fungi are the major component of the rhizosphere of plants and may affect the incidence and severity of root diseases (Table 8.1).

8.2.1 CEREAL CROPS

Kumar et al. (1993) attempted biocontrol of wheat root rots using mixed cultures of *Trichoderma viride* and *Glomus epigaeus* in glasshouse. A mixed culture of *T. viride* and *G. epigaeus* protected wheat plants from all the three pathogens tested. Root rot caused by *Bipolaris sorokiniana* was reduced to 12.2% with the antagonist (A) + mycorrhiza (M) + pathogen (P) treatment as a result of increased mycorrhization of roots. Similar reductions were obtained in the population of *F. avenaceum* and *F. javanicum* when various combinations of symbiont and antagonist were used. Siddiqui and Singh (2005) reported that inoculation of *G. mosseae* in wheat reduced the percentage of leaf infection.

TABLE 8.1 Effect of VAM On Soil Borne Fungal Pathogens.

Pathogen	Host Plant	Effects in mycorrhizal plants*			Reference
		Disease damage	Infection	Reproduction	
Fusarium solani	Bean	−	−	:	Muchovej et al. (1991)
Fusarium solani	Bean	−	:	:	Goncalves et al. (1991)
Fusarium moniliforme Phythium vexans Rhizoctonia spp.	Cardamom	−	:	:	Thomas et al. (1994)
Fusarium oxysporum f.sp ciceri	Chickpea	−	:	:	Rao and Krishmappa (1995)
		−	:	:	Siddiqui and Singh (2004)
Sclerotium rolfsii	Chickpea		:	:	Jayaraman and Kumar (1995)
	cv. BG-256	+	:		
	cv. JG-62	−	:		
Macrophomina phaseolina	Chickpea	−	−	−	Akhtar and Siddiqui (2007)
Sclerotium rolfsii	Chili	−	−	−	Sreenivasa et al. (1992)
Sclerotium rolfsii	Chili	−	:	:	Sreenivasa (1994)
Fusarium oxysporum	Cowpea	−	−	:	Sundaresan et al. (1993)
Rhizoctonia solani	Cowpea	−	−	:	Abdel-Fattah and Shabana (2002)
Fusarium oxysporum f.sp. *vasinfectum*	Cotton	−	:	:	Hu and Gui (1991)
Verticillium dahlia	Cotton	−	:	:	Liu et al. (1995)
Pythium ultimum	Cucumber	−	:	:	Rosendahl and Rosendahl (1990)
Pythium splendens or *Rhizoctonia solani*	Cucumber	−	:	:	Kobayashi (1991)
Fusarium oxysporum	Cucumber	−	:	:	Hao et al. (2005)
Fusarium oxysporum f.sp. *cumini*	Cumin	−	:	:	Champawat (1991)

TABLE 8.1 *(Continued)*

Pathogen	Host Plant	Effects in mycorrhizal plants*			Reference
		Disease damage	Infection	Reproduction	
Verticillium dahlia	Eggplant	—	⋮	⋮	Maisubaro et al. (1995)
Rhizoctonia solani	Fababean	—	⋮	—	Ahemed et al. (1994)
Botrytis fabae	Fababean	—	⋮	—	Rabie (1998)
Sclerotium rolfsii	Groundnut	—	⋮	⋮	Kulkarni et al. (1997)
Fusarium solani	Lettuce	⋮	⋮	—	McAllister et al. (1994)
Verticillium albo–atrum	Lucerne	—	⋮	—	Hwang et al. (1992)
Fusarium oxysporum f.sp. *medicaginis*					
Rhizoctonia solani	Maize	—	⋮	—	Khadge et al. (1990)
Helminthosporium maydis	Maize	×	⋮	⋮	Chhabra et al. (1992)
Fusarium moniliforme		—	⋮	⋮	
Acremonium kiliense		×	⋮	⋮	
Fusarium solani	Maize	⋮	⋮	—	McAllister et al. (1994)
Macrophomina phaseolina	Mungbean	—	—	⋮	Jalali et al. (1991)
Pyrenochaeta terrestris	Onion	×	⋮	⋮	Batcho et al. (1994)
Sclerotium cepivorum	Onion	—	⋮	⋮	Torres-Barragan et al. (1996)
Aphanomyces euteiches	Pea	—	⋮	⋮	Kjoller and Rosendahl (1996)
					Bodker et al. (1998)
					Thygesen et al. (2004)
Fusarium solani f.sp. *piperis*	Pepper	—	⋮	⋮	Chu et al. (1997)
Phytophthora capsici		—	⋮	—	Ozgonen and Erkilic (2007)

TABLE 8.1 *(Continued)*

Pathogen	Host Plant	Effects in mycorrhizal plants*			Reference
		Disease damage	Infection	Reproduction	
Fusarium udum	Pigeonpea	–	...	...	Siddiqui and Mahmood (1995)
Fusarium udum	Pigeonpea	–	...	...	Siddiqui and Mahmood (1996)
Rhizoctonia solani	Potato	...	–	...	Iqbal et al. (1990)
Fusarium sambucinum	Potato	–	...	...	Niemira et al. (1996)
Rhizoctonia solani	Rice	–	...	...	Baby and Manibhushanrao (1996)
Plasmopara helianthi	Sunflower	...	–	–	Toshi et al. (1993)
Thielviopis basicola	Tobacco	–	–	–	Giovannetti et al. (1991)
Sclerotium rolfsii	Tomato	–	...	...	Kichadi and Sreenivasa (1998)
Fusarium oxysporum f.sp. *lycopersici*	Tomato	–	...	...	Ozgonen et al. (1999)
					Bhagwati et al. (2000)
					Akkopru and Demir (2005)
Rhizoctonia solani	Tomato	–	...	...	Berta et al. (2005)
Bipolaris sorokiniana	Wheat	–	...	–	Kumar et al. (1993)
Fusarium avenaceum		–	...	–	
Fusarium javanicum		–	...	–	

*+ = increased, – = decreased, x = no effect, ... = not reported.

Hashemi et al. (2013) studied the biological control of *B. sorokiniana* can be an effective method for control of root rot of wheat. For this purpose, the biocontrol potential of mycorrhiza (*G. fasciculatum*), *Pseudomonas fluorescens* sh4, and a mixture of the two was evaluated for the common root rot in greenhouse condition (22–25°C). Seven wheat seeds variety Chamran were sown in 200 ml polyethylene pots. Half of them were inoculated by *G. fasciculatum* and another half remained uninoculated. They were irrigated by Hogland's nutrient solution. After 35 days, the plantlets were transferred to the polyethylene pots containing 1 kg autoclaved soil. Depending on the different treatments, the pots were inoculated or not, with causal agent of disease *B. sorokiniana*, *P. fluorescens* or mixture of two. Experiment was arranged as a factorial test using completely randomized design with eight treatments and four replications. The plants were irrigated normally and harvested 5 weeks after transplanting of the plantlets. The roots were washed with care and noted for disease index on the roots and fresh weights of roots and shoots were recorded. The results showed that the presence of mycorrhiza, pseudomonas, and the mixture of them around and on the roots of plants, inoculated by *B. sorokiniana*, decreased significantly the severity of disease at the rate of 43, 50.75, and 78.75%, respectively, in comparison with controls (no mycorrhiza and no pseudomonas). Comparable results were noted for the fresh weight of roots and aerial parts of plants in the same treatments. It could be concluded that application of mycorrhiza *G. fasciculatum*, *P. fluorescens*, and a mixture of the two on wheat not only decreases the severity of disease caused by *B. sorokiniana* but also increases the weights of both Bipolaris infected and noninfected plant. This increase in yield was significantly higher in the treatments inoculated with the mixture of mycorrhiza and pseudomonas as compared to the treatments inoculated with only one of each biocontrol agent.

Arabi et al. (2013) studied to assess the biocontrol efficacy of VAM against barley common root rot caused by *Cochliobolus sativus*. Mycorrhization of barley was achieved by growing the plants in expanded clay mixed with 10% (v/v) VAM fungus inoculum in pots experiments. Large differences in disease reactions were observed among genotypes and among treatments. VAM treatments significantly reduced the percentage of disease severity in infected barley plants and increased significantly root biomass, which could be attributed to enhance nutrients uptake, via an increase in the absorbing surface area. It can be concluded that the application of VAM as a biocontrol agent played an important role in plant resistance and exhibited greater potential to protect barley plants against *C. sativus*.

In case of rice, Baby and Manibhushanrao (1996) observed the effect of various organic soil amendents on AM fungal activity on rice plants under greenhouse and field conditions with reference to sheath blight caused by *R. solani*. Organic amendments increased AM spore density, percent infection, and intensity of infection, while disease was decreased. Mycorrhizal formation and sporulation were higher in healthy rice plants than in infected plants. It was found that root dry weight of rice was not affected by *Rhizoctonia solani* in mycorrhizal plants, but the pathogen caused 29% loss in root dry weight in nonmycorrhizal plants (Khadge et al., 1990).

In case maize, Chhabra et al. (1992) observed that the mycorrhizal plants were resistant to *F. moniliforme* but response to the other pathogens was unaffected. The total phenol content in mycorrhizal maize roots was higher than the nonmycorrhizal roots, and it was suggested that this may be responsible for the increased resistance against *F. moniliforme*. McAllister et al. (1994) studied the effects of inoculation by *T. koningii* and *F. solani* on maize with or without colonization by *G. mosseae* in a greenhouse trial. Plant dry weight of nonarbuscular mycorrhizal inoculated maize was unaffected by the presence of *T. koningii* and *F. solani*. In contrast, *T. koningii* decreased plant dry weight and VAM colonization when inoculated into the rhizosphere before or at the same time as *G. mosseae*. In addition, the *T. koningii* population was considerably reduced when *G. mosseae* was inoculated 2 weeks before the saprophytic fungus. Babu et al. (1998) reported that mycorrhizal fungi *G. fasciculatum*, Azosprillum, and Phosphobacterium, when applied alone or in different combinations in field, reduced the population of *P. zeae* and recorded the highest cob yield of maize. Veerabhadrasamy and Garampalli (2011) used three species of AM (*G. fasciculatum*, *G. mosseae*, and *A. laevis*) were used as bioagents to manage black budle disease of maize caused by *C. acremonium*. The results revealed that colonization of AM fungi in root system of host reduce the percentage of disease incidence (0.00%) was recorded, whereas in the pots inoculated with *A. levis* and *G. mosseae*, 16.66% of disease incidence was recorded, and the pots treated with pathogen shows 66.66% of disease incidence compared to control. Among the three arbuscular mycorrhiza, *G. fasciculatum* improved to be more effective in managing the disease followed by *G. mosseae* and *A. levis*; in addition, all the three arbuscular mycorrhizal fungi enhanced the plant growth when they are used alone as inoculums as compared to dual inoculation with the *C. acremonium* and overall control. This suggests that AM fungi, if used, can serve dual purpose. It can be used as biocontrol agent as it shows negative antagonistic interaction soil borne plant pathogens and is used as growth promoter because of the ability to supply macro and micro nutrients to the host plants.

8.2.2 LEGUME CROPS

Rao and Krishnapa (1995) conducted a field experiment to study the management of *Meloidogyne incognita* + *F. oxysporium* wilt complex in chickpea cv. Annegiri. The results indicated that integration of soil solarization, VA mycorrhizal fungus (*G. fasciculatum*), inoculation (12 g/hill), and seed treatment with Carbosulfan (3%w/w) were highly effective in reducing population levels of both pathogens and increasing chickpea grain yield significantly. Jayaraman and Kumar (1995) studied the response of chickpea varieties BG-256 and JG-62 to infection by *Sclerotinia rolfsii* while interacting with *G. fasciculatum*, difolatan, and metalaxyl. In both the varieties, the pathogen reduced yield and growth of plants. The pathogen suppressed mycorrhizal growth formation in both varieties. Singh and Singh (1995) studied the effect of *G. aggregatum* inoculation and P amendments on *Fusarium oxysporum* f.sp. *ciceri* causing chickpea wilt. A total of 80% and 20% plants were found healthy and showing yellowing, respectively, at both 0 and 40 mg P/kg soil after 45 days of sowing in the presence of mycorrhizae and wilt pathogen. All plants, inoculated with *F. oxysporum* f.sp. *ciceri* alone, were found to have wilted after 45 days of sowing at all three levels of P. On the other hand, none of the plants showed wilt symptom in the soil inoculated with VAM fungus. The observations revealed that the presence of mycorrhizal fungi significantly delayed infection, as well as reduced its percentage. The wilting process in mycorrhizal chickpea plants was found to be significantly slow as compared to nonmycorhizal plants.

Kumar et al. (2004) reported that dual inoculations with mycorrhiza and test pathogens increased the seed germination, plant height, number of pods, seed weight, and biomass production as compared to inoculations with pathogen alone. Mycorrhizal inoculation suppressed the incidence of wilt and root-rot disease by 54% and 62%, respectively. Mycorrhizal inoculation resulted in better colonization of roots with VAM and thereby improved uptake of nutrients in chickpea plants. Siddiqui and Singh (2004) also reported the reduction in disease severity of wilt in chickpea by the AM inoculations.

Akhtar and Siddiqui (2007) found reduction in disease severity of root rot in chickpea by inoculation with the *G. intraradices*. They also reported in 2008 that combined inoculation of AM fungus with *Psedomonas* straita and Rhizoctonia caused reduction in the root rot of chickpea. Singh et al. (2013)

showed the status of mycorrhization in three test varieties of chickpea. They reported that when AM colonization was assessed under the influence of soil quality, better results were obtained with unsterilized soil in comparison to sterile soil. Chickpea variety ICC 11322 showed the best result against the Fusarium wilt and chickpea variety ICC4951 was susceptible against the Fusarium wilt. *Acaulospora spinosa* showed best results with JG 74, *G. mosseae* with ICC 4951, and *G. fasciculatum* showed best result with ICC 11322. The lowest percentage mycorrhizal colonization was found on plants with the most severe disease symptoms.

Ahemed et al. (1994) carried out a pot experiment to study the effect of VAM fungi and/or *Rhizobium leguminosarum* on the root-rot disease of fababean caused by *R. solani*. The number of uninoculated seedlings grown in infested soil that survived was 80%. In a similar study (Rabie, 1998), infection of fababeans with *Botrytis fabae* caused significant decrease in growth vigor, total nitrogen content, number of nodules, and nutrient accumulation. In contrast, dual inoculation of *R. leguminosarum* and *G. mosseae* increased all of the above parameters.

Kjoller and Rosendahl (1996) quantified fungal enzyme activities in an interaction between *G. intrartadices* and *Aphanomycetes euteiches* in pea roots. The plants preinoculated with *G. intraradices* showed no symptoms of severe root rot even though the pathogen was present and active in those plants. It was concluded that plants preinoculated with *G. intraradices* were more tolerant of infection with *A. euteiches* than nonmycorrhizal plants. *A. euteiches* enzyme activities in mycorrhizal plants were different to those in nonmycorrhizal plants. Thygesen et al. (2004) observed that inoculation with *G. intraradices* caused reduction in disease severity by *Aphanomyces euteiches*. Siddiqui and Mahamood (1995) used *G. mosseae, T. harzianum, V. chlamydosporium* alone, and in combination for the management of wilt disease complex of pigeonpea caused by *H. cajani* and *F. udum*. Treatment of plants inoculated with pathogens increased plant height, shoot dry weight, number of nodules, and reduced nematode multiplication and wilting index. Simultaneous use of biocontrol agents against pathogens gave better control than their individual application. *T. harzianum* had an adverse effect on root colonization by *G. mosseae*. The highest reduction in nematode multiplication was observed when all three biocontrol agents were used together.

Mikhaleel et al. (2002) evaluated the efficacy of *G. aggregatum* and *Bacillus subtilis* individually or in combination with *Bradyrhizobium japonicum* as biocontrol agent for reducing root-rot disease caused by *F. oxysporum* in soyabean plants. In noninoculated plants, infestation of soil with *F. oxysporum* drastically reduced plant growth parameters compared to noninfested one. However, the detrimental effects caused by *F. oxysporum* infestation were less detected in VAM and/or *B. subtilis* inoculated treatments, being the least pronounced in the VAM+ *B. subtilis* treatment. Results pointed out that the observed polyphyletic effect of mycorrhizal inoculation was not related with plant nutrition, but also reduction of disease severity.

Jalali et al. (1991) revealed that mycorrhizal inoculation significantly restricted pathogen spread in host root tissues of mungbean. Disease incidence was reduced from 77.9% in pathogen inoculated to 13.3% in mycorrhiza + pathogen-inoculated plants. Dual inoculation with mycorrhiza and pathogen resulted in increased total dry matter production, nitrogen, phosphorus, and potassium contents compared with pathogen-inoculated plants. Chandra et al. (1995) evaluated the efficacy of a number of VAM fungi recovered from agricultural fields raising leguminous crops. Interaction between VAM fungi and pathogen was examined by inoculating the crops with selected VAM fungi alone as well as in combination with selected isolate of pathogen in sterilized as well as unsterilized soil. When the inoculum of VAM fungi was added along with that of pathogen, a reduction in rate of mortality and population of pathogen was recorded. Kumar et al. (2007) reported that dual inoculation of *G. fasciculatum* and *R. bataticola* increased the seed germination, yield per plant, and biomass production compared with the inoculation of *R. bataticola* alone. Mycorrhizal inoculation reduced the incidence of root rot by 52%. The efficient colonization of roots by VAM facilitated the reduction of root rot incidence. Mycorrhiza-inoculated plants also exhibited significant increase in the uptake of N, P, K, and Zn. Sankaranarayanan and Sundarababu (1998) observed effect of *Rhizobium* on the interaction of VAM and root knot nematode on blackgram. The combination of Rhizobium and VAM resulted in the least nematode population density and maximum spore, mycorrhizal colonization, total nitrogen, and phosphorus content of blackgram.

8.2.3   VEGETABLE CROPS

Kichadi and Sreenivasa (1998) recorded interaction effects of *G. fasciculatum* and *T. harzianum* on *S. rolfsii* in presence of biogas spent slurry in tomato. Disease severity was least in dual inoculated plants. The interaction effects of the fungi not only increased plant growth and yield but also improved P nutrition. It was observed that when seedlings of tomato were inoculated with *G mosseae*, it showed better growth and nutrient uptake. At the same time, these were resistant against infection caused by *Verticillum dahliae*. Tomato plants inoculated with *G. etunicatum* caused reduction in disease severity (Ozgonen et al., 1999; Bhagwati et al., 2000). The early wilt symptoms caused by *F. oxysporium* on tomato appeared 8–10 days earlier in mycorrhizal plants. Two months later, however, the disease's severity was significantly reduced in these plants. Between the two species tested, *G. etunicatum* was more effective than *G. mosseae* (Sharma and Johri, 2002). Further, significant reduction in wilt severity in tomato was reported by *G. intraradices* inoculation (Akkopru and Demir, 2005). Further, Berta et al. (2005) found that *G. intraradices* inoculation significantly decreased epiphytotic and parasitic growth of pathogen in tomato. Fiorilli et al. (2011) evaluated whether AM symbiosis modifies the response of tomato plants to the attack of the necrotrophic pathogen Botrytis cinerea. Leaves of tomato plants, colonized or not by the AM fungus *G. mosseae*, were infected with *B. cinerea*. A higher disease index in control plants (60.3%) compared to mycorrhizal plants (37.5%) was observed. To assess the potential involvement of salicylic acid (SA), jasmonic acid (JA), and abscisic acid (ABA) in this response, the levels of these hormones were also measured in the leaves of mycorrhizal and nonmycorrhizal plants. While JA was not detected and no differences were observed in the SA content between the two biological conditions, a statistically significant lower content of ABA was detected in mycorrhizal vs control plants. Our results show that AM symbiosis reduces disease severity in tomato plants infected by *B. cinerea* and suggest that ABA is one component of the AM-induced lower susceptibility to *B. cinerea*. Manila and Nelson (2014) conducted a pot experiment in tomato to study the biochemical changes; the protective enzyme activities and disease resistance against the tomato wilt pathogen *F. oxysporum* f. sp. *lycopersici*. Two dominant species of AM fungi *G. fasciculatum* and *A. laevis* were isolated, mass multiplied, and used for further studies. Tomato plants were grown in plastic pots filled with sterile soils and inoculated with AM fungi *G. fasciculatum* and *A. laevis*. The effect of the interaction between the AM fungi and pathogen *F. oxysporum* f. sp. *lycopersici* on tomato plants was monitored regularly. Mycorrhizal colonization significantly increased the mineral nutrient concentration, chlorophyll, protein, amino acids, starch, sugars, and phenolic content. Among the AM fungi, *A. laevis* proved to be the more effective strain compared to *G. fasciculatum*. Reddy et al. (2006) showed that dual inoculations significantly restricted the progression of the pathogen in the root tissues of both the genotypes of tomatoes tested (PED and Arkavikas). Mycorrhizal inoculations not only reduced the percentage of disease incidence but also increased the fruit yield and fruit weight. It is difficult to generalize the interaction of pathogen–host–mycorrhiza, but effective AM fungus can be used in controlling disease biologically.

Influence of VAM and *Aspergillus niger* as deterrents against *R. solani* in potato was recorded by Iqbal et al. (1990). Infection by *R. solani* was greatly reduced in plants preinoculated with VAM fungi. Combined infection by *A. niger* and VAM fungi increased the rate of mycorrhizal infection, improved plant growth, and suppressed infection by *R. solani*. Niemira et al. (1996) reported potato postharvest dry rot (*F. sambucinum*) developed significantly less (20–90% reduction) in the medium with *G. intraradices* when inoculated with *F. sambucinum* compared with those grown without the fungus. Disease suppression also occurred in greenhouse, despite only of trace levels only of arbuscular mycorrhizal colonization of the plants, and no evidence of enhanced plant P nutrition.

Maisubaro et al. (1995) reported growth enhancement and Verticillium wilt control by VAM fungus inoculation in egg plant. The appearance of wilt was delayed and reduced by VAM fungus infection, and the effects were more apparent in *G. margarita*. VAM fungus infected plants yielded more fruits than the noninoculated plants. The incidence of fruit malformation caused by wilt was smaller, especially in *G. etunicatum* infected plants. Observations revealed that secondary cell wall of cortical cells in the third branched roots were thicker in VAM fungus-infected plants than in noninoculated plants. Rao et al. (1998a) conducted screen house experiments to integrate the use of VAM (*G. mosseae*) with a nematophagus fungus *Paecilomyces lilacinus*, for the management of *M. incognita* on egg plants. Both biological control agents did not affect each other's colonization on the roots resulting in an additive effect of both on the management of *M. incognita*. The final nematode population was significantly less in the treatment where *G. mosseae* and *P. lilacinus* were integrated. Rao et al. (1998b) conducted experiments to study the feasibility of interaction of *G. fasciculatum* and *R. communis* cake for the management of *M. incognita* on eggplant. Seedlings colonized with mycorrhiza were least infected by *M. incognita* when transplanted in soil which was amended with castor cake. Futher, significant increases in colonization of *G. fasciculatum* on roots of eggplant and chlamydospores densities of mycorrhiza in this treatment indicated favorable effects of caster cake amendment on the growth of *G. fasciculatum*. Aparajita and Phukan (2003) observed interaction of *G. fasciculatum* and *M. incognita* in brinjal. In their studies, mycorrhizal plants showed relatively higher growth than nonmycorrhizal plants.

In investigations of Batcho et al. (1994), onion plantlets inoculated with *Acaulospora* spp., *G. leptotichum*, *G. mosseae*, and *Glomus* spp. were cultivated in soil infected by *Pyrenochaeta terrestris*. Most roots, VAM inoculated or not, were infected with *P. terrestris* and *Fusarium* spp. The protective effect of mycorrhizas against pink root disease was not evident. Torres-Barragan et al. (1996) used *Glomus* spp. to control onion white rot under field conditions. Inoculations delayed onion white rot epidemics by 2 weeks and provided significant protection against the disease for 11 weeks after transplanting. Mycorrhizal plants showed an increase of 22% in yield, regardless of the presence of *S. cepivorum*. Kucukyumuk et al. (2014) studied the three levels of zinc fertilization (0, 5, and 10 mg/kg), and an arbuscular mycorrhizal (AM) fungus (*G. intraradices*) was tested for their potential to control *Pythium deliense* on inoculated cucumber seedlings. Plant Zn, N, P, K, Mg, Ca, Fe, Mn, Cu contents, dry and fresh weights of plant and roots, and disease severity were determined in the study. Resistance to Pythium rot was determined with the application of mycorrhiza with increasing doses of zinc. Zinc and mycorrhizal fungus applications had significant effects on plant nutrition except for K and Cu. Although the highest N and P concentrations were noted under Zn0 conditions, the values obtained under Zn1 and Zn2 conditions showed differences depending on *G. intraradices* and *P. deliense* treatments. Leaf Ca concentration reached up to the highest level with Zn2GI0Pd1 treatment, and the lowest Ca content was recorded under GI0Pd0 for all Zn applications. Lower level of zinc together with GIPd0 applications resulted in the highest leaf Mg concentration. The highest micronutrient concentrations were analyzed on cucumber plants grown under Zn deficient conditions without GI but with *P. deliense*. Plant dry weight, root fresh, and root dry weights were higher in cucumber plants challenged with AM fungus and *P. deliense* under zinc applied conditions. It was observed that certain rates of zinc and mycorrhiza-based treatments had positive effects on disease factors by suppressing Pythium rot and can be used for biological control. In studies of Vidhyasekaran (2004), onion pink rot caused by *P. terrestris* and tomato root rot caused by *Thielaviopsis basicola* were controlled by mycorrhizal fungi.

Rosendahl and Rosendahl (1990) studied the role of vesicular arbuscular mycorrhiza in controlling damping-off in cucumber caused by *P. ultimum*. VAM inoculation, before or at the same time as inoculation of the pathogen, increased survival of seedlings.

8.2.4 OILSEED CROPS

Interaction between *G. mosseae* and *Plasmopara helianthi* in sunflower plants susceptible and resistant to downy mildew was studied by Toshi et al. (1993). Mycorrhizal plants preinoculated with *G. mosseae* at the highest inoculum concentration showed well-developed mycorrhizal root coloniza-tion and decrease in root infection by *P. helianthi*. Histochemical reactions of incompatibility performed in roots of susceptible and resistant plants showed that no defense response to *P. helianthi* was induced by mycorrhizal infection. Kulkarni et al. (1997) studied that addition of mycorrhizal fungi nullified the effect of the pathogen. *G. fasciculatum* was the most effective and worked best when they were added individually.

8.2.5 OTHER CROPS

Liu et al. (1995) reported that inoculum of cotton with VAM fungi improved seedling growth, increased yield and reduced the incidence and disease index of *Verticillium dahliae* induced wilt. Kobra et al. (2011) studied the interaction of arbuscular mycorrhizal fungi (*G. etunicatum*, *G. intraradices*, and *G. versiforme*) with a wilt-causing *V. dahliae* in cotton. It was found that establishment by arbuscular mycorrhizal fungi reduced disease index. In diseased cotton plants colonized by *G. etunicatum*, the disease index was less than other diseased mycorrhizal and nonmycorrhizal ones. In diseased cotton plants, chlorophyll content was lower than others. Three *Glomus* species significantly increased content of sugar and protein in shoot and root. Pathogen-infected plants had higher proline concentration in shoot and root than healthy plants. On the other hand, the increased content of proline as stress sensor showed that Verticillium accelerates senescence and reduces yield. These results suggest that the beneficial effects of mycor-rhiza can alleviate the pathogenesis effects of *V. dahliae* partly, and also, there is a competitive interaction between the pathogenic and symbiotic fungi.

Doley and Kaur (2012) used AM fungi for the biological control of soil-borne plant pathogen *Macrophomina phaseolina* in groundnut plant. For this, investigation pot culture technique was followed. Soil-based mixture of AM fungi (*G. fasciculatum*) was inoculated into the root of groundnut. In results, the colonization by mycorrhizal fungi significantly resulted into decreased incidence of disease caused by *M. phaseolina*. The growth of groundnut showed marked increase due to mycorrhizal colonization, viz. shoot and root length, fresh and dry weight, leaf, nodule, and pod number. In presence of pathogen, mycorrhizal dependency was significantly higher, but degree of colonization went down. The content of chlorophyll was found to be increased significantly due to inoculation of AM fungi. The various biochemical and defense-related enzyme activities were investigated, and the results obtained showed significant increase in their activities due to pathogen as well as AM fungi inoculation. But, highest activities were recorded where both pathogens as well AM fungi were involved. Thus, inoculation of AM fungi showed great biocontrol ability as well as growth promoter. Moreover, it showed their efficacy in inhibition of damaging effect caused by pathogen *M. phaseolina*.

Hemavati and Thippeswamy (2014) studied the efficacy of arbuscular mycorrhizal fungus (*A. lacunosa*) was evaluated for the biological control of soil-borne plant pathogen *Cercospora arachidicola* in groundnut plant. For this investigation, pot culture technique was followed. Soil-based mixture of AM fungi, *A. lacunosa* was inoculated onto the root of groundnut plant. In results, the colonization by AM fungi significantly resulted into decreased incidence of disease caused by *C. arachidicola*. The growth of groundnut showed marked increase in the shoot and root length, fresh and dry weight due to AM fungi colonization.

Subhashini (1990) reported that inoculation with *G. fasciculatum* reduced the attacks of *Pythium* in tobacco nurseries. Studies on histological, physiological, and biochemical interactions between VAM and *Thielaviopsis basicola* in tobacco plants were attempted by Giovannetti et al. (1991). Investigations indicated that mycorrhizal plants obtained by preinoculation with *G. monosporum* showed a better tolerance to *T. basicola* than nonmycorrhizal. Root infection and chlamydospore formation by *T. basicola* was slower and at a lower level in mycorrhizal plants inoculated with pathogen as compared to pathogen alone inoculated plants.

Thomas et al. (1994) investigated the possibility of using *G. fasciculatum* in control of damping-off of cardamom caused by *F. moniliforme* (*Gibberella fujikuroi*), *P. vexas*, and *Rhizoctonia* spp. Plants inoculated with *G. fujikuroi* only had a disease score 2.0 in the disease scale of 0–3. Uninoculated plants had a score of 0. Plants inoculated with *G. fujikuroi* and *G. fasciculatum* had a score of 0.5, demonstrating that *G. fasciculatum* could reduce the severity of disease. Plants inoculated with *G. fasciculatum* harbored more organisms in their rhizosphere with properties antagonistic to *G. fujikuroi*.

In cumin, the interaction between VAM fungi and *F. oxysporum* f.sp. *cumini* was observed by Champawat (1991). Cumin plants were inoculated with four efficient VAM fungi alone or in association with *F. oxysporum* f.sp. *cumini*. In P-deficient sandy loam soil, VAM enhanced nutrient uptake and reduced wilt disease severity. The interaction between VA-mycorrhizal fungi and *S. rolfsii* in chili (*Capsicum annum* L.) was studied by Sreenivasa et al. (1992) under greenhouse conditions. A significant inhibition of sclerotial bodies was observed in plants inoculated with both *G. fasciculatum* and

G. macrocarpum. Dual inoculation of VAM fungi suppressed the pathogen more effectively than single inoculation.

Alejo-Iturvide et al. (2008) studied the role of mycorrhizal inoculation in chili plants challenged with *Phytophthora capsici* to see the hypersensitive response. In the treatment without mycorrhiza (T3) and with mycorrhiza (T4), visible disorders were detected 2 days after inoculation with *P. capsici*, but in the next days, T3 plants rapidly developed 25% more necrotic lesions on the leaves than T4 plants. From these findings, it is concluded that mycorrhizal colonization contributes significantly in maintaining the redox balance during oxidative stress, but the exact mechanism is still uncertain. Sivaprasad et al. (1999) studied the foot rot of black pepper and rhizome rot of ginger as influenced by vesicular arbuscular mycorrhizal (VAM) fungi. Effect of inoculation with different VAM fungi on foot rot disease of black pepper and rhizome rot of ginger was tested under field conditions. Black pepper cuttings inoculated with *G. monosporum* at nursery level and transplanted to the field significantly reduced the foot rot incidence from 100% infection and 83% mortality observed in control to 33%. Ozgonen and Erkilic (2007) found that AM fungi significantly increased plant growth and reduced disease severity caused by *Phytophtora capsici*.

Interactions of VAM fungi and two wilt pathogens of lucerne, *V. alboatrum*, or *F. oxysporum* f.sp. *medicaginis*, were studied by Hwang et al. (1992) under controlled conditions. There were four treatments of mycorrhizal fungi and two levels of pathogen inoculum. Shoot dry weight of lucerne plants inoculated with VA mycorhizal fungi significantly exceeded those of the nonmycorhizal plants. Inoculation with *V. albo-atrum* or *F. oxysporum* f.sp. *medicaginis* significantly reduced the shoot dry weight of lucerne. Seedlings inoculated with VAM fungi had a lower incidence of wilt than in the nonmycorrhizal ones. Propagule number of both pathogens was lower in the soil inoculated with VAM fungi than in the non mycorrhizal soil.

McAllister et al. (1994) studied the effects of inoculation by *T. koningii* and *F. solani* on lettuce with or without colonization by *G. mosseae* in a greenhouse trial. Plant dry weight of nonarbuscular mycorrhizal inoculated lettuce were unaffected by the presence of *T. koningii* and *F. solani*. In contrast, *T. koningii* decreased plant dry weight and VAM colonization when inoculated into the rhizosphere before or at the same time as *G. mosseae*. In addition, the *T. koningii* population was considerably reduced when *G. mosseae* was inoculated 2 weeks before the saprophytic fungus. *F. solani* had effect on VAM colonization of lettuce roots, which was similar to that of *T. koningii*. Babu et al. (1998) reported that mycorrhizal fungi *G. fasciculatum* and Azosprillum and Phosphobacterium, when applied alone or in different combinations in soil in field, reduced the population of *P. zeae* and recorded the highest cob yield of maize.

8.3 INTERACTION OF VAM WITH NEMATODE PATHOGENS

Plant parasitic nematodes and mycorrhizal fungi are commonly found inhabiting the same soil and colonizing roots of their host plants. These two groups of microorganisms exert characteristics, but opposite effects on plant health. The parasitization of plants by nematodes (mainly endoparasitic) can be influenced by the establishment of a VA-mycorrhiza. Pinochet et al. (1996) reviewed interactions between migratory endoparasitic nematodes and AM fungi in perennial crops. The potential role of mycorrhizal fungi as biological agent for control of nematode plant disease has recently received considerable attention. The damage due to nematode diseases is generally reduced in mycorrhizal plants. Establishment of VAM in plants generally confers resistance to nematode parasitism or adversely affects nematode reproduction. However, sometimes, root colonization by reproduction by VAM fungi may have no effect on nematode reproduction or promote it. The interactions between VAM and soil-borne nematode pathogens reported from 1990 to date are summarized in Table 8.2. A comprehensive review of recent literature on VAM–soil borne nematode pathogen interactions, including mechanisms involved and factors affecting interactions, are presented here.

8.3.1 LEGUME CROPS

Oyekanmi et al. (2007) studied the interaction of two soyabean genotypes through root knot nematode and application of *B. japonicum*, *T. pseudokoningii*, and *G. mosseae*. They reported that these organisms provide equal or better protection against root knot nematode damage than the synthetic nematicide carbofuran. Interaction of *G. fasciculatum* with *M. incognita* on blackgram (*Vigna mungo*) was recorded by Sankaranarayanan and Sundarababu (1997a). Inoculation of blackgram with *G. fasciculatum* 15 and 20 days earlier than inoculation with *M. incognita* controlled the nematode population and also increased biomass production. When *G. fasciculatum* was applied 5 days earlier than the nematode, VA mycorrhizal colonization and spore production were reduced. Sankaranarayanan and Sundarababu (1997b) observed the effect of oil cakes and nematicided on the growth of blackgram inoculated with VAM fungus (*G. fasciculatum*) and root-knot nematode

TABLE 8.2   Effect of VAM on Soil-Borne Nematode Pathogens.

Pathogen	Host plant	Effects in mycorrhizal plants*			Reference
		Disease damage	Infection	Reproduction	
Meloidogyne incognita	Bean	–	⋮	⋮	Osman et al. (1990)
Meloidogyne incognita	Bean	–	⋮	–	Osman et al. (1991)
Meloidogyne incognita	Blackgram	–	⋮	–	Sankaranarayanan and Sundarababu (1997a)
Meloidogyne incognita	Black gram	–	⋮	–	Sankaranarayanan and Sundarababu (1997b)
Meloidogyne incognita	Blackgram	–	⋮	–	Sankaranarayanan and Sundarababu (1998)
Meloidogyne incognita	Black gram	–	⋮	–	Mahanta et al. (2002)
Meloidogyne incognita	Blackgram	–	–	–	Mohanta and Phukan (2004)
Pratylenchus coffeae	Coffee	–	⋮	⋮	Vaast et al. (1998)
Pratylenchus coffeae *Meloidogyne konaensis*	Coffee	–	⋮	⋮	Vaast (1997)
Meloidogyne exigua	Coffee	–	⋮	–	Rivas Platero and Andrade (1998)
Meloidogyne incognita	Cotton	⋮	–	–	Saleh and Sikora (1989)
Meloidogyne incognita	Cowpea	–	–	–	Ahemed and Alsayed (1991)
Meloidogyne incognita	Cowpea	–	⋮	⋮	Santhi and Sundarababu (1995)
Meloidogyne incognita	Cowpea	⋮	⋮	–	Santhi and Sundarababu (1995)
Meloidogyne incognita *Tylenchorhynchus vulgaris*	Lucerne	–	⋮	–	Jain et al. (1998b)
Heterodera cajani	Pigeonpea	⋮	⋮	–	Siddiqui and Mahmood (1995)
Heterodera cajani	Pigeonpea	⋮	⋮	–	Siddiqui and Mahmood (1996)

TABLE 8.2 (Continued)

Pathogen	Host plant	Effects in mycorrhizal plants*			Reference
		Disease damage	Infection	Reproduction	
Meloidogyne incognita	*Piper nigrum*	−	−	−	Sivaprasad et al. (1992)
Meloidogyne incognita	Tobacco	∷	−	∷	Subhashini (1990)
Meloidogyne incognita	Tobacco	∷	−	−	Krishna Prasad (1991)
Meloidogyne incognita	Tomato	−	∷	∷	Neelima et al. (1991)
Meloidogyne javanica	Tomato	−	∷	∷	Sundarababu et al. (1993)
Meloidogyne incognita	Tomato	−	∷	−	Sundarababu et al. (1993)
Meloidogyne javanica	Tomato	−	−	∷	Al-Raddad (1995)
Meloidogyne incognita	Tomato	−	∷	∷	Mishra and Shukla (1996)
Meloidogyne incognita	Tomato	−	∷	−	Rao et al. (1997)
Meloidogyne incognita	Tomato	−	−	∷	Mishra and Sukula (1997)
Meloidogyne arabicida	Tomato	−	∷	−	Rivas-Platero et al. (1998)
Meloidogyne incognita	Tomato	∷	−	−	Rao and Gowen (1998)
Meloidogyne incognita	Tomato	−	∷	−	Reddy et al. (1998)
Meloidogyne incognita	Tomato	−	∷	−	Nagesh et al. (1999)
Meloidogyne incognita	Tomato	∷	−	∷	Ling and Jia (2000)
Meloidogyne incognita	Tomato	∷	∷	−	Bhagawati et al. (2000)
Meloidogyne incognita	Tomato	−	∷	−	Kantharaju et al. (2002)
Meloidogyne incognita	Tomato			−	Pradhan et al. (2003)
Meloidogyne incognita	Tomato			−	Kantharaju et al. (2005)

TABLE 8.2 *(Continued)*

Pathogen	Host plant	Effects in mycorrhizal plants*			Reference
		Disease damage	Infection	Reproduction	
Meloidogyne incognita	Tomato	–			Shreenivasa et al. (2007)
Meloidogyne incognita	Tomato	–			Siddiqui and Akhtar (2008)
Pratylenchus zeae	Maize	...	...	–	Babu et al. (1998)
Meloidogyne incognita	Berseem	–	...	–	Jain et al. (1998a)
Tylenchorhynchus vulgaris					
Meloidogyne incognita	Okra	...	...	–	Jothi et al. (2000)
Meloidogyne incognita	Okra	–	...	–	Sharma et al. (2003)
Meloidogyne incognita	Egg plant	...	...	–	Rao et al. (1998a)
Meloidogyne incognita	Egg plant	–	...	...	Rao et al. (1998b)
Meloidogyne incognita	Brinjal	–	...	–	Aparajita and Phukan (2003)

* – = decreased, ×= no effect, ... = not reported.

(*M. incognita*) in pot studies. Pot soil was either treated with *Azadirachta indica, Madhuca longifolia, Ricinus communis,* or groundnut oilcakes, 15 days prior to sowing of blackgram or inoculated with mycorrhizal fungi at 10 g inoculum/plant, 22 days prior to sowing. After 37 days, nematodes were inoculated at one juvenile/cc soil. One week following nematode application phorate 10 g was applied. All treatments were effective but the greatest reduction in nematode population was found in oil cake treatments.

A further study of Sankaranarayanan and Sundarababu (1997b) recorded the effect of leaf extracts on growth of blackgram inoculated with *G. fasciculatum* and *M. incognita* in pot experiments. There was significant increase in all the growth parameters of blackgram due to the addition of leaf extracts of *Calotropis procera, Targetes erecta, Catharanthus rosea,* and *Bougainvillea spectabilis.* Leaf extracts recorded the highest spore population and mycorrhizal colonization than the nematodes, but the latter recorded less gall index and nematode populations. Sankaranaryanan and Sundarababu (1998) observed the effect of *Rhizobium* on the interaction of VAM and root knot nematode on blackgram. The combination of *Rhizobium* and VAM resulted in the least nematode population density and maximum spore, mycorrhizal colonization, total nitrogen, and phosphorus content of blackgram. Mahanta and Phukan (2004) recorded effect of *G. fasciculatum* on penetration, development, and survival of *M. incognita* on blackgram. The plants inoculated with VAM + nematode after 10 days and VAM and nematode after 15 days showed significantly reduced second-stage juvenile penetration and ultimately delayed the development of different stages of *M. incognita* compared with the plants inoculated with VAM and nematode simultaneously and nematode alone. Significant decreases in nematode penetration, delayed development, number of galls, egg masses, eggs per mass, and final nematode after 15 days followed by the plants inoculated with VAM + nematode after 10 days.

Osman et al. (1990) inoculated bean seedlings in pots with *Glomus* spp. and *M. incognita* either alone or simultaneously. Gall index and final nematode populations were significantly increased with simultaneous inoculations. There was a significant decrease when nematodes were inoculated 15 and 39 days after the fungus. In a similar study (Osman et al., 1991), when the bean plants were infected with *M. incognita* 15 or 30 days after mycorrhizal infection, the development of nematode was inhibited. Application at the time reduced fungal development. Pandey et al. (2005) studied the management of root knot nematode (*M. incognita)* and *F. oxysporium* f.sp. *ciceri* using ecofriendly methods. The highest shoot length, root weight, chlorophyll content, number of pods/plant, 100 seed weight, percent mycorrhizal colonization, and chlamydospore population were obtained when combination of all the treatments were used. Similarly, number of galls/plant, egg mass per plant, nematode soil population and wilt percent was lowest with *G. fasciculatum + T. viride + P. lilacinus* + neem oilseed cake.

Ahemed and Alsayed (1991) evaluated influence of *G. macrocarpum* on the penetration and development of *M. incognita* in cowpea in glass house tests. *G. macrocarpum* significantly reduced nematode numbers in cowpea roots and soils. Similarly, a reduction in number of arbuscules was observed in the presence of *M. incognita*. It was suggested that VAM fungi-infected roots stimulated plant growth by improving host nutrition and minimizing the damage caused by nematode perhaps by change in root exudates causing fewer nematodes to be attracted to and penetrate the plant roots.

Santhi and Sundarababu (1995b) recorded effect of phosphorus on the interaction of VAM fungi with *M. incognita* on cowpea in a pot experiment. Plants with VAM were more resistant to *M. incognita* than those without. Positive correlation was observed between the phosphorus levels and the nematode population, whereas negative correlation between phosphorus levels and the VAM spore population and colonization. Siddiqui and Mahmood (1995) used VAM fungus (*Gigaspora margarita*) and two biocontrol agents for the management of the nematode (*Heterodera cajani*) and the fungus (*Fusarium udum*). All the three management agents, alone or in combination, applied to plants inoculated with pathogens increased shoot dry weight, number of nodules, phosphorus content, and reduced nematode multiplication and wilting index. Simultaneous use of the biocontrol agents and the vesicular arbuscular fungus gave better control of the disease complex than did their individual application. In a further study (Siddiqiue and Mahmood, 1996), *G. mosseae, T. harzianum,* and *V. chlamydosporium* were used alone and in combination for the management of wilt disease complex of pigeonpea. Treatment of plants inoculated with pathogens increased plant length, shoot dry weight, number of nodules, and reduced nematode multiplication and wilting index. Simultaneous use of biocontrol agents against pathogens gave better control than their individual application. *T. harzianum* had an adverse effect on root colonization by *G. mosseae*. The highest reduction in nematode multiplication was observed when all the three biocontrol agents were used together.

Jain et al. (1998a) investigated the influence of the *G. fasciculatum* on *M. incognita* and *Tylenchorhynchus vulgaris* infecting berseem (*Trifolium alexandrium*). Gall formation by *M. incognita* and multiplication of both the nematodes was reduced by 28–50%. Conversely, the prior presence of nematodes reduced the percent root colonization by VAM fungi. Jain et al.

(1998b) reported that preinoculation of lucerne plants with *G. fasciculatum* 2 weeks before introduction of either *M. incognita* or *Tylenchorhynchus vulgaris* significantly reduced the adverse effects of nematodes on total biomass of the plant. Similarly, gall formation and also the multiplication of both nematode species were considerably reduced. In contrast, prior presence of the nematodes reduced mycorrhizal colonization.

8.3.2 VEGETABLE CROPS

Interaction between *G. fasciculatum*, *M. incognita*, and fungicide in tomato was noted by Mishra and Shukla (1996). Preestablishment of *G. fasciculatum* to nemadote increased plant growth decreased the size and number of galls and improved NPK (nitrogen, phosphorus, and potassium) uptake compared with plants inoculated with nematode alone or preinoculated with nematode to VAM. Fungicidal combinations with VAM and nematode producted an overall decrease in plant growth. However, simultaneous applications of Carbendazim + *G. fasciculatum* improved plant growth. Reduced uptake of NPK was recorded in all the treatment combinations of VAM and nematode when combined with fungicides. In interaction between *G. fascuculatum* and *M. incognita* on tomato, the time of inoculation also mattered. Simultaneous inoculations of *G. fasciculatum* with *M. incognita* caused the greatest reduction in number and size of the nematode-induced root galls.

Mittal et al. (1991) concluded that presence of *G. fasciculatum* alone in the roots was inhibitory to the formation of nematode galls. Sundarababu et al. (1993a) also reported that VAM-enhanced plant growth reduced nematode population and increased yields. Rao and Gowen (1998) attempted biomanagement of *M. nicognita* on tomato by integrating *G. deserticola* and *Pasteuria penetrans*. The combination significantly reduced the number of egg masses in root systems and increased the parasitization of females by *P. penetrans*. Root colonization by *G. deserticola* was not affected by *P. penetrans*. It was suggested that *G. deserticola* and *P. penetrans* can prove useful for sustainable management of *M. incognita* on tomatoes. Sundarbabu et al. (1993b) studied influence of time of inoculation on interaction between vesicular arbusular mycorrhiza and *M. javanica* on tomato. In their studies, prior establishment of mycorrhiza could offset the damaging effect of the nematode and improve plant growth. Interaction of *G. mosseae* and *P. lilacinus* on *M. javanica* in tomato was evaluated by Al-Raddad (1995). Inoculation of plants with VAM suppressed gall index and average number of galls/root system by 52% and 66%, respectively, compared with seedings inoculated with *M. javanica* alone. The presence of *G. mosseae* and *P. lilacinus*, together or separately in the layer manure, completely inhibited root infection by nematode. Mycorrhizal colonization was not affected by the layer manure treatment or by root inoculation with *P. lilacinus*. Rivas-Platero et al. (1998) observed interaction between *Glomus* spp. and *M. arabicida* in tomato. *M. arabicida* multiplication rate was reduced 50.5% by *Glomus* spp. and the gall index in tomato roots by 13%. *Glomus* spore production was stimulated by *M. arabicida*. The phosphorus content was higher in the *M. arabicida* + *Glomus* pots than *Glomus* alone ones. Nagesh et al. (1999) studied correlation between *G. fasciculatum* spore density root colonization and *M. incognita* infection on tomato. The nematode populations, egg masses/plant, and root gall index were negatively correlated with spore density. Further, at least 2 spores/g soils were required for 50% root colonization under pot conditions. However, under nursery conditions, 4 spores/g soils were required to obtain 50% root colonization which reduced nematode populations in root by 59 and 61%, respectively.

In studies of Yanling and Zhengjia (2000), *G. mosseae* had an inhibitory effect on *M. incognita* infection in tomatoes in Hubaei, China. Bhagawati et al. (2000) reported that mustard cake and *G. etunicatum* were equally effective in reducing the damage caused by *M. incognita* and/or *F. oxysporum* f.sp. *lycopersici* in tomato. When mustard cake and VAM were applied together, the combined efficacy was much greater than single treatments. The population of pathogens was reduced. Kantharaju et al. (2002) evaluated eight isolates of *G. fasciculatum* against root knot nematode (*M. incognita*) race-1 on tomato. Nematode population in soil, number of galls, and egg masses per root system were lowest in Shimoga Banana isolate followed by Shimoga Areca isolate. Masadeh et al. (2004) studied that *G. intraradices* reduced the number of galls cv. Hildares, but biocontrol of nematode was not achieved in cv. Tiptop. Kantharaju et al. (2005) found that *G. fasciculaum* reduced the nematode population on tomato, but SBI-G.f. isolate was most effective compared with CTI-G.f. isolate. Shreenivasa et al. (2007) reported that *G. fasciculatum* inoculation reduced nematode population and number of galls on tomato. Siddiqui and Akhtar (2007) found that AM (*G. mosseae*) inoculation was superior in reducing galling and nematode multiplication compared with *G. margarita* on tomato. Siddiqui and Akhtar (2008) found that inoculation with AM fungi (*G. intraradices*) with *T. harzianum* caused 37% increase in the growth of nematode inoculated plants. Sharma and Trivedi (1997) studied the concomitant effect of *P. lilacinus* and vesicular arbuscular mycorrhizal fungi on root-knot nematode infected okra.

Combined application of *G. fasciculatum* and *G. mosseae*, 2 weeks prior to inoculation with *M. incognita*, resulted in increased fresh and dry weight of root, shoot, and okra fruit as compared to only nematode-treated plants in pot experiments. The combined effect of the mycorrhizal fungi and *P. lilacinus* gave maximum reduction in nematode galls. Jothi et al. (2000) attempted management of *M. incognita* on okra by integrating nonhost and endomycorrhiza. In their studies, preinoculation with VAM enhanced growth and yield of okra and significantly reduced the population of *M. incognita* (from 994 to 240 nematodes/200 g soil).

The effects of plant growth promoting microbes and vesicular arbuscular mycorrhiza were studied by Sharma et al. (2003) on root knot nematodes (*M. incognita*) infecting okra. Microbes were applied in the top soil in earthen pots alone or in combination with phorate. The maximum reduction in number of galls was obtained with VAM followed by *Azotobater* and *Azosprillum* alone. Combined application of phorate and microbes increased the root-knot galling. Similar trend was noticed with egg mass and eggs/egg mass count. Sharma et al. (2003) reported that *G. mosseae* inoculation reduces the galling on okra. Rao et al. (1998b) conducted experiments to study the feasibility of integration of *G. fasiculatum* and *R. communis* cake for the management of *M. incognita* on eggplant. Seedlings colonized with mycorrhiza were least infected by *M. incognita* when transplanted in soil which was amended with caster cake. Aparajita and Phukan (2003) observed interaction of *G. fasciculatum* and *M. incognita* in brinjal. In their studies, mycorrhizal plants showed relatively higher growth than nonmycorrhizal plants. The treatment with VAM at 600 spores/pot + *M. incognita* was best for reducing galls, egg masses, and nematode population. Nehra (2004) reported that *G. fasciculatum* and *G. mosseae* inoculation in ginger caused reduction in nematode population, but the greatest reduction was recorded by the *G. fasciculatum* inoculation.

8.3.3 OTHER CROPS

Saleh and Sikora (1989) reported that *G. fasciculatum* did not affect plant growth or nematode population density 30 days after simultaneous inoculation at planting but caused significant increases in plant growth and a significant reduction in *M. incognita* population density and reproduction 50 days after inoculation. The fungicides did not alter nematode population on nonmycorrhizal plants. The results demonstrated that fungicides concentration and time of application are important factors influencing the antagonistic interaction between endomycorrhizal fungi and root-knot nematodes on cotton. Vaast et al. (1998) recorded influence of root-lesion nematode, *Pratylenchus coffeae* and two arbuscular mycorrhizal fungi on coffee. The late mycorrhizal inoculation (added simultaneously with nematode) did not enhance coffee tolerance to *P. coffeae*. In the presence of *P. coffeae*, late mycorrhizal plants were P deficient during the entire experiment, and their foliar P concentration remained as low as that of nonmycorrhizal plants. After 7.5 months, nematodes decreased mycorrhizal colonization of late plants and their biomass. These studies confirmed earlier observations that an early VAM inoculation gives better results than simultaneous inoculation with nematodes in coffee. The interactions between mycoorhizal fungi (*Entrophospora colombiana* and *G. margarita*) and *M. exigua* in coffee were observed by Rivas Platero and Andrade (1998). The mycorrhizal fungi reduced the multiplication rate of *M. exigua*. Both the fungi reduced the gall index by 20% compared to nematode alone. The nematode stimulated sporulation of the fungi.

Babu et al. (1998) reported that mycorrhizal fungi and *Azosprillium* and phosphobacterium when applied alone or in different combination in soil field reduced the population of *Pretylenchus zeae* and recorded the highest cob yield of maize. Subhashini (1990) studied the role of VA mycorrhiza in controlling certain root diseases of tobacco. Inoculation with *G. fasciculatum* reduced the intensity of infection by root knot nematode but had no effect on orobanche infestation. Influence of a vesicular arbuscular mycorrhiza on the development and reproduction of root knot nematode in tobacco was recorded by Krishna Prasad (1991). The nematode infestation of seedlings, number of galls, and egg masses per infested seeding were significantly reduced by VAM inoculation.

8.4   FUTURE PROSPECT

A detailed analysis of mycorrhizae, a comprehensive bibliographic database of mycorrhizal research showed that much research on AM has focused on nutrient dynamics, inoculum production, mycorrhiza formation, and morphology and physiology of both arbuscular mycorrhizas and ectomycorrhizas. There has been a shift of attention from ectomycorrhiza to vesicular–arbuscular mycorrhiza since 1970. Very little work has been done on the kinds of mycorrhizas. Barriers to progress in some areas, such as our inability to grow the Glomales in asexinic culture, have stimulated large number of methodological papers over the past 15 years, whereas the genetics and molecular biology, especially of AM, have been neglected. Ecological studies were more numerous than any other kind. Several studies have certainly revealed that mycorrhiza induced resistance in plants, whereas in some fewer cases, either no effect or an increase in the disease severity has been observed. The critical evaluation of AM as a biocontrol agent is therefore one of the challenging areas, and it holds a great promise in our future research endeavors. The effect of vesicular–arbuscular mycorrhizal fungi on plant diseases has not been mostly demonstrated under field conditions, although there are some promising results that support the conclusions of the greenhouse studies. Study on parasitic ability of AM fungi on plant parasitic nematodes is required so that proof of successful parasitism of AM would explore the possibility of managing the nematode population directly acting as biocontrol agent. Few available reports on integration of AM with other biocontrol agents have given promising results. This aspect and utilization of arbuscular mycorrhiza as a factor in integrated plant protection remains to be studied in detail. Arbuscular myrrohizae has to be considered a part of the environment and of the plant itself, which is able to influence its disease resistance. As interactions between the host plants, arbuscular mycorrhizal fungus, soil, and climate will determine the mycorrhizal effect on plant growth, research efforts need to be directed toward finding appropriate host–fungus combinations adapted to well-define soil types and climatic conditions. The management practices enhancing the combined effect of surviving and introduced fungi need to be developed. In order to establish biocontrol efficacy of mycorrhizal fungi, there are many inherent obstacles to overcome. These include strain selection, mass production of mycorrhizal inoculants, storage, the relationship between inoculum potential and infectivity of AM fungi, methods of application, and the effects of nutrients on mycorrhizal responses. For mass production of AM fungi, protoplast fusion with soil culture microorganisms which are harmless to plant growth may be attempted in future. Taxonomic consideration at molecular level warrants immediate attention. Investigation of new technologies is needed, especially genetic manipulation of fungal symbionts to induce new biotypes for improved performance, adaptation of native microbial agents, new disease control requirements, and improved compatibility with chemical fungicides and nematicides. Side effects of pesticides used in normal farming systems on nontarget mycorrhizal endophytes also need to be evaluated more critically for developing viable plant protection strategies. Only fragmentary information is available on the mechanisms involved in the suppression of the severity of soil-borne diseases by AM, endophytes. Although the improvement of plant nutrition, compensation for pathogen damage and competition for photosynthesis or colonization/infection sites have been claimed to play a protective role in AM symbiosis, information is scarce, fragmentary, or even controversial, particularly concerning other mechanisms. Such mechanisms include anatomical or morphological AM induced changes in the root system, microbial changes in rhizosphere populations of mycorrhizal plants and local elicitation of plant defense mechanisms by AM fungi. Information needs to be gathered on these mechanisms, if these fungi are to be fully utilized for meaningful integrated crop disease management. Future research should be directed toward correcting the various imbalances revealed by this analysis.

8.5 CONCLUSION

The potential role of mycorrhizal fungi biocontrol agent for the control of fungal and nematode plant diseases has recently attracted considerable attention. VA-mycorrhizal infections generally inhibit or sometimes increase and occasionally have no effect on disease caused by fungal pathogens. The damage due to nematode diseases has also generally decreased in mycorrhizal plants. Establishment of VAM in plants generally confers resistance to nematode parasitism or adversely affects nematode reproduction. Although the improvement of plant nutrition, compensation for pathogen damage and competition for photosynthates or infection sites have been claimed to play a protective role in VAM symbiosis. Information is scare, fragmentary, or even controversial, particularly concerning other mechanisms such as anatomical or morphological and biochemical VAM-induced changes in the root system, microbial change in the rhizosphere populations of mycorrhizal plants, and local elicitation of plant defense mechanisms by VAM fungi. In the present chapter, interactions between VAM and phytopathogens (fungi and nematode) reported during past 2 decades are reviewed.

KEYWORDS

- **arbuscular mycorrhiza**
- **biocontrol**
- **plant disease**
- **rhizosphere**

REFERENCES

Abdel-Fattah, G. M.; Shabana, Y. M. Efficacy of Arbuscularmycorrhizal Fungus (*Glomus clarum*) in Protection of Cowpea Plants from Root Rot Pathogen *Rhizoctoni asolani*. *J. Plant Dis. Protect.* **2002**, *109*(2), 207–215.

Al-Raddad, A. M. Interaction of *Glomus mosseae* and *Paecilomyces lilacinus* on *Meloidogyne incognita* of Tomato. *Mycorrhiza.* **1995**, *5*(3), 233–236.

Agrios, G. N. Plant Pathology; 5th edn., Elsevier-Academic: San Diego, C.A. 2005; p 922.

Ahemed, M. A.; Saleh, E. A.; E1-Fallol, A. A. The Role of Biofertilizers in Suppression of Rhizoctonia Root Rot Disease of Broad Bean. *Ann. Agric. Sci.* **1994**, *39*(1), 379–395.

Ahemed, S. S.; Alsayed, A. A. Interaction Between the Vesicular Arbuscular Mycorrhiza *Glomus macrocarpus* and *Meloidogyne incognita* Infecting Cowpea. *Ann. Agric. Sci.* **1991**, *29*(4), 1765–1772.

Akhtar, M. S.; Siddiqui, Z. A. Effects of *Glomus fasciculatum* and *Rhizobium spp.* on the Growth and Root-Rot Disease Complex of Chickpea. *Arch. Phytopathol. Plant Protect.* **2007**, *40*, 37–43.

Akkopru, A.; Demir, S. Biocontrol of Fusarium Wilt in Tomato Caused by *Fusarium oxysporum* f. spp. *lycopersici* by AMF *Glomus intraradices* and Some Rhizobacteria. *J. Pyhtopathol.* **2005**, *153*, 544–550.

Alejo-Iturvide, F.; Marquez-Lucio, M. A.; Morales-Ramirez, I.; Vazquez-Garciduenas, M. S.; Olalde-Portugal, V. Mycorrhizal Protection of Chili Plants Challenged by *Phytophthora capsici*. *Eur. J. Plant Pathol.* **2008**, *120*(1), 13–20.

Aparajita B.; Phukan, P. N. Effect of Interaction of *Glomus fasciculatum* and *Meloidogyne incognita* on Growth of Brinjal. *Ann. Plant Protect. Sci.* **2003**, *11*(2), 352–354.

Arabi, M. I. E.; Kanacri, S.; Ayoubi, Z.; Jawhar, M. Mycorrhizal Application as a Biocontrol Agent against Common Root Rot of Barley. *Res. Biotechnol.* **2013**, *4*(4), 7–12.

Arihawa, J.; Karasawa, T. Effect of Previous Crops on Arbuscular Mycorrhizal Formation and Growth of Succeeding Maize. *Soil Sci. Plant Nutr.* **2000**, *46*, 43–51.

Azcon-Aguilar, C.; Barea, J. M. Interaction Between Mycorrhizal Fungi and Other Rhizosphere Microorganisms. In *Mycorrhizal Functioning: An Integrative Plant-fungal Process*; Allen, M.F., Ed.; Chapman and Hall: Lodon, UK, 1992; pp 163–198.

Azcon-Aguilar, C.; Barea, J. M. Arbuscular Mycorrhizas and Biological Control of Soil-Borne Plant Pathogens—An Overview of Mechanisms Involved. *Mycorrhiza.* **1996**, *6*(6), 457–464.

Babu, R. S.; Sankaranarayan, C.; Joshi, G. Management of *Pratylenchus zeae* on Maize by Biofertilizers and VAM. *Indian J. Nematol.* **1998**, *28*(1), 77–80.

Baby, U. I.; Manibhushanrao, K. Influence of Organic Amendments on Arbuscular Mycorrhizal Fungi in Relation to Rice Sheath Blight Disease. *Mycorrhiza.* **1996**, *6*(3), 201–206.

Bagyaraj, D. J. Ecology of Vesicular Arbuscular Mycorrhizae. In *Hand Book of Applied Mycology Vol. I*, Arora, D. K., Rai, B., Mukerji, K. G., Knudsen, G. R., Eds.; Marcel Dekker: New York, 1991; pp 3–34.

Batcho, M.; Kane, A.; Thiam, M. L.; Duponnois, R.; Ducousso, M. Effect of Four Endomycorrhizal Inocula and Basamid on the Development of Onion Roots (*Allium cepa* L.) Cultivated in Soil Infected by *Pyrenochaeta terrestris* (Hansen) Gorenz, Walker and Larson. *Bulletin de I' Institut Fondamental d' Afrique Noire, Serie A, Sciences Naturelles.* **1994**, *47*, 11–32.

Berg, G.; Grosch, R.; Scherwinski, K. Risk Assessment for Microbial Antagonists: Are There Effect on Non-Target Organisms. *Gesunde Pflanzen.* **2007**, *59*, 107–117.

Berta, G.; Sampo, S.; Gamalero, E.; Musasa, N.; Lemanceau, P. Suppression of Rhizoctonia Root-Rot of Tomato By *Glomus mosseae* BEG 12 and *Pseudomonas fluorescens* A6RI is Associated With Their Effect on the Pathogen Growth and on The Root Morphogenesis. *Eur. J. Plant Pathol.* **2005**, *111*, 279–288.

Bhagwati, B.; Goswami, B. K.; Singh, C. S. Management of Disease Complex of Tomato Caused by *Meloidogyne incognita* and *Fusarium oxysporum* f. spp. *lycopersici* through Bioagent. *Indian J. Nematol.* **2000**, *30*(1), 16–21.

Bodker, L.; Kjoller, R.; Rosendahl, S. Effect of Phosphate and the Arbuscular Mycorrhizal Fungus *Glomus intraradices* on Disease Severtity of Root Rot of Peas (*Pisum sativum*) caused by *Aphanomyces euteiches*. *Mycorrhiza*. **1998**, *8*(3), 169–174.

Brundrett, M. C. Coevolution of Roots and Mycorrhizas of Land Plants. *New Phytol.* **2002**, *154*, 275–304.

Champawat, R. S. Interaction between Vesicular Arbuscular Mycorrhizal Fungi, and *Fusarium oxysporum* f. spp. *cumini* and their Effects on Cumin. *Proc. Indian Sci. Acad., B, Biol. Sci.* **1991**, *57*(1), 59–62.

Chandra, S.; Kehri, H. K. VA Mycorrhiza, A New Biotechnological Tool as Biocontrol Agent. In *Biotechnological Approaches for the integrated Management of Crop Diseases*; Mayee, C. D., Manoharachari, C., Tilak, K. V. B. R., Mukadam, D. S., Deshpande, J., Eds.; Daya Publishing Home: Delhi, India. 2004; pp 200–207.

Chandra, S.; Kehri, H. K.; Maheshwari, S. Macrophomina Dry Root Rot of Leguminous Crops and Its Management through VAM Biopesticides. In *Detection of Plant Pathogens and Their Management*; Verma, J. P., Verma, A., Kumar, D., Eds.; Angkor Publishers (P) Ltd.: New Delhi. 1995; pp 357–364.

Chhabra, M, L.; Bhatnager, M. K.; Sharma, M. P. Influence of a Vesicular Arbuscular (VA) Mycorrhizal Fungus on Important Diseases of Maize. *Indian Phytopathol.* **1992**, *45*(2), 235–236.

Chu, E. Y.; Endo, T.; Sten, R. L. B.; Albuqeroue, F. C. Evalution of Arbuscular Mycorrhizal Fungi Inoculation on the Incidence of *Fusarium* Root Rot of Blackpepper. *Fitologia Brasileira*. **1997**, *22*(2), 205–208.

Doley, K.; Jite, P. K. Effect of Arbuscular Mycorrhizal Fungi on Growth of Groundnut and Disease Caused by *Macrophomina phaseolina*. *J. Exp. Sci.* **2012**, *3*(9), 46–50.

Douds, D. D.; Galvez, L.; Franke-Snyder, M.; Reider, C.; Drinkwater, L. E. Effect of Compost Addition and Crop Rotation Point Upon VAM Fungi. *Agric. Ecosyst. Environ.* **1997**, *65*, 257–266.

Fiorilli, V.; Catoni, M.; Francia, D.; Cardinale, F.; Lanfranco1, L. The Arbuscular Mycorrhizal Symbiosis Reduces Disease Severity in Tomato Plants Infected by *Botrytis cinerea*. *J. Plant Pathol.* **2011**, *93*(1), 237–242.

Fitter, A. H.; Garbaye, J. Interaction between Mycorrhizal Fungi and Other Soil Organisms. *Plant Soil.* **1994**, *159*(1), 123–132.

Giovannetti, M.; Tosi, L.; Tore, G. D.; Zazzerini, A. Histological, Physiological and Biochemical Interactions Between Vesicular–Arbucular Mycorrhizae and *Thielaviopsis basicola* in Tobacco Plants. *J. Phytopathol.* **1991**, *131*(4), 265–274.

Goncalves, E. J.; Muchovej, J. J.; Muchovej, R. M. C. Effect of Kind and Method of Fungicidal Treatment of Bean Seed on Infections by the VA-Mycorrhizal Fungus *Glomus macrocarpum* and by the Pathogenic Fungus *Fusarium solani*. 1. Fungal and Plant Parameters. *Plant and Soil.* **1991**, *132*(1), 41–46.

Grosch, R.; Lottmann, J.; Faltin, F.; Berg, G. Use of Bacterial Antagonists to Control Diseases Caused by *Rhizoctonia solani*. *Gesunde Pflanzen*. **2005**, *57*, 149–157.

Hao, Z. P.; Christie, P.; Qin, L.; Wamg, C. X.; Li, X. L. Control of Fusarium Wilt of Cucumber Seedlings by Inoculation with an Arbuscular Mycorrhizal Fungus. *J. Plant Nutr.* **2005**, *28*(11), 1961–1974.

Hashemi, S. G.; Rouhani, H.; Tarighi, S. Study of Interaction Between Mycorrhiza *Glomus fasciculatum* and *Pseudomonas fluorescens* on Control of Common Root Rot of Wheat Caused by Bipolaris sorokiniana. *J. Plant Protect.* **2013**, *27*(2), 13.

Hemavati, C.; Thippeswamy, B. Effect of Arbuscular Mycorrhizal Fungi, *Acaluspora lacunospora* on Growth of Groundnut Disease Caused by *Cercospora arachidicola*. *Int. J. Res. Appl., Nat. Soc. Sci.* **2014**, *2*(4), 57–60.

Hetrick, B. A. D.; Wilson, G. W. T.; Todd, T. C. Mycorrhizal Response to Wheat Cultivars: Relationship to Phosphorus. *Can. J. Bot.* **1996**, *74*, 19–25.

Hock, B.; Verma, A. Mycorrhiza Structure, Function, Molecular Biology and Biotechnology. Springer: Berlin, 1995.

Hu, Z. J.; Gui, X. D. Pretransplant Inoculation with VA Mycorrhizal Fungi and Fusarium Blight of Cotton. *Soil Biol. Biochem.* **1991**, *23*(2), 201–203.

Hwang, S. F.; Chang, K. F.; Chakravarty, P. Effects of Vesicular–Abuscular Mycorrhizal Fungi on the Development of Verticullum and Fusarium Wilts of Alfalfa. *Plant Dis.* **1992**, *76*(3), 239–249.

Iqbal, S. H.; Shahbaz Rana, Khalid, A. N.; Khan, M. The Influence of a Vesicular–Abuscular Mycorrhiaza (VAM) and *Aspergillus niger* as Deterrents against *Rhizoctonia solani* in Potatoes. *Sarhad J. Agric.* **1990**, *6*(5), 481–484.

Jain, R. K.; Hasan, N.; Singh, R. K.; Pandey, P. N. Influence of the Endomycorrhizal Fungus, *Glomus fasciculatum* on *Meloidogyne incognita* and *Tylenchorhynchus vulgaris* Infecting Berseem. *Indian J. Nematol.* **1998a**, *28*(1), 48–51.

Jain, R. K.; Hasan, N.; Singh, R. K.; Pandey, P. N. Interaction Between Plant Parasitic Nematodes, *Meloidagyne incognita, Tylenchorhynchus vulgaris* and VAM Fungi, *Glomus fasciculatum* on Lucerne. *Ann. Plant Protect. Sci.* **1998b**, *6*(1), 37–40.

Jalali, B. L.; Chhabra, M. L.; Singh, R. P. Interaction Between Vesicular Arbuscular Mycorrhizal Endophyte and *Macrophomina phaselina* in Mungbean. *Indian Phytopathol.* **1991**, *43*(4), 527–530.

Jayaraman, J.; Kumar, D. VAM Fungi–Pahogen–Fungicide Interactions in Gram. *Indian Phytopathol.* **1995**, *48*(3), 294–299.

Jothi, G.; Mani, M. P.; Rajeshwari, Sundarababu, R. Management of *Meloidogyne incognita* on Okra by Integrating Non-Host and Endomycorrhiza. *Curr. Nematol.* **2000**, *11*(1–2), 25–28.

Kantharaju, V.; Babu, N. P.; Krishnappa, K.; Reddy, B. M. R. Biodiversity in Indigenous Isolates of VAM, *Glomus fasciculatum* against *Meloidogyne incognita* race-1 on Tomato. In : *Proceedings of National Symposium on Biodiversity and Management of Nematodes in Cropping Systems for Sustainable Agriculture, Jaipur, India,* 11–13 Nov., 2002, 2002; pp 9–14.

Kantharaju, V.; Krishnapp, K.; Ravichandra, N. G.; Karuna, K. Management of Root Knot Nematode, *Meloidogyne incognita* on Tomato by Using Indigenous Isolates of AM Fungi, *Glomus fasciculatum. Indian J. Nematol.* **2005**, *35*, 32–36.

Khadge, B. R.; llage, L. L.; Mew, T. W. Interaction Study of *Glomus mosseae* and *Rhizoctonia solani.* In *Trends in Mycorrhizal Research, Proceedings of the National conference on Mycorrhiza;* Jalali, B. L., Hari Chand, Eds.; Haryana Agricultural University: Hisar-125004, India, 1990; pp 94–95.

Kichadi, S. N.; Sreenivasa, M. M. Interaction Effects of *Glomus fasciculatum* and *Trichoderma harzianum* on *Sclerotium rolfsii* in Presence of Biogas Spent Slurry in Tomato. *Karnataka J. Agric. Sci.* **1998**, *11*(2), 419–422.

Kjoller, R.; Rosendahl, S. The Presence of Arbuscular Mycorrhizal Fungus *Glomus intraradices* Influences Enzymatic Activities of the Root Pathogen Aphanomyces *euteiches* in Pea Roots. *Mycorrhiza.* **1996**, *6*(6), 487–491.

Kobayashi, N. Biological Control of Soil Borne Diseases with VAM Fungi and Charcoal Compost. In *Biological Control of Plant Diseases, Proceedings of the International Seminar "Biological Control of Plant Diseases and Virus Vectors*. Food and Fertilizer Technology Centre for the Asian and Pacific Region: Taipei, Taiwan, 1991; pp 153–160.

Kobra, N.; Jalil, K.; Youbert, G. Arbuscular Mycorrhizal Fungi and Biological Control of Verticillium-Wilted Cotton Plants. *Arch. Phytopathol. Plant Protect.* **2011**, *44*(10), 933–942.

Krishna Prasad, K. S. Influence of Vesicular Arbuscular Mycorrhiza on the Development and Reproduction of Root-Knot Nematode Affecting Flue Cured Tobacco. *Agro-Asian J. Nematol.* **1991**, *1*(2), 130–134.

Kucukyumuk, Z.; Ozgonen, H.; Erdal, I.; Eraslan, F. Effect of Zinc and *Glomus intraradices* on Control of *Pythium deliense*, Plant Growth Parameters and Nutrient Concentrations of Cucumber. *Not Bot Horti Agrobo.* **2014**, *42*(1), 138–142.

Kulkarni, S. A.; Kulkarni, S.; Sreenivas, M. N. Interaction Between Vesicular–Arbucular (VA) Mycorrhizae and *Sclerotium rolfsii* Sacc. In Groundnut. *Karnataka J. Agric. Sci.* **1997**, *10*(3), 919–921.

Kumar, C. P. C.; Garibova, I. V.; Vellickanov, L. L.; Durinina, E. P. Biocontrol of Wheat Root Rots Using Mixed Cultures of *Trichoderma viride* and *Glomus epigaeus*. *Indian J. Plant Protect.* **1993**, *21*(2), 145–148.

Kumar, R. Studies on Host-Pathogen-Mycorrhizal Interaction in Chickpea and Mungbean. Ph.D. Dissertation submitted to CCS Haryana Agriculture University Hisar, 1998; pp 185.

Kumar, R.; Jalali, B. L.; Hari Chand. Interaction Between VA-Mycorrhizal Fungi and Soil-Borne Plant Pathogens of Chickpea. *Legume Res.* **2004**, *27*(1), 19–26.

Kumar, R.; Jalali, B. L.; Mehta, S. K. Management of Root Rot (*Rhizoctonia bataticola*) by Mycorrhiza (*Glomus fasciculatum*) in Mungbean. *Ann. Biol.* **2007**, *13*(1), 45–48.

Liu, R. J.; Li, H. F.; Shen, C. Y.; Chiu, W. F. Detection of Pathogenesis-Related Proteins in Cotton Plants. *Physiol. Mol. Plant Pathol.* **1995**, *47*(6), 357–363.

Mahanta, B.; Phukan, P. N. Comparative Efficacy of *Glomus fasciculatum*, Neem Cake and Carbofuran for the Management of *Meloidgyne incognita* on *Vigna mungo*. *Ann. Plant Protect. Sci.* **2004**, *12*(2), 377–379.

Mahanta, B.; Phukan, P. N.; Mahanta, B. Effect of *Glomus fasciculatum* and *Meloidogyne incognita* on Blackgram. *Ann. Plant Protect. Sci.* **2002**, *10*(2), 407–408.

Maisubaro, Y. I.; Tamura, H.; Harada, T. Growth Enhancement and Verticillium wilt Control by Vesicular–Arbuscular Mycorrhizal Fungus Inoculation in Egg Plant. *J. Jpn. Soc. Hortic. Sci.* **1995**, *64*(3), 555–561.

Manila, S.; Nelson, R. Biochemical Changes Induced in Tomato as a Result of Arbuscular Mycorrhizal Fungal Colonization and Tomato Wilt Pathogen Infection. *Asian J. Plant Sci. Res.* **2014**, *4*(1), 62–68.

Masadeh, B.; von Alten, H.; Grunewaldt-Stoecker, G.; Sikora, R. A. Biocontrol of Root Knot Nematodes using the Arbuscular Mycorrhizal Fungus *Glomus intraradices* and the Antagonistic *Trichoderma viridae* in Two Tomato Cultivars Differing in Their Suitability as Hosts for the Nematodes. *J. Plant Dis. Protect.* **2004**, *111*, 322–333.

McAllister, C. B.; Garcia-Romera, I.; Godeas, A.; Ocampo, J. A. Interactions between *Trichderma koningi*, *Fusarium solani* and *Glomus mosseae*: Effects on Plant Growth, Arbucular Mycorrhizas and the Saprophyte Inoculants. *Soil Biol. Boichem.* **1994**, *26*(10), 1363–1367.

McGonigle, T. P.; Miller, M. H. The Inconsistent Effect of Soil Disturbance on Colonisation of Roots by Arbuscular Mycorrhizal Fungi: A Test of The Inoculum Density Hypothesis. *Appl. Soil Ecol.* **2000**, *14*, 147–153.

Mikhaleel, F. T.; Sherief, F. A.; Rizk, R. K.; Abdalla, F. M. Efficacy of *Glomus aggregatum* and *Bacillus subtilis* as Biocontrol Agent for Reducing Fusarium Root Rot In Soyabean Plants. *Egypt. J. Agric. Res.* **2002**, *80*(3), 987–999.

Mishra, A.; Shukla, B. N. Interaction between *Glomus fasciculatum*, *Meloidogyne incognita* and Fungicide in tomato. *Indian J. Mycol. Plant Pathol.* **1996**, *26*(1), 38–44.

Mishra, A.; Shukla, B. N. Interaction between *Glomus fasciculatum* and *Meloidogyne incognita* on Tomato. *J. Mycol. Plant Pathol.* **1997**, *27*(2), 199–202.

Mittal, N.; Sharma, M.; Saxsena, G.; Mukerji, K. G. Effect of VA Mycorrhiza on Gall Formation in Tomato Roots. *Plant Cell Incomp. Newslett.* **1991**, *23*, 39–43.

Mohanty, K. C.; Sahoo, N. K. Prospects of Mycorrhizae as Potential Nematode Antagonist. In *Advances in Nematology*; Parveen Trivedi, C., Ed.; Scientific Publishers: India, 2003; pp 167–180.

Mosse, B. Mycorrhiza in a Sustainable Agriculture. *Biol. Agric. Horticult.* **1986**, *3*, 191–209.

Muchovej, J. J.; Muchovej, R. M. C.; Goncalves, E. J. Effect of Kind and Method of Fungicidal Treatment of Bean Seed on Infection by The VA-Mycorrhizal Fungus *Glomus macrocarpum* and by the Pathogenic Fungus *Fusarium solani*. II. Temporal Spatial Relationships. *Plant Soil.* **1991**, *132*(1), 47–51.

Nagesh, M.; Reddy, P. P.; Kumar, M. V. V.; Nagaraju, B. M. Studies on Correlation between *Glomus fasciculatum* Spore Density, Root Colonization and *Meloidogyne incognita* Infection on *Lycopersicon esculentum*. *Zeitschrift fur Pflanzenkrankheiten und Pflanzenschutz.* **1999**, *106*, 82–87.

Nehra, S. VAM Fungi and Organic Amendments in the Management of *Meloidogyne incognita* Infected Ginger. *J. Indian Bot. Soc.* **2004**, *83*, 90–97.

Niemira, B. A.; Hammerschmidt, R.; Safir, G. R. Post Harvest Suppression of Potato Dry Rot (*Fusarium sambucinum*) in Prenuclear Minitubers by Arbuscular Mycorrhizal Fungal Inoculum. *Am. Potato J.* **1996**, *73*(11), 509–515.

Norouzi K.; Khara J.; Ghosta Y. Arbuscular Mycorrhizal Fungi and Biological Control of Verticillium-Wilted Cotton Plants. *Arch. Phytopathol. Plant Protect.* **2011**, *44*(10), 933–942.

Ocampo, J. A.; Martin, J.; Hayman, D. S. Influence on Plant Interactions on Vesicular Arbuscular Mycorrhizal Infections. I. Host and Non-Host Plants Grown Together. *New Phytopathol.* **1980**, *84*, 27–35.

Osman, H. A.; Korayem, A. M.; Ameen, H. H.; Badr-Eldin, S. M. S. Interaction of Root-Knot Nematode and Mycorrhizal Fungi on Common Bean (*Phaseolus vulgaris* L.). B, *Eitrage Zur Tropischen Landwirtchhaft und Veterinarmedizin.* **1991**, *29*(3), 341–346.

Osman, H. A.; Korayem, A. M.; Ameen, H. H.; Badr-Eldin, S. M. S. Interaction of Root-Knot Nematode and Mycorrhizal Fungi on Common Bean (*Phaseolus vulgaris* L.). *Anzeiger fur Schadlingskunde Pflanzenschutz Umweltschutz.* **1990**, *63*(1), 129–131.

Oyekanmi, E. O.; Coyne, D. L.; Fagad, O. E.; Osonubi, O. Improving Root Knot Nematode on Two Soyabean Genotypes Through The Application of *Bradyrhizobium japonicum*, *Trichoderma psedokoningii* and *Glomus mosseae* in Full Factorial Combinations. *Crop Protect.* **2007**, *26*(7), 1006–1012.

Ozgonen, H.; Erkilic, A. Growth Enhancement and Phytophthora Blight (*Phytophthora capsici* L.) Control by Arbuscular Mycorrhizal Fungal Inoculation in Pepper. *Crop Protect.* **2007**, *26*, 1682–1688.

Ozgonen, H.; Bicici, M.; Erkilic, A. The Effect of Salicylic Acid and Endomycorrhizal Fungus G. *intraradices* on Plant Development of Tomato and Fusarium Wilt Caused by *Fusarium oxysporum* f. spp. *lycopersici*. *Turky J. Agricult. Forage.* **1999**, *25*, 25–29.

Pandey, R. K.; Goswami, B. K.; Satyendra, S. Management of Root Knot Nematode and Fusarium Wilt Disease Complex by Fungal Bioagents, Neem Oilseed Cake and/or VA Mycorrhiza On Chickpea. *Int. Chickpea Pigeonpea Newslett.* **2005**, *12*, 32–34.

Pathak, D. V.; Singh, S.; Saini, R. S.; Sharma, J.R. Impact of Bio-Inoculants on Seed Germination and Plant Growth of Guava (*Psidium guajava* L.), *Haryana. J. Horticult. Sci.* **2009**, *38*, 26–28.

Pinochet, J.; Calvet, C.; Camrubi, A.; Fernandez, C. Interactions between Migratory Endoparasitic Nematodes and Arbuscular Mycorrhizal Fungi in Perennial Crops: A Review. *Plant Soil.* **1996**, *185*(2), 183–190.

Pradhan, A.; Ganguly, A. K.; Singh, C. S. Influence of *Glomus fasciculatum* on Meloidogyne Incognita Infected Tomato. *Ann. Plant Protect. Sci.* **2003**, *11*, 346–348.

Quarles, W. Plant Disease Biocontrol and VAM Fungi. *IPM Practitioner.* **1999**, *21*(4), 1–9.

Rabie, G. H. Indution of Fungal Disease Resistance in *Vicia faba* by Dual Inoculation with *Rhizobium leguminosarum* and Vesicular–Arbuscular Mycorrhizal Fungi. *Mycopathologia.* **1998**, *141*(3), 159–166.

Rao, M. S.; Gowen, S. R. Bio-Management of *Meloidogyne incognita* on Tomato by Integrating *Glomus deserticola* and *Pasteuria penetrans. Zeitschrift Fur Pfulnzenkrans kheiten und Pflanzenschutz.* **1998**, *105*(1), 49–52.

Rao, M. S.; Reddy, P. P.; Mohandas, M. S. Bio-intensive Management of *Meloidogyne incognita* on Egg Plant by Integrating *Paecilomyces liiacinus* and *Glomus mosseae. Nematologia Mediterranea.* **1998a**, *26*(2), 213–216.

Rao, M. S.; Reddy, P. P.; Sukhada, M.; Nagesh, M.; Pankaj. Management of Root-Knot Nematode on Egg Plant by Integrating Endomycorrhiza (*Glomus fasciculatum*) and Castor (*Ricinus communis*) Cake. *Nematol. Mediterranea.* **1998b**, *26*(2), 217–219.

Rao, M. S.; Somasekhar, N.; Nagesh, M. Management of Root-Knot Nematodes, *Meloidogyne incognita* in Tomato Nursery by Integration of Endomycorrhiza, *Glomus fasciculatum* with Caster Cake. *Pest Manage. Horticult. Ecosyst.* **1997**, *3*(1), 31–35.

Rao, V. K.; Krishnappa, K. Integrated Management of *Meloidogyne incognita Fusarium oxysporum f.* spp. *ciceri* Wilt Disease Complex in Chickpea. *Int. J. Pest Manage.* **1995**, *41*(4), 234–237.

Reddy, B. N.; Raghavender, C. R.; Sreevani, A. Approach for Enhancing Mycorrhiza-Mediated Disease Resistance of Tomato Damping off. *Indian Phytopathol.* **2006**, *59*(3), 299–304.

Reddy, P. P.; Nagesh, M.; Divappa, V.; Kumar, M. V. V. Management of *Meloidogyne incognita* on Tomato by Integrating Endomycorrhiza, *Glomus mosseae* with Oil Cakes under Nursery and Field Condition. *Zeitschrift fur Pfaunzenkrantkeiten und Pflanzenschutz.* **1998**, *105*(1), 53–57.

Redecker, D.; Morton, J. B.; Bruns, T. D. Ancestral Lineages of Arbuscular Mycorrhizal Fungi (*Glomales*). *Mol. Phylogeny Evol.* **2000**, *14*, 276–284.

Rivas Platero, G. G.; Andrade, J. C. Interaction of Mycorrhizal Fungi and *Meloidogyne exigua* in Coffee. *Manejo Integrado de Plagas.* **1998**, *49*, 68–72.

Rivas Platero, G. G.; Miranda, T. R.; Andrade, J. C. Vesicular Arbuscular Fungus, *Glomus* spp. Integration with *Meloidogyne arabicida* in Tomato. *Manejo Integrado de Plagas.* **1998**, *47*, 41–43.

Rosendahl, C. N.; Rosendahl, S. The Role of Vesicular Arbuscular Mycorrhiza in Controlling Damping-Off and Growth Reduction in Cucumber caused by *Pythium ultimum*. *Symbiosis.* **1990**, *9*(1–3), 363–366.

Sahay, N. S.; Sudha, Singh, A.; Verma, A. Trends in Endomycorrhizal Research. *Indian J. Exp. Biol.* **1998**, *36*(11), 1069–1086.

Saleh, H. M.; Sikora, R. A. Effect of Quintozen, Benomyl and Carbendazim on the Interaction Between the Endomycorrhizal Fungus *Glomus fasciculatum* and the Root-Knot Nematode *Meloidogyne incognita* on Cotton. *Nematologica.* **1989**, *34*(4), 432–442.

Sankaranarayanan, C.; Sundarababu, R. Effect of *Rhizobium* on the Interaction of Vesicular–Arbuscular Mycorrhizae and Root-Knot Nematode on Blackgram (*Vigna mungo*). *Nematol. Mediterranea.* **1998**, *26*(2), 195–198.

Sankaranarayanan, C.; Sundarababu, R. Effect of Oil Cakes and Nematicides on the Growth of Blackgram (*Vigna mungo*) Inoculated with VAM Fungus (*Glomus fasciculatum*) and Root-Knot Nematode (*Meloidogyne incognita*). *Indian J. Nematol.* **1997a**, *26*(2), 144–147.

Sankaranarayanan, C.; Sundarababu, R. Effect of Leaf Extracts on Growth of Blackgram Inoculated With Vesicular Arbuscular Mycorrhiza (*Glomus fasciculatum*) and Root Knot Nematode (*Meloidogyne incognita*). *Indian J. Nematol.* **1997b**, *27*(1), 128–130.

Santhi, A.; Sundarababu, R. Effect of Phosphorus on the Interaction of Vesicular–Arbuscular Mycorrhizal Fungi with *Meloidogyne incognita* on Cowpea. *Nematol. Mediterranea.* **1995a**, *23*(2), 263–265.

Santhi, A.; Sundarababu, R. Effect of Three Species of VAM viz *Glomus fasciculatum, G. versiforme, G. etunicatum* and Root Knot Nematode *Meloidogyne incognita* on Cowpea Growing in Different Types of Soil. *Int. J. Trop. Plant Dis.* **1995b**, *13*(1), 63–68.

Schussler A (2005). http://www.tudarmstadt.de/fb/bio/bot/schuessler/amphylo/amphylogeny. html (accessed 19-Oct-2005).

Schussler, A.; Schwarzott, D.; Walker, C. A New Fungal Phylum, the Glomeromycota, Phylogeny And Evolution. *Mycol. Res.* **2001**, *105*, 1413–1421.

Sharma, A. K.; Johri, B. N. Arbuscular Mycorrhizae, Interactions in Plants, Rhizosphere and Soils. Oxford IBH Publication Co: New Delhi, 2002; pp. 311.

Sharma, H.; Pankaj, K.; Mishra, S. D. Effect of Plant Growth Promoter Microbes on the Root-Knot Nematode *Meloidogyne incognita* on Okra. *Curr. Nematol.* **2003**, *14*(1/2), 57–60.

Sharma, W.; Trivedi, P. C. Cocomitant Effect of *Paecilomyces lilacinus* and Vesicular–Arbuscular Mycorrhizal Fungi on Root-Knot Nematode Infected Okra. *Ann. Plant Protect. Sci.* **1997**, *5*, 70–74.

Shreenivasa, K. R.; Krishnappa, K.; Ravindandra, N. G. Interaction Effects of Arbuscular Mycorrhizal Fungus *Glomus fasciculatum* and Root Knot Nematode, *Meloidogyne incognita* on Growth and Phosphorus Uptake of Tomato. *Karnataka J. Agric. Sci.* **2007**, *20*, 57–61.

Siddiqui, Z. A.; Akhtar, M. S. Effects of AM Fungi and Organic Fertilizers on the Reproduction of the Nematode *Meloidogyne incognita* and On The Growth and Water Loss of Tomato. *Biol. Fertility Soils.* **2007**, *43*, 603–609.

Siddiqui, Z. A.; Akhtar, M. S. Synergistic Effect of Antagonistic Fungi and a Plant Growth Promoting Rhizobacterium, An Arbuscular Mycorrhizal Fungus, or Composed Cow Manure on the Populations of *Meloidogyne incognita* and Growth of Tomato. *Biol. Sci. Technol.* **2008**, *18*, 279–290.

Siddiqui, Z. A.; Mahmood, I. Some Observations on the Management of Wilt Disease Complex of Pigeonpea by Treatment With a Vesicular Arbuscular Fungus and Biocontrol Agents for Nematodes. *Bioresour. Technol.* **1995**, *54*(3), 227–230.

Siddiqui, Z. A.; Mahmood, I. Biological Control of *Heterodera cajani* and *Fusarium udum* on Pigeonpea by *Glomus mosseae, Trichoderma harzianum* and *Vericillium hamydosporium. Israel J. Plant Sci.* **1996**, *44*(1), 49–56.

Siddiqui, Z. A.; Singh, L. P. Effects of Soil Inoculants on the Growth, Transpiration and Wilt Disease of Chickpea. *J. Plant Dis. Protect.* **2004**, *111*, 151–157.

Siddiqui, Z. A.; Singh, L. P. Effects of Fly Ash and Soil Micro-Organisms on Plant Growth, Photosynthetic Pigments and Leaf Blight of Wheat. *J. Plant Dis. Protect.* **2005**, *112*, 146–155.

Singh, P. K.; Singh, M.; Agnihotri, V. K.; Vyas, D. Arbuscular Mycorrhizal Fungi: Biocontrol against *Fusarium* Wilt of Chickpea. *Int. J. Sci. Res. Pub.* **2013**, *3*(1), 1–5.

Singh, R. S.; Singh, D. Effect of *Glomus aggregatum* Inoculation and P-Amendments on *Fusarium oxysporum f.* spp. *ciceri* Causing Chickpea Wilt. In *Mycorrhizae: Biofertilizers for the Future*; Adholeya, A., Singh, S., Eds.; TERI: New Delhi, 1995; pp 119–123.

Sivaprasad, P.; Suneetha, S.; Joseph, P. J.; Sulochana, K. K.; Rajmohan, K. Influence of Arbuscular Mycorrhizal Fungi (AMF) on Establishment and Growth of Tissue Culture Plantlets of Banana and Alocasia. In *National Conference on Mycorrhiza,* Barkatulla University, Bhopal, 1999; pp 12.

Sivaprasad, R.; Jacob, A.; Sulochana, K. K.; Visalak-Syhy, A.; George, B. Growth, Root-Knot Nematode Infestation and Phosphorus Nutrition in *Piper nigrum* (L.) as Influenced by Vesicular Arbucular Mycorrhizae. In Proceedings of the 3rd International Conference on Plant Protection in the Tropics, Genting Highlands, Malaysia, (Eds. F.A.C. Ooi., G.S. Lim and P.S. Teng), Kuala Lumpur, Malaysia, *Malaysian Plant Protect. Soc.* 1992, *6*, 34–37.

Smith, S. E.; Read, D. J. *Mycorrhizal Symbiosis*, 2nd edn, Academic: London, 1997.

Sreenivasa, M. N.; Bagyaraj, D. J. Use of Pesticides for Mass Production of Vesicular–Arbuscular Mycorrhizal Inoculum. *Plants Soil.* **1989**, *119*, 127–132.

Sreenivasa, M. N. Biological Deterrent Activities of VA Mycorrhiza and *Trichoderma harzianum* on *Sclerotium rolfsii* at Different P-Levels in Chilli. *Environ. Ecol.* **1994**, *12*(2), 319–321.

Sreenivasa, M. N.; Nirmalnath, P. J.; Kulkarni, S. Interaction Between VA-Mycorrhizal Fungi and *Sclerotium rolfsii* in Chilli (*Capsicum annum* L.). *Zentralblatt fur Mikrobiologie.* **1992**, *147*(8), 509–512.

Subhashini, D. V. The Role of VA Mycorrhiza in Controlling Certain Root Disease of Tobacco. In *Trends in Mycorrhizal Research. Proceedings of the National Conference on Mycorrhiza*; Jalali, B. L., Hari Chand Eds.; Haryana Agricultural University: Hisar, 1990; pp 102.

Sundarababu, R.; Sankaranarayanan, C.; Vadivelu, S. Interaction of Mycorrhiza Species with *Meloidogyne incognita* on Tomato. *Indian J. Nematol.* **1993a**, *23*(1), 121–123.

Sundarababu, R.; Sankaranayanan, C.; Santhi, A. Interaction Between Vesicular–Arbucular Mycrrhiza and *Meloidogyne incognita* on Tomato as Influenced by Time of Inoculation. *Indian J. Nematol.* **1993b**, *23*(1), 125–126.

Sundaresan, P.; Raja, N. U.; Gunasekaran, P. Interaction and Accumulation of Phytoalexins in Cowpea Roots Infected with a Mycorrhizal Fungus *Glomus fasciculatum* and Their Resistance to Fusarium Wilt Disease. *J. Biosci.* **1993**, *18*(2), 291–301.

Tahat, M. M.; Sijam, K.; Othman, R. Mycorrhizal Fungi as a Biocontrol Agent. *Plant Pathol. J.* **2010**, *9*(4), 198–207.

Thomas, L.; Mallesha, B. C.; Bagyaraj, D. J. Biological Control of Damping-Off of Cardamom by the VA Mycorrhizal Fungus, *Glomus fasciculatum. Microbiol. Res.* **1994**, *149*(4), 413–417.

Thygesen, K.; Larsen, J.; Bodker, L. Arbuscular Mycorrhizal Fungi Reduce Development of Pea Root-Rot Caused by *Aphanomyces euteiches* Using Oospores as Pathogen Inoculum. *Eur. J. Plant Pathol.* **2004**, *110*, 411–419.

Torres-Barragan, A.; Zavaleta-Mejia, E.; Gonzalez-Chavez, C.; Ferrera-Cerrato, R. The Use of Arbuscular Mycorrhizae to Control Onion White Rot (*Sclerotium cepivorum* Berk.) under Fields Conditions. *Mycorrhiza.* **1996**, *6*(4), 253–257.

Toshi, L.; Giovannetti, M.; Zazzerini, A.; Sbrana, C. Interactions between *Plasmopara helianthi* and Arbuscular Mycorrhizal Fungi in Sunflower Seedling Susceptible and Resistant to Downy Mildew. *Phytopathol. Mediterranea.* **1993**, *32*(2), 106–114.

Vaast, P. Interaction of Endomycorrhiza of *Arabica coffee* and the Chemical Control of Nematodes *Pratylenchus caffeae* and *Meloidogyne konaensis*. In *Dix-Septieme colleque Scientifique International sur le Café, Naiobi.* Association Scientifique Internationale du cafe: Paris, France, 1997; pp 564–571.

Vaast, P.; Caswell-Chen, E. P.; Zasoski, R. J. Influences of a Root-Lesion Nematode, *Pratylenchus coffeae*, and Two Arbuscular Mycorrhizal Fungi, *Acaulospora mellea* and *Glomus clarum* on coffee (*Coffea arabica* L.). *Biol. Fertility Soils.* **1998**, *26*(2), 130–135.

Veerabhadrasamy, A. L.; Garampalli, R. H. Effect of Arbuscular Mycorrhizal Fungi in the Management of Black Bundle Disease of Maize Caused by *Cephalosporium acremonium*. *Sci. Res. Rep.* **2011**, *1*(12), 96–100.

Vidhyasekaran, P. *Encyclopedia of Plant Pathology.* Viva Books Private Limited: New Delhi, 2004, pp 619.

Vyas, S. C. Arbuncular Mycorrhizal Fungi and Agricultural Interaction in Sustainable Agriculture. In *Proceedings National Symposium on Sustainable Agriculture in Subhumid Zone*; Dasgupta, M. K., Ghosh, D. C., Das Gupta, D., Majumdar, D. K., Chattopadhyay, G. N., Ganguli, P. K., Munsi, P. S., Bhatta-Charya, D., Eds.; Institute of Agriculture: Sriniketan, West Bengal, 1995; pp 376–380.

Wood, T.; Cummings, B. Biotechnology and The Future of VAM Commercialization. In *Mycorrhizal Functioning: An Integrative Plant Fungal Process*; Allen, M. F., Ed.; Chapman and Hall: London, U.K., 1992 pp 468–487.

Yanling, W.; Zhengjia, H. Effect of VA Mycorrhiza on Nematode Infection in Tomato. *J. Hauzhong Agric. Univ.* **2000**, *19*(1), 25–28.

CHAPTER 9

SIGNIFICANCE OF MICROBIAL BIOAGENTS IN SUSTAINABLE AGRO-ECOSYSTEM: AN OVERVIEW

SATISH K. SAIN and ABHAY K. PANDEY*

Plant Health Management Division, National Institute of Plant Health Management, Department of Agriculture and Cooperation, Ministry of Agriculture, Government of India, Rajendranagar, Hyderabad 500030, Telangana, India

Corresponding author. E-mail: abhaykumarpandey.ku@gmail.com

CONTENTS

ABSTRACT

Agriculture is the lifeline of farmers; however, crop damage by various pests during plantation/vegetative stage affects the production yield. To get the sustainable agriculture, microbial control is an alternative approach for the management of diseases and insect-pests. In biological control, antagonists and entomopathogens interactions may be helpful to select more fruitful microbial bioagents. The mechanisms involved between entomopathogens and antagonists are antibiosis and competition; mycoparasitism, cell wall degrading enzymes, and induced resistance and hypovirulence/crossprotection are described in the present chapter. Further, the role of different microbial agents such as *Trichoderma, Pseudomonas, Verticillium (Lecanicillium), Bauveria, Noumeria, Metarrhizium, Paecilomyces, Hirsutella, Baculoviruses, Bacillius, Glomus,* and *Entomopathogenic nematodes* in the management of different diseases and insect-pests has been elaborated here. The finding of the scholars revealed that the future view of microbes-based control of the diseases and insect-pests is bright and promising with the growing demand for microbial products among the growers. It is possible to use the biological control as an effective strategy to manage the plant diseases and insect-pests, increase yield, and protect the environment, and bioresources for sustainable agro-ecosystem.

9.1   INTRODUCTION

Agriculture and its allied sectors are indisputably the biggest livelihood providers in India, particularly in the vast rural areas. For the holistic rural development, sustainable agriculture, in terms of food security, rural employment, and eco-friendly technologies like soil conservation, natural resource management, and biodiversity protection, is essential. From time to time, Indian agriculture and allied activities have witnessed green, white, yellow, and blue revolution. However, crop productions for human consumption are at risk due to several diseases and insect-pests. Crop protection measures are regarded as helpful means for the reduction of incidence of these crop losses (Oerke, 2006). There are several tools of crop protection which have been implemented together for the management of such pests including cultural, mechanical, physical, biological, and chemical. These tools together have been compiled and used in the form integrated pest management approach. However, the indiscriminate use of chemical pesticides has created several complications, like pesticide resistance, pest resurgence, toxic residues on food, water, air and soil, abolition of natural enemies, and ecosystem disruption (De Waard et al., 1993). On the other hand, microbial agents such as microbial pesticides offer a very good alternative to manage the insect-pests and diseases in an eco-friendly approach. Because, mostly they occur naturally, have high specificity to target pests, no or little adverse effect on beneficial insects, resistance development to them is slow or less common, they have no unknown environmental hazards, have less residual activity and are effective against insecticide/fungicide resistance species of insects. The role microbial pesticides are regarded as potential and reliable tools in Integrated Pest Management Programme (IPM) to manage pests due to the aforesaid reasons. It is the durable and self-sustaining cure process for managing invasive plants. The present chapter describes the management of fungal, bacterial pathogens, plant parasitic nematodes, and insect-pests by application of microbial bioagents.

9.2 MICROBIAL BIOAGENTS

Biological control or microbial biocontrol is the use of invasive or native natural enemies like parasitoids and predators, parasites, and pathogens to reduce the population of pests below a desired level. In another word, biological control is the process in which the pests are controlled by using other living organisms, and microbes involve in such type of controlling mechanism are called microbial bioagents. These microbial bioagents include bacteria, fungi, virus, and nematodes. Their antagonistic activity against pests and pathogens has led to their mass production which is being used in both protected and field crops for managing the various diseases and crop pests. Their mode of action is found to be extremely varied which directly start a toxic biological process or suppress the pathogens by competition or parasitism. Sometimes, in the plants, they induce resistance factors. Yet microbial bioagents can be used in several field and orchard crops; however, they are not extensively used on grapevine, despite their large potential as a replacement for chemical pesticides recognize to their low impact on the environmental their human health safety, and, very relevant, the fact they do not induce pesticide resistance. The most successful microbe is *Bacillus thuringiensis* (Bt)—a bacterium which is found to act as potential insecticide because of its toxin producing nature and *Ampelomyces quisqualis*, a fungal antagonist which is active against the harmful group of pathogens, powdery mildew causing agents. Similarly, *Trichoderma* and *Pseudomonas* are found to be effective in management of various soil-borne diseases as described in this chapter.

9.3   NEED OF MICROBIAL BIOAGENTS IN SUSTAINABLE AGRO-ECOSYSTEM

Beyond good agricultural practices, growers are found to be totally depending on chemical fertilizers and pesticides. However, indiscriminate use and misuse of commercial pesticides has led to significant changes in people's approaches toward the use of safer and eco-friendly pesticides in agriculture system. This has happened due to its nonbiodegradability, side effects on the environment, and beneficial organisms. The increase in the population of insect-pests and diseases resulted in the increased use of toxic chemicals for their management. Ultimately, because of indiscriminate use of chemical pesticides, the number of species resistant to pesticides and fungicides has increased. In recent years, after validation of the general agreement of trade and tariff of World Trade Organization, more emphasis is given to the use of eco-friendly pesticides for crop production in view of their least toxic nature, low levels of disease resistance, and low residue problems. However, biological controls should be implemented with other pest management methods because under different conditions, different methods are effective at different times.

9.4   TYPES OF INTERACTIONS CONTRIBUTING TO BIOCONTROL AGENTS

Studying the phenomenon of biological control in plant diseases through the interactions between antagonists and pathogens may allow us to select the more potent microbial agents and to manipulate it in the soil environment to create favorable conditions for fruitful biocontrol. In the past few decades, number of studies has been carried out to know the molecular mechanism of bioagents, particularly in plant disease control. Predominantly, molecular approaches are useful in determining the relative contributions of different genetic traits in complicated processes. These mechanism of microbial biocontrol may involve and be divided into antibiosis, competition, myco-parasitism, cell-wall-degrading enzymes, induced resistance, and hypoviru-lence/cross protection. However, these mechanisms of biological control are probably never mutually exclusive. The basic mechanisms are described in the following sections.

9.4.1 COMPETITION

To proliferate and survive in the natural habitats, microorganisms are found to compete for several components like space, nutrients, minerals, etc. for their growth and multiplication in both rhizosphere as well as phyllosphere. This mechanism plays a very important role in the microbial management of *Fusarium, Rhizoctonia, Sclerotium, Phytophthora*, and *Pythium* by some strains of fluorescent *Pseudomonas*. Fungi having more mass of mycelial growth and having highest number of propagules have found to be more advantageous in competitive mechanisms. The microbial agents have a more effectual uptake system for the substance than do the pathogens (Nelson, 1990). For example, in alkaline soils, iron competition may be a limiting factor for microbial growth. Some bacteria, especially fluorescent *Pseudomonads* produce siderophores that have very high affinities for iron and deprive the pathogens from iron nutrition, thus resulting in decrease in diseases and increase of crop yields (Loper and Buyer, 1991).

9.4.2 ANTIBIOSIS

Antibiosis plays a vital role in biological control. In this way of mechanism, the antagonism is facilitated by certain enzymes, lytic agents, metabolites, volatile compounds, or other toxic substances. In the antibiosis, the metabolites are secreted by underground parts of plants, soil microorganism, plant residues, etc. It happens when the pathogen is inhibited or killed by metabolic products of the antagonists (lytic agents, enzymes, volatile compounds, etc.) (Sharma et al., 2013). For example, chaetomin is produced by *Chaetomium globosum*, peptaibols are produced by *Trichoderma harzianum*, and pyrones are produced by *Trichoderma* spp. (Schirmbock et al., 1994). Gliotoxin produced by *Trichoderma virens* is found to inhibit *Pythium* a causal agent of damping-off of cotton seedlings (Wilhite et al., 1994). Among bacteria, *Agrobacterium radiobacter* strain K84 is found to produce antibiotic agrocin 84 which is one of the best described examples of microbial control to control crown gall caused by virulent *Agrobacterium tumefaciens* strains (Kerr, 1989).

9.4.3 MYCOPARASITISM/HYPERPARASITISM

When one fungus is parasitic on the other fungus, the fact mycoparasitism occurs. The parasitized fungus is regarded as hypoparasite, whereas the parasiting ones are hyperparasite. During this mechanism, while invading pathogens, the antagonist secretes some enzymes like chitinases, celluloses, glucanases, and lytic enzymes. There are many mycoparasites occurring on a wide range of plant pathogenic fungi and some of them are playing a noteworthy role in disease control (Lo et al., 1998), for example, *Sphaerellopsis filum* is parasite of rust fungi, *Puccinia*, and *Uromyces* (Sundheim and Tronsmo, 1988). Weindling (1932) observed *Trichoderma viride* parasitizing on the hyphae of *Rhizoctonia solani* and suggested that inoculating soil with *Trichoderma* spores is found to control damping off of citrus seedling (Weindling and Fawcett, 1936). Similarly, cellwall-degrading enzymes have also been shown to be involved in the inhibition of pathogenic fungi. The gene(s) for some of those enzymes have been isolated and identified (Harman et al., 1993).

9.4.4 CELL-WALL DEGRADING ENZYMES: LYTIC ENZYMES

The complete or partial damage of a cell by enzymes is regarded as lysis, and this is of two types, endolysis and exolysis. Destruction of the cell wall does not occur in endolysis, whereas in exolysis (heterolysis), destruction of cell is found to occur by the enzymes chitinases, cellulases, etc. Microbes are also found to produce the extracellular hydrolytic enzymes which have a key role in the inhibition of plant pathogenic fungi. Researchers have also demonstrated the role of chitinase or ß-1,3-glucanase in in-vitro lysis of fungal cell walls (Harman et al., 1993). Genetic proof regarding the significance of the aforesaid enzymes in microbial control has showed that a chitinase (ChiA) deficient mutant of *Serratia marcescens* will have less inhibitory activity against fungal germ tube elongation and also reduced microbial control of pea *Fusarium* wilt seedling (Lam and Gaffney, 1993). Similarly, several species of transgenic plants containing the gene for endochitinase from *T. harzianum* have been investigated (Harman et al., 1994). In these, transgenic plants increased resistance against plant pathogenic fungi have been recorded (Lorito et al., 1993) which indicates that these enzymes have a significant role in the microbial control, and this ability of some microbes may further be improved by alteration with chitinolytic enzymes.

Entomopathogenic fungi have difference mechanism of pathogenicity than viruses and bacteria because the entomopathogenic fungi infect insects by penetrating the insect cuticle (Richard et al., 2010). They degrade the protein, chitin, and lipids of the host cuticle and allow penetration of fungal hyphae (Cho et al., 2006). Moreover exoenzymes, during in-vitro and in-vivo conditions the *Beauveria bassiana* is reported to secrete some toxin proteins and metabolites. These types of protein include Cyclosporins A and C as well as cyclic peptides. The toxins such as beauvericin, enniatins, oosporein, and bassianolide produced by *B. bassiana* have potent insecticidal properties (Vey et al., 2001), Bassiacridin is another toxin produced by *B. bassiana* has significant toxic activity against pests (Enrique and Alain, 2004). However, the main issue which limits the market share of these entomopathogenic fungi is relating to the mycoinsecticides property in relation to the direct proportional to insect killing speed (St Leger and Wang, 2009). Based on this method, to explain the most relevant cuticle degrading enzymes, several schools of thought on entomopathogenic fungi virulence factor have been studied (St Leger and Wang, 2009) because in engineered strains, over-expression occur in more fungi that are more poisonous toward insects (Fang et al., 2005).

9.4.5 INDUCED SYSTEMIC RESISTANCE (ISR)

Induced systemic resistance (ISR) is the capacity of a microbial agent to induce the defense mechanisms in the plants which led to occurrence of systemic resistance to a number of pathogens. ISR in plants occurs when the plants are inoculated with weak pathogens, nonpathogens, or microbial agents. Among these, the microbial agents are found to play an important role in ISR by stimulating the physical and mechanical strength of the host plant cell wall. These microbial agents are also found to change the physiological and biochemical reactions of host during which synthesis of defense chemicals/metabolites against target pathogens is found to occur. Several pathogens related (PR) proteins (chitinase, ß-1,3-glucanse), chalcone synthase, phenylalanine ammonialyase, peroxidase, phenolics, callose, lignin, and phytoalexins are found to accumulate during defense reaction in the host plant. In the management of several plant diseases by inducing plant systemic resistance, many strains of plant growth promoting rhizobacteria (PGPR) have been found to show their potential nature (Alstrom, 1995). Maurhofer et al. (1994) reported that strain CHA96 from *Trichoderma* species could induce PR-proteins (e.g., endochitinases and ß-1,3-glucanases) in the intercellular fluid of leaves and thus could increase resistance to Tobacco Necrosis Virus (Liu et al., 1995). In addition, against the *Fusarium* wilt in radish, the lipopolysaccharide with the Oantigenicside chain produced by strain WCS374 of *Pseudomonas fluorescens* is involved for initiation of systemic resistance (Leeman et al., 1996). Similarly, *Trichoderma* species are also found to induce systemic resistance in cauliflower and tomato against *Sclerotinia sclerotiorum* a stalk rot pathogen (Sharma and Sain, 2004) and against downy mildew pathogen in cauliflower (Sharma et al., 2004).

9.4.6 PLANT GROWTH PROMOTION

Microbial biocontrol agents also produce growth hormones like, auxins, cytokinin, gibberellins, etc. These hormones suppress the deleterious pathogens and promote the growth of plants and simultaneously increase the yield. The studies on mechanism of growth promotion indicates that PGPR promotes plant growth and vigor directly by producing growth regulators or indirectly by stimulating nutrient uptake, by producing siderophores or antibiotics to protect plant from soil-borne pathogens or deleterious rhizosphere microorganisms. *Pseudomonas* spp. is reported to increase plant growth by producing gibberellins like substances, mineralizing phosphates. PGPR like *Pseudomonas* spp. and *Bacillus* spp. have been reported to enhance plant growth in cauliflower, cabbage, cluster bean, soybean, etc. (Sain, 2010; Sain and Gour, 2009, Sain et al., 2007). Similarly, *Trichoderma* species enhance rate of seed germination, plant vigor and yield in several crops (Sharma and Sain 2004; Sharma et al., 2004).

9.4.7 HYPOVIRULENCE/CROSS PROTECTION

The most mycoviruses (mostly dsRNA (ribonucleic acid) viruses) prime "secret lives;" some of the dsRNAs reduce the plant pathogenic ability of their fungal hosts (plant pathogens). As the plant pathogens are attacked with dsRNAs, their disease-causing ability will be reduced, and this feature is known as hypovirulence. The hypovirulence property of some fungal isolates has attracted attention for managing several fungal diseases in crop plants. However, presently, a little or limited strategies are available for the control of these diseases ecologically (Nuss, 2005). Features of three interacting-trophic levels (virus, fungus, and tree) and the environment are determining aspects for the success or failure of hypovirulence strategy. Fungal vegetative incompatibility constrains virus transmission, but this factor alone is a poor predictor of biological control. For the use of hypovirulance, the best example is in controlling the chestnut blight fungus, *Cryphonectria parasitica*, in Europe and in Michigan in the United States. However, epidemiological dynamics of the hypovirulance system to determine the crucial factors regulating the establishment of hypovirulence in chestnut forests need to be studied further (Milgroom and Cortesi, 2004). However, the recent development of an infectious cDNA (complementary DNA)-based reverse genetics system for members of the Hypoviridae mycovirus family has enabled the analysis of basic aspects of this fascinating virus–fungus–plant interaction, including virus–host interactions, the mechanisms underlying fungal pathogenesis, fungal signaling pathways, and the evolution of RNA silencing. Such systems also provide a means for engineering mycoviruses for enhanced biocontrol potential (Nuss, 2005). In 1929, the mechanism of crossprotection was first defined by H. H. McKinney to have prospective for biological control of plant viral diseases. In Brazil, citrus tristeza virus disease was controlled by inoculating the citrus trees with a mild strain of the same virus, which then protects the trees against the more severe strains (Costa and Muller, 1980). Citrus tristeza virus (CTV) strains have also been found to be effective in recovering the shoots and rootstock of sour orange (Wallace and Drake, 1976). Moreover, two strains of cucumber mosaic virus (CMV), that is, CMV-S and CMV-P, were found to differ in three features of value, that is, one strain was found as milder (CMV-S) than the other (CMV-P) in tobacco, tomato, and squash. The mild strain protected plants of all these three hosts from second strain and prevented the visions accumulation and the challenge strain of dsRNAs (Dodds et al., 1985).

9.5 APPLICATION OF MICROBIAL BIOAGENTS

Microbial pesticides are applied through different way including the most common like seed, seedling treatment, and soil application. However, there are some conditions which should be fulfilled during the application of microbial to achieve the successful results:

- Legally registration is required for all microbial biopesticides at least at country level for each crop against each pest. However, biopesticides produced by farmers for own consumption at his on-farm production unit, registration is not required.
- Farmers and advisers should be trained in the appropriate application techniques for microbial agents.
- There should be availability of efficiently products with aggressive/efficacious strain.
- There should be availability of efficient strains selected against the main pathogens and pests.

- Suitable ecological conditions should be indicated for application.
- The appropriate soil and ecosystem management should be applied for better results.
- For commercialization, proper registration procedures and regulations should be followed.

9.6 FORMULATION CHALLENGES

From the lab-to-land conditions, the microbial bioagents that have successfully made the complex transition, the record for transfer of bio-control agents from on-station field experiments to on-farm use, have showed extremely poor results. Most of the bioagents that perform excellently under controlled conditions exhibit little or no effect under field conditions. For the microbial agents, a major difficulty appears when their formulated products are applied in large scale for the disease management without much research work on their field performance. Possibility of formulation of inoculum occur under the experimental conditions within 24–48 h of planting, whereas microbial bioagents must survive better during transported on seeds or in formulated packages. Future research should have a vital role in developing the formulation of microbial bioagents to facilitate the better transportation and storage quality. In the bioformulation, a gain has been made by supplementing the stabilized nutrients for longer shelf life of microbes in their resting stages, but more research is needed in this area to explore the acre age and crop types in which microbial pesticide can be used dependably.

9.7 MICROBIAL BIOAGENTS AGAINST IN DISEASE AND PESTS CONTROL

Biological control of diseases and insect-pests by the use of fungal antagonists and entomopathogens, respectively, can be fruitfully utilized particularly within the frame task of integrated pest management. The use of microbial bioagents offer an ecologically sound and effective method for managing the pest problems because they are not harmful to the ecosystem as well as for human health. Some of the most commonly used microbial bioagents species of *Trichoderma*, *Pseudomonas*, *Bacillus*, etc. are contributing an important role in sustainable agriculture are describe here.

9.7.1   TRICHODERMA

Trichoderma spp. is used for biocontrol of several plant pathogens. The potential use of the *Trichoderma* species as a biocontrol agent has been suggested more than 70 years ago by Weindling (1932). He first time explained the parasitic nature of this genus against soil-borne fungal or bacterial pathogens. The mechanisms projected to describe the biocontrol of plant pathogens by using *Trichoderma* species are presumptive. The described mechanisms for biocontrol are lysis, antibiosis, mycoparasitism, and competition (Papavizas and Lumsden, 1980; Cook and Baker, 1983). These mechanisms may act alone or in combination. *Trichoderma* species are also effective against various bacterial pathogenic species. They produce among 40 different metabolites of *T. harzianum* and ciprofloxacin and norfloxacin in cultures of *T. viride* which are antibacterial in nature. However, their efficient interaction with the host plant needs to be accompanied by production of secondary metabolites and cell wall degrading enzymes (Lorito et al., 2010).

In recent years, *Trichoderma* is found be strongly affecting the plant development and biochemistry (Mach et al., 1993). Some antagonistic effects include nutrient competition, mechanical barriers, or pH changes. Perusal of research literature has indicated the fungal inhibition of various other plant pathogens. By the conventional method, *Trichoderma* species are isolated from the different sources and have been used in the production of several enzymes as well as in biological control (Almeida et al., 2007; Lopes et al., 2012). *T. viride*, *Trichoderma hamatum*, and *Trichoderma koningii* are found to exhibit pronounce antagonist activity against *R. solani*, *Fusarium oxysporum*, *F. oxysporum* f. sp. *lycopersici*, *Pythium aphanidermatum*, *Phytophthora infestance*, *Colletrotrichim capsici*, and *Macrophomina phaseolina* (Mukhopadhyay, 1987; Lewis et al., 1990; Arora, 1990; Gaikwad et al., 1999; Yadav and Tripathi, 1999; Sharma and Sain 2005; Sharma et al., 2005). The cellulose content is decreased when *Trichoderma* spp. is used in decomposing bark, which leads to the activation of chitinase genes of *Trichoderma*, which finally produce the chitinase to parasitize *R. solani* (Benhamou and Chet, 1997). Similarly, *T. harzianum* isolates are strong antagonists against fungal species like *Aspergillus flavus*, *Aspergillus candidus*, *Aspergillus fumigates*, *Aspergillus niger*, *Aspergillus terreus*, *Fusarium graminarium*, *Fusarium semitectum*, *Cladosporium*, and *Rhizopus* (Jegathambigai et al., 2009). *T. harzianum* and *Trichoderma asperellum* are reported to be the most effective antagonists against the *Fusarium solani*, *R. solani*, and *S. sclerotiorum*. Both bioagents when multiplied in the broth medium produce and secrete several enzymes like chitinase, phosphatase, protease, β-1,3-glucanase, NAGAse, and alginate lyase (Qualhato et al., 2013). The minimum inhibitory concentration of *T. harzianum* on fungal isolates *A. terreus*, *A. fumigates*, *Aspergillus clavatus* ranges from 100 to 150 µl/ml and for bacterial isolates *Staphylococcus aureus*, *Escherichia coli* and *Klebsiella* ranges from 50 to 100 µl/ml of media. However, some *A. niger*, *A. clavatus*, and *Proteus* isolates are found to be resistant to antimicrobial activity of *T. harzianum* extract (Leelavathi et al., 2014).

Several isolates of *T. viride*, isolated from different regions of Allahabad district, are found to be efficacious exhibiting significant growth inhibition (GI) against *R. solani* (70% GI), *Sclerotium rolfsii* (68.2% GI), *M. phaseolina* (70% GI), *Alternaria alternata* (73.3% GI), *F. solani* (69.3% GI), and *C. capsici* (70.1% GI) of green gram (*Vigna radiata*). However, 20% concentration is found to be more effective exhibiting 100% GI against all the fungal species (Mishra et al., 2011). Some species of *Trichoderma* like *T. viride*, *T. harzianum*, *T. koningii*, *Trichoderma pseudokoningii*, and *T. virens* have been shown remarkable variance against two fungal pathogens *Alternaria spinaciae* (cause of leaf spot) and *F. oxysporum* f. sp. *spinaciae* (cause of wilt) in dual culture. Among these, *T. virens* resulted in 50% reduction of fungal growth (Bhale et al., 2012). Similarly, *T. harzianum* exhibited good bioefficacy even against ominivorus disease causing fungi, including *S. rolfsii* (Benhamou and Chet, 1996) and *S. sclerotiorum* (Inbar et al., 1996). Henis et al. (1983) reported that the different isolates of *T. harzianum* parasitized on *S. rolfsii* with varying percentages of inhibition. *Armillaria* rot in tea has been effectively managed by several *Trichoderma* species such as *T. harzianum*, *T. koningii*, and *Trichoderma longibrachiatum* (Osando and Waudo, 1994). In addition, *T. koningii* has been successful in management of corn rot caused by *F. solani* (McAllister et al., 1994). Among five species of this genus, *T. viride* had greater inhibition of *S. rolfsii* than *T. harzianum*, *T. longibrachiatum*, *Trichoderma paraseramosum*, and *Trichoderma hamatum* (Shaigan et al., 2008).

In addition, *T. viride* is found to be extremely effective against *R. solani*, *S. rolfsii*, and *S. sclerotiorum* in comparison with *T. harzianum* (Amin et al., 2010). The mycelial growth of *S. sclerotiorum* is inhibited by culture filtrates of *T. viride* because of antibiotic production (Kapil and Kapoor, 2005); and volatile secondary metabolites play a key role not only in mycoparasitism by *T. harzianum* and *Trichoderma atroviride*, but also in their interactions with tomato and canola seedlings growth (Vinale et al., 2008). Volatile and nonvolatile compounds from *Trichoderma* spp. are found to be effective against *C. capsici*, a causal agent of anthracnose disease in bell peppers (Ajith and Lakshmidevi, 2010). *T. atroviride* (T-15603.1) stain has been found to be more effective in managing basidomycetes fungi like *Ganoderma adspersum*, *Ganoderma lipsiense*, *Inonotus hispidus*, *Polyporus squamosus*, and ascomycete like *Kretzschmaria deusta*. Wilt of safflower (*F. oxysporum* f. sp. *carthami*) has been effectively managed by *Trichoderma* species (Waghmare and Kurundkar, 2011). *Trichoderma* isolates such as TCVSI-1, -3, -4, -5, -8, and -10 exhibited significant antagonistic activity against *Fusarium moniliformae*—a pathogen causing wilt disease in sugarcane with the Percent Inhibition of Mycelial Growth recorded high of TCVSI-1 (39.70), TCVSI-1 (48.86), TCVSI-1 (58.78), and TCVSI-3 (86.26), followed by TCVSI-5 (33.59), TCVSI-3 (38.17), TCVSI-3 (53.44), and TCVSI-5 (82.44) in 25, 50, 75, and 100% concentrations, respectively (Gawade et al., 2012). The *Trichoderma* formulation T-15603.1 also induces the growth of several wood decaying fungi of urban sites up to 82.3%. RAPD-PCR (Random Amplified Polymorphic DNA-polymerase chain reaction) results show that when 0.1% urea and 0.2% glucose combined with a humidity storing gel as a carrier substance enhanced conidial formulation, germination rate, and conidial viability which lead to the T-15603.1 establishment in the wood substrate (Schubert et al., 2008).

9.7.2   VERTICILLIUM (LECANICILLIUM)

This genus has broad spectrum hosts, and it has been isolated from variety of substrates and commercialized as the biopesticides, that is, Mycotal which exhibits pathogenic activity toward thrips, whiteflies and verticillin toward mites, aphids, and whiteflies. Genera also exhibits pronounce efficacy against various plant pathogenic fungi such as powdery mildews fungus (Miller et al., 2004), green molds (Benhamou and Brodeur, 2000), and *Pythium* (Benhamou and Brodeur, 2001).

Some isolates of some *Lecanicillium* species which produce chitinase enzyme are found to be mycoparasitic to powdery mildew fungal mycelia and conidia by penetrating the hyphae in it (Askary et al., 1997). Soil application of *Verticillium chlamydosporium* has shown significant control of *Heterodora avenae* (Kerry *et. al*, 1984) and *Meloidogyne arenuria* (Neal) ChitWood (Godoy et al., 1983) in pots. *Lecanicillium* spp. are able to pierce and colonize uredial sori of *Puccinia coronata* (Leinhos and Buchenauer, 1992). The change in host cells in *Penicillium digitatum* is also recorded prior to contact by the *Lecanicillium* spp. (Benhamou and Brodeur, 2000). Besides mycoparasitism by *Lecanicellium* in *Penicillium ultimatum*, it is found that the mode of action is linked to host plant tissue colonization which triggers the plant defense reactions (Benhamou and Brodeur, 2001). ISR is also reported in cucumber roots when treated by spores of *Lecanicillium muscarium*.

Fewer lesions have been observed to be with decreased severity when plants were inoculated with *L. muscarium* as compared with noninoculated plants. Furthermore, treatment of roots with *L. muscarium* results in reduced wilt incidence and other soil-borne disease in tomato (*Verticillium dahlia*), Japanese radish (*Verticillium dahliae*), and melon (*F. oxysporum* f.sp. *melonis*) (Kusunoki et al., 2006; Koike et al., 2007). For instance, several species of this bioagent like *Lecanicillium psalliotae*, *Lecanicillium antillanum*, and other species are also reported to be ovicidal nature against the root-knot nematode (*Meloidogyne incognita*) eggs (Nguyen et al., 2007). Similarly, *V. chlamydosporium* isolates reduced 80% population of *Meloidogyne arenaria* in tomato plants (Leij and Kerry, 1991). *Lecanicillium* spp. is also found to infect cysts and eggs of *Heterodera glycines* (soybean cyst nematode) and reduce their population in a laboratory and green house conditions (Meyer et al., 1997). *Lecanicillium attenuatum* and *Lecanicillium longisporum* are found to be virulent against aphids as well as inhibited the *Sphaerotheca fuliginea* mycelial growth and spore production (Askary et al., 1998; Kim et al., 2008).

In addition, *Lecanicillium lecanii* is primarily identified as soft-scale insect pathogens (Zare and Gams, 2001) and *Lecanicillium nodulosum* as pathogen for mites (Zare and Gams, 2001). Combination of arachid oil (peanut oil 0.5%) and two invert emulsions using either Sunspray 6N or paraffin oil (0.5%) with *Lecanicillium Lacanii* is found to be more effective for the management of cucumber powdery mildew pathogen *S. fuliginea* (Verhaar et al., 1999). In greenhouse testing, the population of cotton aphid is decreased by 60% by the initial application of *Verticillium lecanii* spores (10^8 spores ml^{-1}) formulated in 1% (w/v) of montmorillonite SCPX-1374 and 1% (w/v) of the wetting agent, that is, EM-APW-2 (Lee et al., 2006). Integrated application of *L. muscarium* and imidacloprid has been reported to result in significant mortality in *Bemisia tabaci* infecting verbena foliage (Cuthbertson et al., 2005).

9.7.3 BEAUVERIA

The genus includes *B. bassiana* (Bals.) Vuill. species is an important microbial pathogen of insects and is base of several microbial insecticides against major arthropod pests in agricultural, forest as well as aquatic ecosystem.

B. bassiana has been recorded 78 and 92% mortality in cotton boll weevil under laboratory conditions in feeding and direct contact method of application, respectively (Wright and Chandler, 1991, 1992). The efficacy of *B. bassiana* against cotton leaf roller under both in-vitro and in-vivo condition showed mortality of the pest within 4 days of application (Ramesh et al., 1999) and also exhibited 100% mortality of clover root weevil (*Sitona lepidus*) adults (Willoughby et al., 1998). A total of 95% of aphid, for example, *Myzus persicae* were found to be infected by this fungal bioagent (Chen et al., 2008) and exhibiting dual properties, that is, natural enemies of pests and plant pathogens during the biological control process (Bonnie et al., 2009). *B. bassiana* and its formulation Botanigard 22WP® shows potential microbial control agent for greenhouse tomato insect-pests and sweet pepper and found to be compatible with natural enemies like *Amblyseius swirskii*, *Aphidius colemani*, *Encarsia formosa*, *Eretmocerus eremicus*, and *Orius insidiosus* (Shipp et al., 2012). *B. bassiana* spray application at 1×10^8 spores/ml concentration works as potent biopesticides by exhibiting LT_{50} values in range of 2.19–3.73 days against *Rhopalosiphum padi*, *Schizaphis graminum*, *Lipaphis erysimi*, and *Brevicoryne brassicae*, as well as not showing any harmful effects to *Coccinella septempunctata* (Akmal et al., 2013; Thungrabeab and Tongma, 2007). While in contrary to this, James and Lighthart (1994) found *Nomuraea rileyi* to be safe for the beneficial insects (*C. septempunctata*) as compared to *B. bassiana*, *Metarhizium anisopliae*, and *Paecilomyces fumosoroseus* application to manage against *Hippodamia convergens*.

The application of oil palm with *B. bassiana* resulted in 98.3% reduction in the number of leaf web worms after 21st day (Rajasekhar and Kalidas, 2012) and also exhibited 90% mortality of the larvae. Similarly, it has been recorded to reduce the diseases caused by several soil-borne plant pathogens like *Fusarium*, *Rhizoctonia*, and *Pythium*. Being eco-friendly, the mechanism involved colonization of the hyphae endophytically resulted in induce systemic resistance (Ownley et al., 2010) with low mammalian toxicity, and no residual toxicity (Copping, 2004) and has been successful as fungal insecticides for aphids (Shah and Pell, 2003). *B. bassiana* is found to be more efficacious if used in combination with *V. lecanii* and *Paecilomyces furnosoroseus* for the management of leaf rust of wheat (Sheroze et al., 2002). Nymphal stage of whiteflies infesting cucurbit crops is also found to be controlled by *B. bassiana* (Altre et al., 1999; James et al., 2003; Ali et al., 2009). Its formulation exhibits significant mortality of *Helicoverpa armigera* up to 60–100% (Ritu et al., 2012) and highest spore suspension (2.4×10^7) has pronounced larvicidal activity with the least pupation (43.33%)

(Malarvannan et al., 2010). In addition, 77% mortality of *Plutella xylostella* larvae, on oilseed rape leaves (Lohse et al., 2014) and 100% larval mortality of Manuka beetles (serious and persistent pests of dairy pastures on Cape Foulwind, Westport) is observed after 8-week postfoliar application of *B. brongniartii* (F636) (Townsend et al., 2010). *B. bassiana causes significant* reduction in adult emergence and longevity of *Corcyra cephalonica* Stainton when the grains are treated with 2.02×10^8 spores/ml; thus, it may play a pivotal role in the IPM (Kaur et al., 2014).

   B. bassiana is reported to secrete some toxin proteins and metabolites including Cyclosporins A and C as well as cyclic peptides. The toxins such as beauvericin, enniatins, oosporein, and bassianolide produced by *B. bassiana* have potent insecticidal properties (Vey et al., 2001; Enrique and Alain, 2004). The virulence of *B. bassiana* for aphids (*M. persicae*) is found to be enhanced due to over-expression of a chitinase gene (Bbchit1) (Fang et al., 2005). Notably, the genes exhibiting pathogenic properties of *B. bassiana* derived the enzymes having the nature of culticle degradation (Fang et al., 2005), a perilipin-like protein which regulates the appressorium turgor pressor, differentiation, and a coat protein (G protein as well as its regulator) having cell protective nature are involved in evading host immune responses (Wang and St. Leger, 2007, Fang et al., 2007).

9.7.4   PAECILOMYCES

P. fumosoroseus (Wise) Brown has been shown to have potent nymphalcidal activity of whiteflies damaging cucurbits (Altre et al., 1999; James et al., 2003; Ali et al., 2009). The fungus is as effective as *Paecilomyces lilacinus* (Thom) Samson against *M. arenaria* but did not appear to survive well in soil (Godoy et al., 1983). *P. lilacinus* and Furadan (Carbofuran) used with oil cakes are found effective against root-rot–root-knot disease complex in mungbean (Ehtesham-ul-Haque et al., 1995) and alone in controlling *M. incognita* in tomato in Sri Lanka (Ekanayake and Jayasundara, 1994). Similarly, isolates of *Paecilomyces variotii*, *P. lilacinus* are also found to have better antagonistic activity than *P. fumosoroseus* against *M. incognita* interms of reduction of number of root galls per root stem and increased shoot and root weights in mungbean (Perveen and Shahzad, 2013). The *P. variotii* also shows good management of root-knot nematodes in the peach orchards (Al-Qasim et al., 2009).

   Formulation of *P. lilacinus* (10 g/kg) is found to be effective for the management of wilt caused by *F. oxysporum* f.sp *lycopersici* in tomato.

Similarly, other formulations of *P. fumosoroseus* are effective against cabbage-heart caterpillar, *Crocidolomia binotalis* Zeller. Second instar larvae are most susceptible to *P. fumosoroseus* and absolute mortality (80–100%) occurred at the concentration of 2×10^7 conidia/ml (Hashim and Ibrahim, 2003). Seed treatment plus soil application recorded significantly increased the shoot length, root length; biomass, and fruit yield (Sivakumar et al., 2008). The EC_{50} for *P. fumosoroseus* against cabbage-heart caterpillar, *C. binotalis* Zeller (Lepidoptera: Pyralidae) is assessed at $1.926 \times 1rY$ conidia/ml in a laboratory bioassay. The mortality for all the treatment bioassay was >70%. Application of the conidia of this bioagent in palm oil Vesawit® has been given the most favorable results against the cabbage-heart caterpillar (Hashim and Ibrahim, 2003).

P. variotii exhibits broad spectrum nature against melon *Fusarium* wilt and tomato bacterial spot (Suárez-Estrella et al., 2013). *P. fumosoroseus* and *Paratoxotus farinosus* (Holm ex S. F. Gray) Brown and Smith are entomo-parasitic to larvae and pupae of apple fruit moth (*Argyresthia conjugella Zell.*). Soil drenching by *P. fumosoroseus suspension* at 1×10^7 conidia per cm² reduced the emergence of *A. conjugella* adults by 70.1% as compared to untreated control (Vänninen and Hokkanen, 1997). Application of *P. lilacinus* mass multiplied on rice is found to be reducing infection of *Rotylenchulus reniformis* in tomato up to 46 and 48% both undergreenhouse and field microplot, respectively (*Walters and Barker, 1994*). They demonstrated that a preplanting soil application of a commercial formulation of *P. lilacinus* strain 251 (PL251) at $2 \times 10^{(5)}$ CFU/g soil is found to be adequate to decrease root galling in tomato caused by root-knot nematode *M. incognita* by 45% along with 69% reduction in egg masses (Kiewnick et al., 2011).

Combined application of *Bradyrhizobium* and *Purpureocillium lilacimus* 10 days after root knot nematode incocualtion have been proved be more effective against *M. incognita* infection compared to application of *Bradyrhizobium* alone in black gram and resulted no significant damage in plant growth. Nitrogenase activities are found to be higher is infested with *M. incognita* as compared with *Bradyrhizobium beibre* and *P. lilacinus* application (Bhat et al., 2012). *P. lilacinus* (2.3×10, sup>9-1) is also effective in controlling cotton aphids in combination with other eco-friendly control practices such as *Azadirachta indica* (10 ml/l) and diatomaceous earth formulation (PyriSec) (3 g/l) (*Wakil* et al., *2012*). *Phidippus carneus* and *P. farinosuss* species of this genus can effectively manage the *Ostrinia nubilalis*, *Sesamia cretica*, and *Chilo agamemnon* under laboratory, green house, and field conditions, and LC_{50} of the *P. carneus* is found to be 149×10^4, 166×10^4, and 176×10^4 spores/ml for *O. nubilalis*, *S. cretica*, and *C. agamemnon*, respectively, and LC$_{50}$ for *P. farinosuss* is 156×10^4, 189×10^4, and 195×10^4 spores/ml (Magda and Said, 2014).

9.7.5 METARHIZIUM

Metarhizium species are one of the important entomopathogens which has shown promising results for managing many soil-borne as well as foliar insect-pests in different crop conditions. Bioassays have been conducted to test isolates of *M. anisopliae* (Metschnikoff) Sorokin against several insect-pests show that *M. anisopliae* isolates have poor to high bioefficacy efficacies (Mazodze and Zvoutete, 1999). Among 28 isolates of *M. anisopliae* studied by Nussenbaum and Lecuona (2012) against the boll weevil (*Anthonomus grandis*) cotton main pest in America, Ma 50 and Ma 20 isolates are found as the most virulent against *A.* grandis at their LC$_{50}$ values 1.13×10^7 and 1.20×10^7 conidia/ml, respectively, which shows the opportunity of the application of *M. anisopliae* against boll weevils. *Metarhizium* brunneum is able to cause 61.7% mortality in ambrosia beetle *Xylosandrus* germanus 6 days after postspray in the field (Castrillo et al., 2011).

Use of *M.* brunneum F52 (90 ml a.i./ha), *M.* brunneum + azadirachtin (45 ml a.i./ha + 742 ml a.i./ha) and *M.* brunneum + spinosad (45 ml a.i./ha + 45 grams a.i./ha) are found to be effective in controlling the sweetpotato weevil, *Cylas* formicarius in both the laboratory and in the field. Combining application of *M.* brunneum with *B. bassiana* is found to have enhanced effectiveness against weevil (*C.* formicarius*) resulted in* reduced tuber damage which leads to the higher production yield (Reddy et al., 2014). *M. anisopliae* also reduced the fruit damage in tomato due to fruit borer by 87.01% in a farmer's field in Jorhat, Assam when applied (Phukon et al., 2014). Field application of BioCane™ the formulation of *M. anisopliae* isolate FI-1045 has been able to reduce incidence of white grub (*Dermolepida* albohirtum*)* (Waterhouse) in sugarcane in Australia (Samson et al., 2006).

The spray application of *M. anisopliae* is found effective against nymphal stage of long-horned grasshopper *Uvarovistia zebra* with 53.3% mortality at its highest concentrations of 2×10^7 spores/ml when ingested by insects using lettuce baits inoculated the mortality caused by *M. anisopliae* is 43% (Mohammadbeigi and Port, 2013). Among 12 selected isolates of *M. anisopliae* collected from different places, M.a (HG5-BD), M.a (HG3-B) and M.a (OM3-BD) were found to be comparatively better effective for controlling, *Leptocorisa acuta* Thunberg (Hemiptera: Alydidae) the rice ear head bug in field condition than others. The efficacy of this bioagent perceived from 7

DAT and extended to its highest peak at 14 DAT. At 10 DAT, *M. anisopliae* has also been caused the *L. acuta* field mortality of 63.6–86.6% (Sivakumar et al., 2008).

9.7.6 NOMURAEA

N. rileyi (Farlow) Samson is also one of the important entomopathogens used as an environmental-friendly biopesticide. However, insecticidal activity of *N. rileyi* to lepidopteran pests indicates that it could be a candidate for further development as a microbial insecticide (Ignoffo, 1981). The earlier experiments indicate that the *N. rileyi* isolates conidial concentration at 2×10^8 conidia/ml showed 77–80% mycosis and mortality of *Spodoptera litura* larvae in 7 days and 79–85% mortality of *H. armigera* in 8 days (Devi et al., 2003). Pathogenesis study of *N. rileyi* against *S. litura* showed that its infection process starts with adhesion of conidia on the insect cuticle (Srisuk-chayakul et al., 2005). Application of *N. rileyi* at 1.25 kg/ha combined with spinosad 0.0045%, *N. rileyi* at 1.25 kg/ha + novaluron 0.004%, *N. rileyi* at 1.25 kg/ ha + methomyl 0.025%, and *N. rileyi* at 1.25 kg/ha + indoxa-carb 0.0015% are found as effective for the control of *S. litura* infesting in groundnut field ecosystem at Junagarh (Kachhadiya et al., 2014). Similarly, pyrethroid cypermethrin and oil-based formulation (Diesel: Sunflower oil; 7:3) of *N. rileyi* N812 are found to be most effective in suppressing the pest population of beet armyworm, *S. litura* in sugarbeet field (Yadav and Deshpande, 2012) while 1.5×10^5 conidia/ml LC_{50} values significantly pathogenic to third instars (9–10 days old) larvae of *S. litura*. Mycoinsecticide produced from *N. rileyi* applied at 30 kg/ha (1×10^8 and 2×10^8 conidia/l) is found to be effective in managing the *H. armigera* (Shekharappa, 2009). LC_{50} of this entomopathogenic fungus against insect-pests of olive (*Bactrocera oleae, Ceratitis capitata*, and *Prays oleae*) under laboratory conditions are recorded to be 142, 145, and 155 spores/ml under laboratory conditions.

The percentage of *P. oleae, C. capitata*, and *B. oleae* infestations have significantly decreased in field plots treated with *N. rileyi* to 10, 13, and 10 individuals as compared to 38, 31, and 33 individuals of the corresponding pests in the control. The treated trees with *N. rileyi* scored the highest weight 3090 kg/ Feddanas compared to 2169 ± 80.53 kg/Feddan among the control trees (Sabbour, 2013). The suspensions of 10 isolates of *N. rileyi* at 10^8 conidia/ml found to be active against third instars of *S. litura*, and resulting in 85–97% mortality with LT_{50} values from 5.5 to 6.6 days (Padanad and Krishnaraj, 2009).

Combined field application of endosolphon at 1.6% and 3.2×10^8 spores/ml of *N. rileyi* found to be effective in decreasing the incidence of *H. armigera*, *S. litura*, and *Trichoplusia ni* in cabbage in Karnataka region (Gopalakrishnan and Mohan, 2002). Concerning the bioefficacy toward *Bemisia tabaci* and *M. persicae* pests in tomato crop, this bioagent has been proved as efficacious at the LC_{50} values are at 103.7×10^4 and 89.1×10^4 spores/ml, respectively, decreases the infestation of the plants by them both under laboratory and field conditions. Interestingly, at the highest lethal concentration (1×10^8 spores/ml), *N. rileyi* shows the safety levels to the predator *Coccinella undecimpunctata* (Matter and Sabbour, 2013). The *N. rileyi* at 1×10^8 and 2×10^8 conidial suspension per liter proved pathogenic to leaf eating caterpillar in soybean in *kharif* (2008 and 2009) at the Dharwad, Karnataka, India and found to be nonpathogenic to other beneficial insects and is found to reduce the larval population, damage to pods and increasing the seed yield (Patil and Abhilash, 2014).

9.7.7 HIRSUTELLA

The genus *Hirsutella* (Hypocreales) exhibited pathogenic nature against several mites, nematodes, and other lepidopteran insects (McCoy, 1996; Jaffe, 2000; Chandler et al., 2000). There are several species under this genus like *Hirsutella citriformis*, *Hirsutella gigantean*, and *Hirsutella nodulosus* infect dipterans, and hemipteran and lepidopterans are found to produce a compact group of erect conidiophores (synnemata), whereas *Hirsutella rhossiliensis* and *Hirsutella minnesotensis* infect plant–parasitic nematodes. Mite-specific *Hirsutella thompsonii* is the most widely studied species of this genus (McCoy, 1981). This fungus is found to be worldwide in distribution on several tetranychid and eriophyoid mite hosts. During the hot and humid weather conditions, it can cause remarkable natural epizootics among the population of mites (e.g., citrusrust mites, blueberry, coconut and tomato mites, etc.) and is regarded as a major natural enemy of the several mite pests (Chandler et al., 2000). *H. thompsonii* produces a single polypeptide chain, insecticidal protein named hirsutellin A which is poised of 130 amino acid residues and cause infection in plant damaging mites (Herrero-Galán et al., 2008).

Hirsutella is a common fungus having a wide host range in the insects (Samson et al. 1988). However, not much work has been reported in the microbial control of insects in field conditions. The widely use species are *H. thompsonii* which has been found to control mites, *H. citriformis* effectively manage the rice brown planthopper, *Nilaparvata lugens* (Stahl) (Rombach et al. 1986) have yet to be accepted as one of the potential means of insect-pest management. The population of psyllid is found to be reduced by *H. citriformis* (Carruthers and Hural, 1990). One of the species of *Hirsutella* such as *H. rhossiliensis* is having nematophagous nature and found to be very effective against soybean cyst nematode (Zhang et al., 2008).

9.7.8 PSEUDOMONAS

Pseudomonas species are one of the most vital bacteria dwelling the rhizosphere of diverse crop plants and have been often reported as microbial biocontrol agent. *Pseudomonas* are Gram-negative rod-shaped bacteria (Palleroni, 1984) and are aerobic, produce exopolysaccharides those generate biofilms (Hassett et al., 2002). Over the years, various isolates of *Pseudomonas* have been assessed as potential biocontrol agents against plant pathogens of soil-borne origin. Many strains of *P. fluorescens* showed potential bioactivity against plant pathogens especially the pathogens found to be associated with plant roots.

For sustainable agriculture, recent analysis on genomic of biocontrol traits and rhizosphere competence will likely lead to the progress of several special tools for potential management of indigenous bioagent as well as a better exploitation of their plant-beneficial properties. *P. fluorescens* strains, viz., CHA0 and Pf-5, have been effective against the pathogenic fungi associated with roots of plant species such as *Pythium* and *Fusarium* and phytophagous nematodes (Haas and Keel, 2003). The inoculation of *P. fluorescens* strains (UP61, UP143, and UP148) and *Rhizobium loti* B816 simultaneously in *Lotus corniculatus* plants, no change in nitrogen fixation has been found and disease rate due to the *Pythium ultimum*, and *R. solani* are found to be reduced (Bagnasco et al., 1998). Similar strains UP61, UP143, and UP148 isolated from Uruguayan soils are also found to be effective for the management of damping-off in birds-foot trefoil (Fuente et al., 2002).

The antagonistic potential of 142 fluorescent isolates of *Pseudomonas* from the maize rhizosphere showed significant antagonism against three plant pathogenic fungi and bacterium, that is, *Ralstonia solanacearum* (Costa et al., 2006). Application of *Pseudomonas putida* WCS358r, *P. fluorescens* WCS374r, *P. fluorescens* WCS417r, and *Pseudomonas aeruginosa* 7NSK2 are found to control *R. solanacearum* causing the bacterial wilt in *Eucalyptus* (Ran et al., 2005). The bacterium is also effective to manage the bacterial wilt in tomato caused by the same pathogen up to 68.4% (Guo et al., 2004). Concerning the antagonistic role of this bioagent (*Pseudomonas fluorescence* and *P. aeruginosa*) against fungi, the bacterium pronounced efficacy against *F. oxysporum, A. niger*, and *A. alternata* pathogens of Castor (*Ricinus communis* L.), Peanut (*Arachis hypogaea* L.), and Mung bean (*Vigna radiata* (L.) R. Wilczek) in Gujrat (Khanuchiya et al., 2012).

9.7.9 BACILLUS

Against the plant bacterial disease, *Bacillus subtilis* and other Bacilli have long been used as bioagents but the detailed mechanisms through which bacteria confer defense are not well understood. *B. subtilis*, a soil bacterium, isolated from the different chili rhizospheres displayed high antagonistic activity against *Colletotrichum gloeosporioides* OGC1. Chili seeds treated with *Bacillus* suspension showed 100% germination index and 65% disease incidence reduction (Ashwini and Srividya, 2014). In a similar study, Saleem and Kandasamy (2002) investigated the in-vitro suppression of fungal growth by *Bacillus* strain BC121 in comparison with a mutant of that strain, which lacks both antagonistic activity and chitinolytic activity. It can be also used as biocontrol agent against *Pseudomonas syringae* pv. *tomato* DC3000 infecting *Arabidopsis* roots (Bais et al., 2003).

The minimum inhibitory concentrations 25 μg/ml and lipopeptide level in roots colonized by *B. subtilis* are found to be sufficient for the inhibition of *P. syringae*. *B. subtilis* (B11), *Bacillus pumilus* (B19), *Bacillus cereus* (B16), *B. cereus* (B17), *Bacillus brevis* (EN63-1), and *Bacillus licheniformis* (EN74-1) exhibit potential inhibitory activity against *Botrytis mali* causing gray mold of apple in Iran (Jamalizadeh et al., 2009). The spread of *B. mali* lesion has found to be inhibited (9–32.2 mm) in apples stored at 40°C when treated by the all isolated of this bioagent while in control lesion are found having 41.6–51.4 mm diameter. At 20°C, the lesion size recorded in antagonistic bacterial treatment bioassay are from 7 to 24.9 mm for treatments with 42.2–46.6 mm in control bioassay.

Bt is a facultative anaerobic, Gram-positive, motile, and spore-forming soil bacterium. The spores have capacity to produce a specific type of substance having insecticidal crystal proteins, that is, Cry and/or Cyt proteins (also known as δ-endotoxins) which have insecticidal activity. The Bt formulations have great potential in IPM programs in agriculture. In genetically modified bacterium, it also acts as main source of genes for transgenic expression to provide better resistance against pests in micorbials and plants as pest control agents. Six isolates of *B. subtilis* from China are found to exhibit >50% biocontrol efficacy against bacterial wilt pathogen *R. solanacearum* in greenhouse tomato (Chen et al., 2013).

For sustainable agriculture and bicontrol, the high specificity of Bt crops provides a new keystone that will allow both to expand during this century (Federici, 2007). However, government policy and regulation limitations are the key factors which influence the availability of BTI (*Bacillus subtilis* I) in the vector control program (Yu et al., 2015). Later on, Cry6 A toxin from Bt is a representative nematicidal crystal protein with a variety of nematicidal properties to free-living nematode *Caenorhabditis elegans* and possesses great potential in plant–parasitic nematode management and construction of transgenic crop with constant resistance to nematode (Yu et al., 2015).

9.7.10   *GLOMUS*

Glomus is the largest genus of arbuscular mycorrhizal (AM) fungi having capacity to form symbiotic relationships (mycorrhizas) with roots of higher plants. During the last 30 years, the ability of AM for management of soil-borne pathogens has been extensively investigated. A total of 85 species have been described, but is currently known as nonmonophyletic. For improving fertility of the soil, there are several *Glomus* species including *Glomus aggregatum*, which are cultivated and sold as inoculant of mycorrhizal origin. By increasing protection against biotic and abiotic stresses, the AM association also improves the health of the plants. The application of this bioagent to the soil also leads to the inhibition of soil-borne bacterial pathogens and bioremediation of polluted soils (Barea et al., 2005). When mycorrhizal establishment occur on the roots, a significant reduction in bacterial diseases are found (Dehne, 1982). Inoculation of the mycorrhiza is also found to be suppressing the growth of root pathogens such as *V. dahlia* (Matsubara et al., 1995), *F. oxysporum* (Stephan et al., 1999), *Phytophthora* species (Trotta et al., 1996), *R. solani* (Stephan et al., 1999; Matloob and Juber, 2013), and *Pythium ultimum* (Calvet et al., 1993) on different crops.

In addition, plant pathogenic nematodes are also found to be suppressed by the inoculation of AM fungi (Stephan et al., 1999; Diedhiou et al., 2003). *R. solani* infection is found to be reduced in shoots and crowns by 60–71.2% in potato plants varieties inoculated with *Glomus etunicatum* (Yao et al., 2002). In-vitro studies showed that strawberry plant roots inoculated with *G. etunicatum* and *Glomus monosporum* reduced *Phytophthora fragariae* sporulation by 67 and 64%, respectively, in 48 h (Norman and Hooker, 2000). AM, *Glomus intraradices* significantly reduced the *R. solani* disease incidence and severity in bean plants under different conditions (Matloob and Juber, 2013), and seed decay, root, and crown rot tomato cause by *R. solani*. The AM treatment also results in positive influence on plant growth and other health indicators like seeds germination, fresh weight, dry weight, roots volume, and fruit weight (Kareem and Hassan, 2014).

9.7.11   ENTOMOPATHOGENIC NEMATODES (EPN)

The nematodes are associated with insects that cause disease to them is referred to as entomopathogenic nematodes (EPN). The potentialities of EPNs have been reported as biological agents against a number of insect-pests and disease. They are found to be effective against development of second generation of adults in the field. About 30 families of nematodes are known to parasitize insects and nematodes. But only seven families, viz., Mermethidae, Allantonematidae, Neotylenchidae, Sphaerularidae, Rhabditidae, Steinernematidae, and Heterorhabditidae possess a definite potential of causing mortality to insects. However, the families, Steinernematidae, and Heterorhabditidae contain most efficient EPNs and are of prime importance as far as biological control of insects is concerned. They possess many features such as broad spectrum of the host, looking the host actively and kill them within 48 h, easy mass production, long-term efficacy, easy application, compatibility with most chemicals, and are environmentally safe.

EPNs are suitable for biocontrol programs because of pathogenic nature, behavior of host searching, and their survivability. EPNs have been efficaciously useful against soil dwelling (as soil application) and aboveground insects (foliar spray) in cryptic. EPNs are exempt from federal and state registration requirements, and they are considered safe to vertebrates and vegetation, and may be applied through several methods, including drip irrigation. A nematode, that is, *Steinernema riobravis* Cabanillas, isolated from the Rio Grand Valley of Texas (Cabanillas et al., 1994), is found to be effective against last instars of *Galleria mellonella* (L.) at 37 °C (Grewal et al., 1994). There are various EPNs commercially available in the markets and are easily mass produced, stored, and transported (Gaugler and Kaya, 1990).

At present, EPNs are also being used in combination with many pesticides (Kaya, 1990), which provide satisfactory results in the integrated pest management system. *Steinernema glaseri* (NC1 strain), *Steinernema scarabaei* (AMK001 strain), *Heterorhabditis bacteriophora* (GPS11 strain), and *H. zealandica* (X1 strain) have also been exhibited pathogenic nature against third-instars scarabs (*Popillia japonica, Anomala orientalis, Cyclocephala borealis*, and *Rhizotrogus majalis*), and late-instar greater wax moth (*G. mellonella*) (Koppenhöfer and Fuzy, 2008). Similarly, *S. scarabaei, H. zealandica*, and *H. bacteriophora* (GPS11 and TF strains) have been found effective against third instars of Asiatic garden beetle (*Maladera castanea*), European chafer (*R. majalis*), Japanese beetle (*P. japonica*), oriental beetle (*Anomala (=Exomala) orientalis*), and northern masked chafer (*C. borealis*), both under laboratory and greenhouse conditions. *H. bacteriophora* and *H. zealandica* have been found to have modest virulence to *P. japonica, A. orientalis, C. borealis*, and *M. castanea*. But low virulence has been observed against *R. majalis, H. zealandica*, whereas *H. bacteriophora* caused only erratic and very low mortality.

In contrast, modest virulence against *C. borealis* has been recorded in *S. scarabaei*, but is highly virulent against *R. majalis, P. japonica, A. orientalis*, and *M. castanea* with *M. castanea* being the least susceptible and *R. majalis* the most susceptible (Koppenhöfer et al., 2006). *S. scarabaei* is again found to be efficacious than *H. bacteriophora* in laboratory, greenhouse, and field experiments against two white grub species, European chafer, (*R. majalis*) and Japanese beetle, (*P. japonica*) with an LC_{50} of 5.5–6.0 and 5.7 infective juveniles per third-instar larva in *R. majalis* and *P. japonica*, respectively (David and Albrecht, 2003). On the contrary, some authors found the low susceptibility of *S. scarabaei and H. bacteriophora* to the three white grub species, viz., *P. japonica, R. majalis*, and *C. borealis* (Ruisheng and Grewal, 2007).

9.7.12   BACULOVIRUSES

More than 1600 viruses have been isolated from more than 1100 insect and mites species. Among these, three families (Ascoviridae, Baculoviridae, and Polydnaviridae) are specific for insect-pests and other arthropods. Among these families, the baculoviruses are the most widely exploited virus group for managing the insect-pests; they are very different from other viruses that infect vertebrates and are considered to be very safe to use. Baculoviruses include two genera; nucleopolyhedroviruses (NPV) and granuloviruses (GV). According to the family Baculoviruses, their pathogenesis mode and replication vary, but ingestion is the always cause of infection. Virions penetrate epithelial cells through the binding to the receptors in the gut. The infection often spreads to the hemocoel and then to essential organs and tissues, particularly fat bodies. In 5–14 days, acute infections occur causing the death of the host. Mass production of NPV and GV can only be done in vivo, but is economically viable for larger hosts like lepidoptera (*Heli-coverpa—HaNPV* and *Spodoptera—SINPV*) and formulation development and method of application are conventional. Approximately, 16 baculoviruses based biopesticides are available commercially for use and some are under development. For example, codling moth granulovirus, CpGV (*Cydia pomonella* Granulovirus) is an effective biopesticide of codling moth caterpillar pests of apples.

9.8   CONCLUSION AND FUTURE PROSPECTS

Enhance in productivity of the crop from the current agricultural systems reaching a plateau in several countries including India and due to excessive and indiscriminate use of chemical fertilizers and pesticides the environmental problems are becoming a matter of great concern. Thus, the microbial-based biological control can be an alternate system, which may contribute an important role in achieving the goal of sustainable agriculture. Microbial biopesticides are likely to play a vital role in IPM in modern and sustainable agriculture for controlling pests of vegetables, fruit crops, cereal crops, forest pests, and pests of domestic and public health importance. Because of their slow active nature, there is a need to develop more effective strategies for their sustainable use in agriculture. Extension workers and farmers need to be educated on their benefits, appropriate application methods, and their proper use. There should be timely availability of effective biopesticides to the farmers. The price of the commercial biopesticides has to be competitive with synthetic chemical pesticides or alternately the government has to provide subsidies for encouraging their use in agriculture to safeguard human health. But the investigation carried out through the need to use this practice in agriculture system has opened attractive biology and led to abundant essential discoveries. The prospect to conduct such basic studies is relying on responsibility to resolve the problems that occur during applied aspects, those prevent the successful exploitation of many microbial bioagents. The future thrust in microbial biocontrol investigation will require future trends in understanding fundamental biology of microbials with the quest for solutions that will make microbial biocontrol integral to the easy, safe, and sustainable management of all agro-ecosystems.

KEYWORDS

- **microbial bioagents**
- **fungal and insect-pests**
- **biocontrol**
- **sustainable agriculture**

REFERENCES

Ajith, P. S.; Lakshmidevi, N. Effect of Volatile and Nonvolatile Compounds from *Trichoderma* spp. Against *Colletotrichum capsiciincitant* of Anthracnose on Bell Peppers. *Nat. Sci.* **2010,** *8,* 265–269.

Akmal, M.; Freed, S.; Malik, M. N.; Gul, H. T. Efficacy of *Beauveria bassiana* (Deuteromycotina: Hypomycetes) Against Different Aphid Species under Laboratory Conditions. *Pak. J. Zool.* **2013,** *45,* 71–78.

Ali, S.; Huang, Z.; Ren, S. X. Media Composition Influences on Growth, Enzyme Activity and Virulence of the Entomopathogen Hyphomycetes *Isaria fumosoroseus. Entomol. Exp. Appl.* **2009,** *131,* 30–38.

Almeida, F. B.; Cerqueira, F. M.; Silva, R. N.; Ulhoa, C. J. Mycoparasitism Studies of *Trichoderma harzianum* Strains Against *Rhizoctonia solani*: Evaluation of Coiling and Hydrolytic Enzyme Production. *Biotechnol. Lett.* **2007,** *29,* 1189–1193.

Al-Qasim, M.; Abu-Gharbieh, W.; Assas, K. Nematophagal Ability of Jordanian Isolates of *Paecilomyces variotii* on the Root-knot Nematode *Meloidogyne javanica. Nematol. Mediter.* **2009,** *37,* 53–57.

Alstrom, S. Evidence of Disease Resistance Inducedby Rhizosphere Pseudomonad against *Pseudomonas syringae* pv. *phaseolicola. J. Genet. Appl. Microbiol.* **1995,** *41,* 315–325.

Altre, J. A.; Vandenberg, J. D.; Cantone, F. A. Pathogenicity of *Paecilomyces fumosoroseus* Isolates to Diamondback Moth, *Plutella xylostella*: Correlation with Spore Size, Germination Speed, and Attachment to Cuticle. *J. Invert. Pathol.* **1999,** *73,* 332–338.

Amin, F.; Razdan, V. K.; Mohid, F. A.; Bhat, K. A.; Banday, S. Potential of *Trichoderma* Species as Biocontrol Agents of Soil Borne Fungal Propagules. *J. Phytopathol.* **2010,** *10,* 38–41.

Arora, R. K. Evaluation of Bioagents for Control of Soil and Tuber Borne Diseases of Potato. *Indian Phytopathol.* **1990,** *52,* 310.

Ashwini, N.; Srividya, S. Potentiality of *Bacillus subtilis* as Biocontrol Agent for Management of Anthracnose Disease of Chilli Caused by *Colletotrichum gloeosporioides* OGC1. *Biotechnology.* **2014,** *4,* 127–136.

Askary, H.; Benhamou, N.; Brodeur, J. Ultrastructural and Cytochemical Investigations of the Antagonistic Effect of *Verticillium lecanii* on Cucumber Powdery Mildew. *Phytopathology.* **1997,** *87,* 359–368.

Askary, H.; Carriere, Y.; Belanger, R. R.; Brodeur, J. Pathogenicity of the Fungus *Verticillium lecanii* to Aphids and Powdery Mildew. *Biocontrol Sci. Technol.* **1998,** *8*(1): 23–32.

Bagnasco, P.; De La Fuente, L.; Gualtieri, G.; Noya, F.; Arias, A. Fluorescent *Pseudomonas* spp. as Biocontrol Agents Against Forage Legume Root Pathogenic Fungi. *Soil Biol. Biochem.* **1998**, *30,* 1317–1322.

Bais, H. P.; Fall, R.; Vivanco, J. M. Biocontrol of *Bacillus subtilis* Against Infection of *Arabidopsis*roots by *Pseudomonas syringae* is Facilitated by Biofilm Formation and Surfactin Production. *Plant Physiol.* 2004, *134,* 307–319.

Barea, J. M.; Pozo, M. J.; Rosario, A.; Aguilar, C. A. Focus Paper: Microbial Co-operation in Therhizosphere. *J. Exp. Bot.* **2005,** 56, 1761–1778.

Benhamou, N.; Brodeur, J. Evidence for Antibiosis and Induced Host Defense Reactions in the Interaction Between *Verticillium lecanii* and *Penecillium digitatum,* the Causal Agent of Green Mold. *Phytopathology.* **2000,** *90,* 932–943.

Benhamou, N.; Brodeur, J. Pre-inoculation of Ri T-DNA Transformed Cucumber Roots with the Mycoparasite, *Verticillium lecanii,* Induces Host Defense Reactions Against *Pythium ultimum* Infection. *Physiol. Mol. Plant Pathol.* **2001,** *58,* 133–146.

Benhamou, N.; Chet, I. Parasitism of Sclerotia of *Sclerotium rolfsii* by *Trichoderma harzianum:* Ultrastructural and Cytochemical Aspects of the Interaction. *Biochem. Cell Biol.* **1996,** *86,* 405–415.

Benhamou, N.; Chet, I. Cellular and Molecular Mechanisms Involved in Theintersection between *Trichoderma harzianum* and *Pythium ultimum. Appl. Environ. Microbiol.* **1997,** *63,* 2095–2099.

Bhale, U. N.; Ambuse, M. G.; Chatage, V. S.; Rajkonda, J. N. Bioefficacy of *Trichoderma* Isolates Against Pathogenic Fungi Inciting Spinach (*Spinacea oleracea* L.). *J. Biopesticide.* **2012,** *5,* 1–6.

Bhat, M. Y.; Hamid, W. A.; Munawar, F. Effect of *Paecilomyces lilacinus* and Plant Growth Promoting Rhizobacteria on *Meloidogyne incognita* Inoculated Black Gram, *Vignamungo* Plants. *J. Biopesticides.* **2012,** *5,* 36.

Bonnie, H.; Kimberly, D.; Fernando, E. Endophytic Fungal Entomopathogens with Activity Against Plant Pathogens: Ecology and Evolution. *BioControl.* **2009,** *55,* 113–128.

Cabanillas, H. R.; Poinar, G. O.; Raulston J. R. *Steinernema riobravis* sp. (Rhabditida: Steinernematidae) from Texas. *Fundam. Appl. Nematol.* **1994,** *17,* 123–131.

Calvet, C.; Pera, J.; Barea, J. M. Growth Response of Marigold (*Tagetes erecta*) to Inoculation with *Glomus mosseae, Trichoderma aureoviride* and *Pythium ultimum* in a Peat-perlite Mixture. *Plant Soil.* **1993,** *148,* 1–6.

Carruthers, R. I.; Hural, K. Fungi as a Naturally Occurring Entomopathogens. In *New Directions in Biological Control: Alternatives for Suppressing Agricultural Pests and Diseases;* Baker, R. R.; Dunn, P. E., Eds; Alan R. Liss, Inc.: New York, **1990,** pp. 115–138.

Castrillo, L. A.; Griggs, M. H.; Ranger, C. M.; Reding, M. E.; Vandenberg, J. D. Virulence of Commercial Strains of *Beauveria bassiana* and *Metarhizium* brunneum (Ascomycota: Hypocreales) Against Adult *Xylosandrus germanus* (Coleoptera: Curculionidae) and Impact on Brood. *Biol. Control.* **2011,** *58,* 121–126.

Chandler, D.; Davidson, G.; Pell, J. K.; Ball, B. V.; Shaw, K.; Sunderland, K. D. Fungal Biocontrol of Acari. *Biocontrol Sci. Technol.* **2000** *10,* 357–384.

Chen, C.; Li, Z. Y.; Feng, M. G. Occurrence of Entomopathogenic Fungi in Migratory Alate Aphids in Yunnan Province of China. *BioControl.* **2008,** *53,* 317–326.

Chen, Y.; Yan, F.; Chai, Y.; Liu, H.; Kolter, R.; Losick, R.; Guo, J. H. Biocontrol of Tomato Wilt Disease by *Bacillus subtilis* Isolates from Natural Environments Depends on Conserved Genes Mediating Biofilm Formation. *Environ. Microbiol.* **2013,** *15,* 848–64.

Cho, E. M.; Boucias, D.; Keyhani, N. O. EST Analysis of cDNA Libraries from the Entomo-pathogenic Fungus *Beauveria* (*Cordyceps*) *bassiana*. II. Fungal Cells Sporulating on Chitin and Producing Oosporein. *Microbiology.* **2006**, *152*, 2855–2864.

Cook, R. J.; Baker, K. F. The Nature and Practice of Biological Control of Plant Pathogens. The American Phytopathological Society: St. Paul, MN, **1983**, p. 539.

Copping, L. G. The Manual of Biocontrol Agents, British Crop Protection Council, Crop Protection. *Crop Prot.* **2004**, *23*, 275–285.

Costa, R.; Gomes, N. C. M.; Peixoto, R. S.; Rumjanek, N.; Berg, G.; Mendonça-Hagler, L. C. S.; Smalla, K. Diversity and Antagonistic Potential of *Pseudomonas* spp. Associated to the Rhizosphere of Maize Grown in a Subtropical Organic Farm. *Soil Biol. Biochem.* **2006**, *38*, 2434–2447.

Cuthbertson, A. G. S.; Walters, K. F. A.; Northing, P. Susceptibility of *Bemisia tabaci* Imma-ture Stages to the Entomopathogenic Fungus *Lecanicillium muscarium* on Tomato and Verbena Foliage. *Mycopathologia.* **2005**, *159*, 23–29.

David, L. C.; Albrecht, M. K. *Steinernema scarabaei*, an Entomopathogenic Nematode for Control of the European Chafer. *Biol. Control.* **2003**, *28*, 379–386.

De Waard, M. A.; Georgopoulos, S. G.; Hollomon, D. W.; Ishii, H.; Leroux, P.; Ragsdale, N. N.; Schwinn, F. J. Chemical Control of Plant Diseases: Problems and Prospects. *Ann. Rev. Phytopathol.* **1993**, *31*, 403–421.

Dehne, H. W. Interaction between Vesicular-Arbuscular Mycorrhizal Fungi and Plant Patho-gens. *Phytopathology.* **1982**, *72*, 1115–1119.

Devi, P. S.; Prasad, Y. G.; Chowdary, D. A.; Rao, L. M.; Balakrishnan, K. Identification of Virulent Isolates of the Entomopathogenic Fungus *Nomuraea rileyi* (F) Samson for the Management of *Helicoverpa armigera* and *Spodoptera litura* (Identification of Virulent Isolates of *N. rileyi*). *Mycopathologia.* **2003**, *156*, 365–73.

Diedhiou, P. M. H. J.; Oerke, E. C.; Dehne, H. W. Effects of Arbuscularmycorrhizal Fungi and a Non-pathogenic *Fusarium oxysporum* on *Meloidogyne incognita* Infestation of Tomato. *Mycorrhiza.* **2003**, *13*, 199–204.

Dodds, J. A.; Lee, S. Q.; Tiffany, M. Cross Protection Between Strains of Cucumber Mosaic Virus: Effect of Host and Type of Inoculum on Accumulation of Virions and Double-stranded RNA of the Challenge Strain. *Virology.* **1985**, *144*, 301–309.

Ehtesham-ul-Haque, S.; Abid, M.; Ghaffar, A. Efficacy of *Bradyrhizobium* spp. and *Paecilo-myces lilacinus* with Oil Cakes in the Control of Root Rot of Mungbean. *Trop. Sci.* **1995**, *35*, 294–299.

Ekanayake, H. M. R. K.; Jayasundara, N. J. Effect of *Paecilomyces lilacinus* and *Beauveria bassiana* in Controlling *Meloidogyne incognita* on Tomato in Sri Lanka. *Nematol. Mediter.* 1994, *22*, 87–88.

Enrique, Q.; Alain V. E. Y. Bassiacridin, A Protein Toxic for Locusts Secreted by the Entomo-pathogenic Fungus *Beauveria bassiana*. *Mycol. Res.* **2004**, *108*, 441–452.

Fang, W.; Leng, B.; Xiao, Y.; Jin, K.; Ma, J.; Fan, Y.; Feng, J.; Yang, X.; Zhang, Y.; Pei, Y. Cloning of *Beauveria bassiana* Chitinase Gene Bbchit1 and its Application to Improve Fungal Strain Virulence. *Appl. Environ. Microbiol.* **2005**, *71*, 363–370.

Fang, W.; Pei, Y.; Bidochka, M. J. A regulator of a G protein Signalling (RGS) Gene, *cag8*, from the Insect-pathogenic Fungus *Metarhizium anisopliae* is Involved in Conidiation, Virulence and Hydrophobin Synthesis. *Microbiol.* **2007**, *153*, 1017–1025.

Federici, B. A. Bacteria as Biological Control Agents for Insects: Economics, Engineering, and Environmental Safety. Novel Biotechnologies for Biocontrol Agent Enhancement and Management. In *NATO Security through Science Series;* Vurro, M., Gressel, J. Eds., Springer: Netherlands, **2007,** pp. 25–51.

Fuente, L. D. A.; Quagliotto, L.; Bajsa, N.; Fabiano, E.; Altier, N.; Arias, A. Inoculation with *Pseudomonas fluorescens* Biocontrol Strains Does Not Affect the Symbiosis between Rhizobia and Forage Legumes. *Soil Biol. Biochem.* **2002,** *34,* 545–548.

Gaikwad, V. T.; Charde, T. H.; Gohalkar, R. T. Antagonistic Effect of *T. viride* and *T. harzianum* on *Fusarium oxysporum* f. sp. *lycopersici. Indian Phytopathol.* **1999,** *52,* 313.

Gaugler, R.; Kaya, H. K. *Entomopathogenic Nematodes in Biological Control.* CRC Press: Boca Raton, 1990.

Gawade, D. B.; Pawar, B. H.; Gawande, S. J.; Vasekar, V. C. Antagonistic Effect of *Trichoderma* Against *Fusarium moniliformae* the Causal of Sugarcane Wilt. *American-Eurasian J. Agric. Environ. Sci.* **2012,** *12,* 1236–1241.

Godoy, G.; Rodriguez-Kabana, R.; Morgan-Jones, G. Fungal Parasites of *Meloidogyne arenaria* Eggs in an Alabama Soil. A Mycological Survey and Greenhouse Studies. *Nematropica.* **1983,** *13,* 201–213.

Gopalakrishnan, C.; Mohan, K. S. Field efficacy of *Nomuraea rileyi* (Farlow) Samson Against Certain Lepidopteran Insect Pests of Cabbage. In *Biological Control of Lepidopteran Pests,* Proceedings of the Symposium of Biological Control of Lepidopteran Pests, July 17–18; Tandon, P. L.; Ballal, C. R.; Jalali, S. K.; Rabindra, R. J., Eds.; Bangalore, India **2003,** 2002, pp. 247–250.

Grewal, P. S.; Selvan, S.; Gaugler, R. Thermal Adaptation of Entomopathogenic Nematodes: Niche Breadthfor Infection, Establishment, and Reproduction. *J. Therm. Biol.* **1994,** *19,* 245–253.

Guo, J.; Qi, H.; Guo, Y.; Ge, H.; Gong, L.; Zhang, L.; Sun, P. Biocontrol of Tomato Wilt by Plant Growth-Promoting Rhizobacteria. *Biol. Control.* **2004,** *29,* 66–72.

Haas, D.; Keel, C. Regulation of Antibiotic Production in Root-colonizing *Pseudomonas* spp. and Relevance for Biological Control of Plant Disease. *Ann. Rev. Phytopathol.* **2003,** 41, 117–153.

Harman, G. E.; Hayes, C. K. *Biologically Based Technologies for Pest Control: Pathogens that are Pests of Agriculture. A Report to the Office of Technology Assessment,* US Congress. **1994,** p.75.

Harman, G. E.; Hayes, C. K.; Lorito, M.; Broadway, R. M.; Di Pietro, A.; Peterbauer, C.; Tronsmo, A. Chitinolytic Enzymes of *Trichoderma harzianum* Purification of Chitobiosidase and Endochitinase. *Phytopathology.* **1993,** *83,* 313–318.

Hashim, N.; Ibrahim, Y. B. *Efficacy of Entomopathogenic Fungi, Paecilomyces fumosoroseus, Beauveria bassiana and Metarhizium anisopliae var. majus Against Crocidolomia Binotalis (Lepidoptera; Pyralidae). Pertanika J. Trop. Agric. Sci.* **2003,** *26,* 103–108.

Hassett, D.; Cuppoletti, J.; Trapnell, B.; Lymar, S.; Rowe, J.; Yoon, S.; Hilliard, G.; Parvatiyar, K.; Kamani, M.; Wozniak, D.; Hwang, S.; McDermott, T.; Ochsner, U. Anaerobic Metabolism and Quorumsensing by *Pseudomonas aeruginosa* Biofilms in Chronically Infected Cystic Fibrosis Airways: Rethinking Antibiotic Treatment Strategies and Drug Targets. *Adv. Drug Deliv. Rev.* **2002,** *54,* 1425–1443.

Henis, Y.; Adams, P. B.; Lewis, J. A.; Papavizas, G. C. Penetration of Sclerotia of *Sclerotium rolfsii* by *Trichoderma* spp. *Phytopathology.* **1983,** *73,* 1043–1046.

Herrero-Galán, E.; Lacadena, J.; Martínez, del Pozo A.; Boucias, D. G.; Olmo, N.; Oñaderra, M.;Gavilanes, J. G. The insecticidal Protein Hirsutellin A from the Mite Fungal Pathogen *Hirsutella thompsonii* is a Ribotoxin. *Proteins.* **2008,** *72,* 217–28.

Ignoffo, C. M. The Fungus *Nomuraearileyi* as a Microbial Insecticide. In *Microbial Control of Pests and Plant Diseases*, 1970–80; Burges H. D., Ed.; Academic Press: London, UK, **1981**, pp 513–538.

Inbar, J.; Mendez, A.; Chet, I. Hyphal Interaction between *Trichoderma harzianum* and *Sclerotinia sclerotiorum* and its Role in Biological Control. *Soil Biol. Biochem.* **1996**, *28*, 757–763.

Jaffe, B. A. Augmentation of Soil with the Nematophagous Fungi *Hirsutella rhossiliensi* and *Arthrobotrys haptyla*. *Phytopathology.* **2000**, *90*, 498–504.

Jamalizadeh, M.; Etebarian, H.; Aminian, H.; Alizadeh, A. Evaluation of *Bacillus* spp. as Potential Biocontrol Agent for Postharvest Gray Mold Control on Golden Delicious Apple in Iran. *J. Plant Prot. Res.* **2009**, *49*, 405–410.

James, R. R.; Lighthart, B.; Susceptibility of the Convergent Lady Beetle (Coleoptera: Coccinellidae) to Four Entomogenous Fungi. *Environ. Entomol.* **1994**, *23*, 190–192.

James, R. R.; Buckner, J. S.; Freeman, T. P. Cuticular Lipids and Silverleaf Whitefly Stage Affect Conidial Germination of *Beauveria bassiana* and *Paecilomyces fumosoroseus*. *J. Invert. Pathol.* **2003**, *84*, 67–74.

Jegathambigai, V.; Wilson, R. S.; Wijeratnam; Wijesundera, R. L. C. *Trichoderma* as a Seed Treatment to Control *Helminthosporium* Leaf Spot Disease of *Chrysalidocarpus lutescens*. *World J. Agric. Sci.* **2009**, *5*, 720–728.

Kachhadiya, N. M.; Kapadia, M. N.; Jethva, D. M. Field Efficacy of *Nomuraea rileyi* (Farlow) Samson alone and in Combination with Insecticides Against *Spodoptera litura* (Fabricius) Infesting Groundnut. *Int. J. Plant Prot.* **2014**, *7*, 143–146.

Kapil, R.; Kapoor, A. S. Management of White Rot of Pea Incited by *Sclerotinia sclerotiorum* using *Trichoderma* spp. and Biopesticides. *Indian Phytopathol.* **2005**, *58*, 10–16.

Kareem, T. A.; Hassan, M. S. Evaluation of *Glomus mosseae* as Biocontrol Agents Against *Rhizoctonia solani* on tomato. *J. Biol. Agric. Healthc.* **2014**, *4*, 15–19.

Kaur, S.; Thakur, A.; Rajput, M. A Laboratory Assessment of the Potential of *Beauveria bassiana* (Balsamo) Vuillemin as a Biocontrol Agent of *Corcyra cephalonica* Stainton (Lepidoptera: Pyralidae). *J. Stored Prod. Res.* **2014**, *59*, 185–189.

Kaya, H. K. Soil Ecology. In *Entomopathogenic Nematodes in Biological Control*; Gaugler, R.; Kaya H. K., Eds.; CRC: Boca Raton, FL, 1990 pp. 93–111.

Kerr, A. Commercial Release of a Genetically Engineered Bacterium for the Control of Crown Gall. *Agric. Sci.* **1989**, *2*, 41–48.

Kerry, B. R.; Simon, A.; Rovira, A. D. Observations on the Introduction of *Verticillium chlamydosporium* and Other Parasitic Fungi into Soil for Control of the Cereal-cyst Nematode, *Heterodera avenue*. *Ann. Appl. Biol.* **1984**, *105*, 509–516.

Khanuchiya, S.; Parabia, F. M.; Patel, M.; Patel, V.; Patel, K.; Gami, B. Effect of *Pseudomonas fluorescence, P. aeruginosa* and *Bacillus subtilis* As Bio Control Agent for Crop Protection. *Cibtech J. Microbiol.* **2012**, *1*, 52–59.

Kiewnick, S.; Neumann, S.; Sikora, R. A.; Frey, J. E. Effect of *Meloidogyne incognita* Inoculum Density and Application Rate of *Paecilomyces lilacinus* Strain 251 on Biocontrol Efficacy and Colonization of Egg Masses Analyzed by Real-time Quantitative PCR. *Phytopathology.* **2011**, *101*, 105–12.

Kim, J. J.; Goettel, M. S.; Gillespie, D. R. Evaluation of *Lecanicillium longisporum*, Vertalec for Simultaneous Suppression of Cotton Aphid, *Aphis gossypii*, and Cucumber Powdery Mildew, *Sphaerotheca fuliginea*, on potted cucumbers. *Biological Control*, 2008. do10.1016/j.biocontrol.2008.02.003.

Koike, M.; Yoshida, S.; Abe, N.; Asano, K. Microbial Pesticide Inhibiting the Outbreak of Plant Disease Damage. US National Phase Appl. No. 11/568,369, 371(c), 2007.

Koppenhöfer, A. M.; Fuzy, E. M. Attraction of Four Entomopathogenic Nematodes to Four White Grub Species. *J. Invert. Pathol.* **2008,** *99,* 227–234.

Koppenhöfer, A. M.; Grewal, P. S.; Fuzy, E. M. Virulence of the Entomopathogenic Nematodes *Heterorhabditis bacteriophora, Heterorhabditis zealandica,* and *Steinernema scarabaei* Against Five White Grub Species (Coleoptera: Scarabaeidae) of Economic Importance in Turfgrass in North America. *Biol. Control.* **2006,** *38,* 397–404.

Kusunoki, K.; Kawai, A.; Aiuchi, D.; Koike, M.; Tani, M.; Kuramochi, K.; Biological Control of *Verticillium* Black-spot of Japanese Radish by Entomopathogenic *Verticillium lecanii* (*Lecanicillium* spp.). *Res. Bull. Obihiro Univ.* **2006,** *27,* 99–107.

Lam, S. T.; Gaffney, T. D. Biological Activities of Bacteria Used in Plant Pathogen Control. In *Biotechnology in Plant Disease Control*; Chet, I., Ed.; John Wiley: New York, **1993,** pp. 291–320.

Lee, J. Y.; Kang, S. W.; Yoon, C. S.; Kim, J. J.; Choi, D. R.; Kim, S. W. *Verticillium lecanii* Spore Formulation Using UV Protectant and Wetting Agent and the Biocontrol of Cotton Aphids. *Biotechnol. Lett.* **2006,** *28,* 1041–1045.

Leelavathi, M. S.; Vani L.; Reena, P. Antimicrobial Activity of *Trichoderma harzianum* Against Bacteria and Fungi. *Int. J. Curr. Microbiol. Appl. Sci.* **2014,** *3,* 96–103.

Leeman, M.; Den Ouden, F. M.; Van Pelt, J. A.; Dirkx, F. P. M.; Steijl, H.; Bakker, P. A. H. M.; Schippers, B. Iron Availability Affects Induction of Systemic Resistance to *Fusarium* Wilt of Radish by *Pseudomonas fluorescens. Phytopathology.* **1996,** *86,* 149–155.

Leij, De A. A. M.; Kerry, B. R. The Nematophagous Fungus *Verticilium chlamydosporizkrn* as a Potential Biological Control Agent for *Meloidogyne arenaria. Frans Revue Nématolol.* **1991,** *14,* 157–164.

Leinhos, G. M. E.; Buchenauer, H. Hyperparasitism of Selected Fungi on Rust Fungi of Cereal. *Z. Pflanzenkr. Pflanzenschutz.* **1992,** *99,* 482–498.

Lewis, J. A.; Barksdale, T. H.; Papavizas, G. C. Greenhouse and Field Management of Tomato Fruit Rot Caused by *Rhizoctonia solani. Crop Prot.* **1990,** *9,* 8.

Liu, L.; Kloepper, J. W.; Tuzun, S. Induction of Systemic Resistance in Cucumber Against Bacterial Angular Leaf Spot by Plant Growth-promoting Rhizobacteria. *Phytopathology.* **1995,** *85,* 843–847.

Lo, C. T.; Nelson, E. B.; Hayes, C. K.; Harman, G. E. Ecological Studies of Transformed *Trichoderma harzianum* Strain 1295-22 in the Rhizosphere and on the Phylloplane of Creeping Bentgrass. *Phytopathology.* **1998,** *88,* 129–136.

Lohse, R.; Jakobs-Schönwandt, D.; Patel, A. V. Screening of Liquid Media and Fermentation of an Endophytic *Beauveria bassiana* strain in a Bioreactor. *AMB Exp.* **2014,** *4,* 47.

Loper, J. E.; Buyer, J. S. Siderophores Inmicrobial Interactions on Plant Surfaces. *Mol. Plant Microbe Interact.* **1991,** *4,* 5–13.

Lopes, F. A, Steindorff, A. S, Geraldine, A. M.; Brandão, R. S.; Monteiro, V. N.; Lobo, M. Jr.; Coelho, A. S.; Ulhoa, C. J.; Silva, R. N. Biochemical and Metabolic Profiles of *Trichoderma* Strains Isolated from Common Bean Crops in the Brazilian Cerrado, and Potential Antagonism Against *Sclerotinia sclerotiorum. Fungal Biol.* **2012,** *116,* 815–824.

Lorito, M.; Hayes, C. K.; Peterbauer, C.; Tronsmo, A.; Klemsdal, S.; Harman, G. E. Antifungal Chitinolytic Enzymes from *Trichoderma harzianum* and *Gliocladium virens*: Purification, Characterization, Biological Activity and Molecular Cloning In *Chitin Enzymology*; Muzzarelli, R. A., Ed.; Eur. Chitin Soc.: Ancona, Italy, **1993,** pp. 383–392.

Lorito, M.; Woo, S. L.; Harman, G. E.; Monte, E. Translational Research on *Trichoderma*: From Omics to the Field. *Ann. Rev. Phytopathol.* **2010**, *48*, 395–417.

Mach, R. L.; Butterweck, A.; Schindler, M. Messner, R.; Herzog, P.; Kubicek C. P. In *Molecular Regulation of Formation of Xylanase (XYN) I and II by Trichoderma reesei, Cellulases and Other Hydrolases: Enzyme Structures, Biochemistry, Genetics and Applications*; Suominen, P.; Reinikainen, T., Ed.; Foundation of Biotechnical and Industrial Fermentation Research: Helsinki, Finland, **1993**, pp. 211–216.

Magda, S.; Said, S. Efficacy of Two Entomopathogenic Fungi Against Corn Pests Under Laboratory and Field Conditions in Egypt. *Eur. J. Acad. Essays.* **2014**, *1*, 1–6.

Malarvannan, S.; Murali, P. D.; Shanthakumar, S. P.; Prabavathy, V. R.; Nair, S. Laboratory Evaluation of the Entomopathogenic Fungi, *Beauveria bassiana* Against the Tobacco Caterpillar, *Spodoptera litura* Fabricius (Noctuidae: Lepidoptera). *J. Biopesticides.* **2010**, *3*, 126–131.

Matloob, A. A. H.; Juber, K. S. Biological Control of Bean Root Rot Disease Caused by *Rhizoctonia solani* Under Green House and Field Conditions. *Agric. Biol. J. N. Am.* **2013**, *4*, 512–519.

Matsubara, Y.; Tamura, H.; Harada, T.; Growth Enhancement and *Verticillium* Wilt Control by Vesicular Arbuscular Mycorrhizal Fungus Inoculation in Eggplant. *J. Jpn. Soc. Hortic. Sci.* **1995**, *64*, 555–561.

Matter, M. M.; Sabbour, M. M. Differential Efficacies of *Nomuraea rileyi* and *Isaria fumosorosea* on Some Serious Pests and the Pests' Efficient Predator Prevailing in Tomato Fields in Egypt. *J. Plant Prot. Res.* **2013**, *53*, 103–109.

Maurhofer, M.; Hase, C.; Meuwly, P.; Metraux, J. P.; Defago, G. Induction of Systemic Resistance to Tobacco Necrosis Virus. *Phytopathology.* **1994**, *84*, 139–146.

Mazodze, R.;Zvoutete, P. Efficacy of *Metarhizium anisopliae* Against *Heteronychus licas* (Scarabaedae: Dynastinae) in Sugarcane in Zimbabwe. *Crop Prot.* **1999**, *18*, 571–575.

McAllister, C. B.; García-Romero, I.; Godeas, A.; Ocampo, J. A. *In vitro* Interactions Between *Trichoderma koningii, Fusarium solani* and *Glomus mosseae*. *Soil Biol. Biochem.* **1994**, *26*, 1369–74.

McCoy, C. W. Fungi: Pest Control by *Hirsutella thompsonii*. In *Microbial Control of Insects, Mites and Plant Diseases*; Burges, H. D. Ed. Academic Press: New York, 1981, pp 499–512.

McCoy, C. W. *Pathogens of Eriophyoid Mites*. In *Eriophyoid Mites–Their Biology, Natural Enemies and Control*; Lindquist, E. E.; Sabelis, M. W.; Bruin, J. Eds.; Elsevier Science B. V.: Amsterdam, **1996**, pp 481–490.

Meyer, S. L. F.; Johnson, G.; Dimock, M.; Fahey, J. W.; Huettel, R. N. Field Efficacy of *Verticillium lecanii*, Sex Pheromone, and Pheromone Analogs as Potential Management Agents for Soybean Cyst Nematode. *J. Nematol.* **1997**, *29*, 282–288.

Milgroom, M. G.; Cortesi, P.; Biological Control of Chestnut Blight with Hypovirulence: A Critical Analysis. *Annu. Rev. Phytopathol.* **2004**, *42*, 311–38.

Miller, T. C.; Gubler, W. D.; Laemmlen, F. F.; Geng, S.; Rizzo, D. M. Potential for Using *Lecanicillium lecanii* for Suppression of Strawberry Powdery Mildew. *Biocontrol Sci. Technol.* **2004**, *14*, 215–220.

Mishra, B. K.; Mishra, R. K.; Mishra, R. C.; Tiwari, A. K.; Yadav, R. S.; Dikshit, A. Biocontrol Efficacy of *Trichoderma viride* Isolates Against Fungal Plant Pathogens Causing Disease in *Vigna radiata* L. *Arch. Appl. Sci. Res.* **2011**, *3*, 361–369.

Mohammadbeigi, A.; Port, G. Efficacy of *Beauveria bassiana* and *Metarhizium anisopliae* Against *Uvarovistia zebra* (Orthoptera: Tettigoniidae) via Contact and Ingestion. *Int. J. Agric. Crop Sci.* **2013**, 5, 138–146.

Mukhopadhyay, A. N. Comparative Antagonistic Properties of *Gliocladium virens* and *Trichoderma harzianum* on *Sclerotium rolfsii* and *Rhizoctonia solani*. *Indian Phytopathol.* **1987**, *40*, 276.

Murali, R. K.; Mohsan, C. H, Arunalakshmi, K.; Padmavathi, J.; Umadevi, K. *Beauveria bassiana* (Bals.)Vill. (Hyphomycetes, Moniliales) in Cotton Pest Management, a Field Trial on Cotton Leaf Roller (*Sylepta derogeta* (Fabricius) Lepidoptera; pyralidae). *J. Entomol. Res.* **1999**, *23*, 267–71.

Nelson, E. B. Exudate Molecules Initiating Fungal Responses to Seeds and Roots. *Plant Soil.* **1990**, *129*, 61–73.

Nguyen, N. V.; Kim, Y. J.; Oh, K. T.; Jung, W. J.; Park; R. D. The Role of Chitinase from *Lecanicillium antillanum* B-3 in Parasitism to Root-knot Nematode *Meloidogyne incognita* eggs. *Biocontrol Sci. Technol.* **2007**, *17*, 1047–1058.

Norman, J. R.; Hooker, J. E.; Sporulation of *Phytophthora fragariae* Shows Greater Stimulation Byexudates of Non-mycorrhizal Strawberry Roots. *Mycol. Res.* **2000**, *104*, 1069–1073.

Nuss, D. L. Hypovirulence: Mycoviruses at the Fungal-plant Interface. *Nat. Rev. Microbiol.* **2005**, *8*, 632–42.

Nussenbaum, A. L.; Lecuona, R. E. Selection of *Beauveria bassiana* Sensulato and *Metarhizium anisopliae* Sensulato Isolates as Microbial Control Agents Against the Boll Weevil (*Anthonomus grandis*) in Argentina. *J. Invert. Pathol.* **2012**, *110*, 1–7.

Oerke, E. C. Crop Losses to Pests. *J. Agric. Sci.* **2006**, *144*, 31–43.

Osando, J. M.; Waudo, S. W. Interaction between *Trichoderma* Species and *Armillaria* Root Rot Fungus of Tea in Kenya. *Int. J. Pest Manag.* **1994**, *40*, 69–74.

Ownley, B. H.; Kimberly, D. G.; Fernando E. V.; Endophytic Fungal Entomopathogens with Activity Against Plant Pathogens: Ecology and Evolution. *BioControl.* **2010**, *55*, 113–128.

Padanad, M. S.; Krishnaraj, P. U. Efficacy of *Nomuraea rileyi* and Spinosad Against Olive Pests Under Laboratory and Field Conditions in Egypt, Pathogenicity of Native Entomopathogenic Fungus *Nomuraea rileyi* Against *Spodoptera litura*. *Plant Health Prog.* **2009**, doi:10.1094/PHP-2009-0807-01-RS.

Palleroni, J. N. (1984) Pseudomonadacea. In *Bergay's Manual Systematic Bacteriology, Vol. 1*; Kreig, N. R.; Holt, J. Eds. Williams and Wilkins: Baltimore, 1984.

Papavizas, G. C.; Lumsden, R. D. Biological Control of Soil-borne Fungal Propagules. *Ann. Rev. Phytopathol.* **1980**, *18*, 389–413.

Patil, R. H. and Abhilash, C. *Nomuraea rileyi (Farlow) Samson: A Bio-pesticide IPM Component for the Management of Leaf Eating Caterpillars in Soybean Ecosystem*. International Conference on Biological, Civil and Environmental Engineering (BCEE-2014) March 17–18, Dubai (UAE), 2014.

Perveen, Z.; Shahzad, S. A Comparative Study of the Efficacy of *Paecilomyces* Species Against Root-Knot Nematode *Meloidogyne incognita*.
Pak. J. Nematol. **2013**, *31*, 125–131.

Phukon, M.; Sarma, I.; Borgohain, R.; Sarma, B.; Goswami, J. Efficacy of *Metarhizium anisopliae, Beauveria bassiana* and Neemoil Against Tomato Fruit Borer, *Helicoverpa armigera* under in Field Condition. *Asian J. Bio Sci.* **2014**, *9*, *151–155.*

Qualhato, T. F.; Lopes, F. A. C.; Steindorff, A. S. Brandão, R. S.; Jesuino, R. S. A.; Ulhoa, C. J. Mycoparasitism Studies of *Trichoderma* Species Against Three Phytopathogenic Fungi: Evaluation of Antagonism and Hydrolytic Enzyme Production. *Biotechnol. Lett.* **2013**, *35*, 1461–1468.

Rajasekhar, P.; Kalidas, P. *Beauveria bassiana*–A Novel Biocontrol Agent Against the Leaf Webworms of Oil Palm. *Curr. Biotica.* **2012**, *6*, 334–341.

Ramesh, K.; Murali Mohan, C.; Arunalakshmi, K.; Padmavathi, J.; Umadevi, K. *Beauveria bassiana* (Bals.) Vill. (Hyphomycetes, Moniliales) in Cotton Pest Management; a Field Trial on Cotton Leaf Roller (*Sylepta derogeta* (Fabricius) Lepidoptere; Pyralidae. *J. Ent. Res.* **1999**, *23*, 267–271.

Ran, L. X.; Liu, C. Y.; Wu, G. J.; Loon L. C. van, Bakker, P. A. H. M. Suppression of Bacterial Wilt in *Eucalyptus urophylla* by Fluorescent *Pseudomonas* spp. in China. *Biol. Control.* **2005**, *32*, 111–120.

Reddy, G. V. P.; Zhao, Z.; Humber, R. A. Laboratory and Field Efficacy of Entomopathogenic Fungi for the Management of the Sweetpotato Weevil, *Cylas formicarius* (Coleoptera: Brentidae). *J. Invert. Pathol.* **2014**, *122*, 10–15.

Richard, J. S.; Neal, T. D.; Karl, J. K.; Michael, R. K. Model Reactions for Insect Cuticle Sclerotization: Participation of Amino Groups in the Cross-linking of *Manduca sexta* Cuticle Protein MsCP36, Insect. *Biochem. Mol. Biol.* **2010**, *40*, 252–258.

Ritu, A.; Anjali, C.; Nidhi, T.; Sheetal, P.; Deepak, B. Biopesticidal Formulation of *Beauveria bassiana* Effective Against Larvae of *Helicoverpa armigera*. *J. Biofertil. Biopesticide.* **2012**, *3*, 120.

Rombach, M. C.; Aguda, R. M.; Shepard, B. M.; Roberts, D. W. Infection of the Brown Planthopper, *Nilaparvata lugens* (Delphacidae: Homoptera) by Field Application of Entomogenous Fungi (Deuteromycotina). *Environ. Entomol.* **1986**, *15*, 1070–1073.

Ruisheng, A.; Grewal, P. S. Differences in the Virulence of *Heterorhabditis bacteriophora* and *Steinernema scarabaei* to Three White Grub Species: The Relative Contribution of the Nematodes and their Symbiotic Bacteria. *Biol. Control.* **2007**, *43*, 310–316.

Sabbour, M. M. Department of Pests and Plant Protection, National Research Centre, Dokki, Cairo, Egypt. *Global J. Biodivers. Sci. Manag.* **2013**, *3*, 228–232.

Sain S. K.; Gour, H. N. Efficacy of Isolated Bacteria in In Vitro Inhibition of *Xanthomonas axonopodis* pv. *cyamopsidis* and Prevention of Bacterial Leaf Blight of Cluster Bean in the Field. *J. Biol. Control.* **2009**, *23*, 421–425.

Sain S. K. Efficacy of Plant Growth Promoting Rhizobacteria in In Vitro Inhibition of *Xanthomonas axonopodis* pv. *glycines* and Prevention of Bacterial Pustules of Soybean in the Field. *J. Biol. Control.* **2010**, *24*, 333–337.

Sain S. K.; Gour H. N.; Sharma, P. Evaluation of Botanicals and PGPRs Against *Xanthomonas campestris* pv. *campestris*, An Incitant of Black Rot of Cauliflower. *J. Eco-Friendly Agric.* **2007**, *2*, 178–182.

Saleem, B.; Kandasamy, U. Antagonism of *Bacillus* Species (strain BC121) Towards *Curvularia lunata*. *Curr. Sci.* **2002**, *82*, 1457–1463.

Samson, P. R.; Staier, T. N.; Bull, J. I. Evaluation of An Application Procedure for *Metarhizium anisopliae* in Sugarcane Ratoons for Control of the White Grub *Dermolepida albohirtum*. *Crop Prot.* **2006**, *25*, 741–747.

Samson, R. A.; Evans, H. C.; Latge, J. P. *Atlas of Entomopathogenic Fungi*. Springer Verlag: Berlin, 1988.

Schirmbock, M.; Lorito, M.; Wang, Y-L.; Hayes, C. K.; Arisan-Atac, I.; Scala, F.; Harman, G. E.; Kubicek, C. Parallel Formation and Synergism of Hydrolyticenzymes and Peptaibol Antibiotics, Molecular Mechanisms Involved in the Antagonistic Action of *Trichoderma harzianum* against Phytopathogenic Fungi. *Appl. Environ. Microbiol.* **1994**, *60*, 4364–4370.

Schubert, M.; Fink S.; Schwarze, F. W. M. R. Evaluation of *Trichoderma* spp. as a Biocontrol Agent Against Wood Decay Fungi in Urban Trees. *Biol. Control.* **2008**, *45*, 111–123.

Shah, P. A.; Pell, J. K. Entomopathogenic Fungi as Biological Control Agents. *Appl. Microbiol. Biotechnol.* **2003**, *61*, 413–423.

Shaigan, S.; Seraji, A.; Moghaddam, S. A. M. Identification and Investigation on Antagonistic Effect of *Trichoderma* spp. on Tea Seedlings White Foot and Root Rot (*Sclerotium rolfsii* Sacc.) In Vitro Condition. *Pak. J. Biol. Sci.* **2008**, *19*, 2346–2350.

Sharma P.; Sain, S. K. Use of Biotic Agents and Abiotic Compounds Against Damping Off of Cauliflower Caused by *Pythium aphanidermatum*. *Indian Phytopathol.* **2005**, *58*, 395–401.

Sharma P.; Kadu, L. N.; Sain, S. K. Biological Management of Die-back and Fruit Rot of Chilli (*Capsicum annum* L.) Caused by *Colletotrichum capsici* (Syd.) Butler and Bisby. *Indian J. Plant Prot.* **2005**, *33*,226–230.

Sharma, A.; Diwevidi, V. D.; Singh, S.; Pawar, K. K.; Jerman, M.; Singh, L. B.; Singh, S.; Srivastawa, D. Biological Control and its Important in Agriculture. *Int. J. Biotechnol. Bioeng. Res.* **2013**, *4*, 175–180.

Sharma, P.; Sain, S. K. Induction of Systemic Resistance in Tomato and Cauliflower by *Trichoderma* Species Against Stalk Rot Pathogen (*Sclerotinia sclerotiorum*). *J. Biol. Control.* **2004**, 18, 21–28.

Sharma, P.; Sain S. K.; Sindhu M.; Kadu L. N. Integrated use of CGA 2045704 and *Trichoderma harzianum* on Downy Mildew Suppression and Enzymatic Activity in Cauliflower. *Ann. Agric. Res.* **2004** 25, 129–134.

Shekharappa. Biological Control of Earhead Caterpillar, *Helicoverpa armigera* Hubner in Sorghum. *J. Plant Prot. Sci.* **2009**, *1*, 69–70.

Sheroze, A.; Rashid, A.; Nasir, M. A.; Shakir, A. S. Evaluation of Some Biocontrol Agents/ Antagonistic Microbes Againstpastule Development of Leaf Rust of Wheat Caused by *Puccinia recondita* f. sp. *tritici* Roberge ex. Desmaz (Erikson and Henn) D. M. Henderson. *Plant Pathol. J.* **2002**, *1*, 51–53.

Shipp, L.; Kapongo, J. P.; Park, H.; Keven, P. Effect of Bee-vectored *Beauveria bassiana* on Greenhouse Beneficials Under Greenhouse Cage Conditions. *Biol. Control.* **2012**, *63*, 135–142.

Sivakumar, T.; Eswaran, A.; Balabaskar, P. Bioefficacy of Antagonists Against for the Management of *Fusarium oxysporum* f.sp. *lycopersici* and *Meloidogyne incognita* Disease Complex of Tomato Under Field Condition. *Plant Arch.* 2008, *8*, 373–377.

Srisukchayakul, P.; Wiwat, C.; Pantuwatana, S. Studies on the Pathogenesis of the Local Isolates of *Nomuraea rileyi* Against *Spodoptera litura*. *Sci. Asia.* **2005**, *31*, 273–276.

St Leger, R. J.; Wang, C. Entomopathogenic Fungi and the Genomic Era, In *Insect Pathogens*: *Molecular Approaches and Techniques*; Stock, S. P.; Vandenberg, J.; Glazer, I.; Boemare, N., Eds.; CABI: Wallingford, UK, **2009**, pp. 366–400.

Stephan, Z. A.; Hassan, M. S.; Abbass, H. I.; Antoan, B. G. Influence of Vesicular–Arbuscularmycorrhizae on Wilt–Root Knot Disease Complex of Eggplant and Tomato Seedlings. *Iraq. J. Agric.* **1999**, *4*, 54–60.

Suárez-Estrella, F.; Arcos-Nievas, López, M. J.; Vargas-García, M. C.; Moreno, J. Biological Control of Plant Pathogens by Microorganisms Isolated from Agro-Industrial Composts. *Biol. Control.* 2013, *67*, 509–515.

Sundheim, L.; Tronsmo, A. Hyperparasites Inbiological Control. In *Biocontrol of Plant Diseases*; Mukerji, K. G.; Garg, K. L., Eds.; CRC Press: Boca Raton, FL, **1988**, pp. 53–69.

Thungrabeab, M.; Tongma, S. Effect of Entomopathogenic Fungi, *Beauveria bassiana* (Balsam) and *Metarhizium anisopliae* (METSCH) on Non-target Insects. *KMITL Sci. Technol. J.* **2007**, 7(S1), 8–12.

Townsend, R. J.; Nelson, T. L.; Jackson T. A. *Beauveria brongniartii*–A Potential Biocontrol Agent for Use Against Manuka Beetle Larvae Damaging Dairy Pastures on Cape Foulwind. *N. Z. Plant Prot.* **2010**, *63*, 224–228.

Trotta, A.; Varese, G. C.; Gnavi, E.; Fusconi, A.; Sampo, S.; Berta, G. Interactions between the Soilborne Root Pathogen *Phytophthora nicotianae* var. *parasitica* and the Arbuscular Mycorrhizal Fungus *Glomus mosseae* in Tomato Plants. *Plant Soil.* **1996,** *185,* 199–209.

Vänninen, I.; Hokkanen, I. Efficacy of Entomopathogenic Fungi and Nematodes Against *Argyresthia conjugella* (*Lep.:Yponomeutidae*). *Entomophaga.* **1997,** *42,* 377–385.

Verhaar, M. A.; Hijwegen, T.; Zadoks, J. C. Improvement of the Efficacy of *Verticillium lecanii* Used in Biocontrol of *Sphaerotheca fuliginea* by Addition of Oil Formulations. *Biocontrol.* **1999,** *44,* 73–87.

Vey, A.; Hoagland, R.; Butt, T. M.. Toxic Metabolites of Fungal Biocontrol Agents, In *Fungi as biocontrol agents*; Butt, T. M.; Jackson, C. W.; Magan, N. Eds.; CAB International: Wallingford, **2001,** pp.311–345.

Vinale, F.; Sivasithamparam, K.; Ghisalberti, E. L.; Marra, R.; Barbetti, M. J.; Li, H. *et al.* A Novel Role for *Trichoderma* Secondary Metabolites in the Interactions with Plants. *Physiol. Mol. Plant Pathol.* **2008,** *72,* 80–86.

Waghmare, S. J.; Kurundkar, B. P. Efficacy of Local Isolates of *Trichoderma* spp. Against *Fusarium oxysporum* f. sp. *carthami* Causing Wilt of Safflower. *Adv. Plant Sci.* **2011,** *24,* 37–38.

Wakil, W.; Ghazanfar, M. U.; Kwon, Y. J.; Ullah, E.; Islam, S.; Ali, K. Testing *Paecilomyces lilacinus*, Diatomaceous Earth and *Azadirachta indica* Alone and in Combination Against Cotton Aphid (*Aphis gossypii* Glover) (Insecta: Homoptera: Aphididae). *Afr. J. Biotechnol.* 2012, *11,* 821–828.

Wallace, J. M.; Drake, R. J. *Progress Report of Studies in California on Preimmunization Against Citrus Tristeza Virus in Budded Citrus Trees.* Proc. 10th Conf. IOCV. IOCV: Riverside, CA, **1976,** pp. 58–62.

Walters, A.; Barker, K. R. Efficacy of *Paecilomyces lilacinus* in Suppressing *Rotylenchulus reniformis* on Tomato. *Suppl. J. Nematol.* **1994,** *26*(4S), 600–605.

Wang, C.; St. Leger, R. J. A Scorpion Neurotoxin Increases the Potency of a Fungal Insecticide. *Nat. Biotechnol.* **2007,** *25,* 1455–1456.

Weindling R. *Trichoderma lignorum* As a Parasite of Other Soil Fungi. *Phytopathology.* **1932,** *22,* 837–845.

Weindling, R.; Fawcett, H. S. Experiments in the Control of *Rhizoctonia* Damping Off of Citrus Seedling. *Hilgardia.* **1936,** *10,* 1–16.

Wilhite, S. E.; Lumsden, R. D.; Straney, D. C. Mutational Analysis of Gliotoxin Production by Thebiocontrol Fungus *Gliocladium virens* in Relation to suppression of *Pythium* Damping-off. *Phytopathology.* **1994,** *84,* 816–821.

Willoughby, B. E.; Glare, T. R.; Kettlewell, F. J.; Nelson, T. L. *Beauveria bassiana* as a Potential Biocontrol Agent Against the Clover Root Weevil, *Sitona Lepidus.* Proc. 51st N. Z. Plant Prot. Conf. **1998,** 9–15.

Wright, J. E.; Chandler, L. D. Laboratory Evaluation of the Entomopathogenic Fungi *Beauveria bassiana* Against Boll Weevil (Coleoptera; Curculionidae). *J. Invert. Pathol.* **1991,** *58,* 448–49.

Wright, J. E.; Chandler, L. D. Development of Biorational Mycoinsecticide *Beauveria bassiana* Conidial Formulation and Its Application Against Boll Weevil Populations (Coleoptera; Curculionidae). *J. Econ. Entomol.* **1992,** *85,* 1130–35.

Yadav, J. P. S.; Tripathi, N. N. Antagonistic Effects of *T. viride* Against *Rhizoctonia solani.* *Indian Phytopathol.* **1999,** *52,* 321.

Yadav, P.; Deshpande, M. V. Control of Beet Armyworm, *Spodoptera litura* (Fabricius) by Entomopathogenic Fungi, *Nomuraea rileyi* N812, *Beauveria bassiana* B3301 and *Metarhizium anisopliae* M34412. *Biopesticide Int.* **2012**, *8,* 107–114.

Yao, M. K.; Tweddell, R. J.; Désilets, H. Effect of Two Vesicular-arbuscular Mycorrhizal Fungi on the Growth of Micropropagated Potato Plantlets and on the Extent of Disease Caused by *Rhizoctonia solani. Mycorrhiza.* **2002,** *12,* 235–242.

Yu, Z.; Xiong, J.; Zhou, Q.; Luo, H.; Hu, S.; Xia, L.; Sun, M.; Li, L.; Yu, Z. The Diverse Nematicidal Properties and Biocontrol Efficacy of *Bacillus thuringiensis* Cry6A Against the Root-knot Nematode *Meloidogyne hapla. J. Invert. Pathol.* **2015,** *125,* 73–80.

Zare, R.; Gams, W. A Revision of *Verticillium* Section Prostrata. IV. The genera *Lecanicillium* and *Simplicillium. Nova Hedwigia.* **2001,** *73,* 1–50.

Zhang L.; Yang E.; Xiang M.; Liu X.; Chen, S. Population Dynamics and Biocontrol Efficacy of the Nematophagous Fungus *Hirsutella rhossiliensis* as Affected by Stage of the Soybean Cyst Nematode. *Biol. Control.* **2008,** *47,* 244–249.

PART IV
Industrial Microbiology and Microbial Biotechnology

CHAPTER 10

MASS PRODUCTION, QUALITY CONTROL, AND SCOPE OF BIOFERTILIZERS IN INDIA

HARSHA N. SHELAT*, R. V. VYAS, and Y. K. JHALA

Department of Agricultural Microbiology & Bio-fertilizers Projects, B. A. College of Agriculture, Anand Agricultural University, Anand 388110, Gujarat, India

Corresponding author. E-mail: hnshelat@aau.in; hnshelat@gmail.com

CONTENTS

ABSTRACT

In India, chemical fertilizers which are based on fossil fuel are mainly imported. Moreover, subsidies on fertilizers impact on GDP of the country. Dependence on chemicals results in loss in soil health. Microorganisms have emerged as the potential alternative for the productivity, reliability, and sustainability of the global food chain. Biofertilzers (carrier based) are paramount over the chemical fertilizers having remarkable effect on the global agriculture productivity last 2 decades. A few limitations of the carrier based biofertilizers are rectified by developing liquid biofertilizers which are having great promise for the cost-effective sustainable agriculture (Vyas et al., 2008). Biofertilizers are produced and sold by private sector, NGOs, State and Central Government units, state agricultural universities (SAUs), and others for the last 3 decades but still their usage is not up to the mark. Quality control guidelines for biofertilizers were included in the Fertilizer Control Order issued by Government of India which if, followed during production and sell of biofertilizers, will ensure supply of good quality biofertilizers to farmers, thereby generating faith in biofertilizers—an alternative to chemical fertilizers. To meet the rising demands for food commodities, efforts are being made by the state and central governments for sufficient agricultural production by popularizing biofertilizers and making them available to the farming community.

10.1   INTRODUCTION

A biofertilizer could be defined as the formulated product containing one or more microorganisms that enhance the nutrient status and the growth and yield of the plants either by replacing soil nutrients, by making nutrients more available to plants, and/or by increasing plant access to nutrients.

India is probably the country with the most complete legal framework related to biofertilizers. Ministry of Agriculture, Department of Agriculture and Cooperation, Government of India, New Delhi, vide their order dated March 24, 2006 included biofertilizers and organic fertilizers under Section 3 of the Essential Commodities Act, 1955 (10 of 1995), in Fertilizer (Control) Order, 1985 (TFCO, 1985). These rules were further amended in respect of applicability, specifications, and testing protocols vide Gazette notification November 2009. In this act, the term biofertilizer means "the product containing carrier based (solid or liquid) living microorganisms which are agriculturally useful in terms of nitrogen fixation, phosphorus solubilization, or nutrient mobilization to increase the productivity of the soil and/or crop." The term is covered under the broad definition of fertilizers, which means "any substance used or intended to be used as a fertilizer of the soil and/or crop." (Vora et al., 2008)

The quality of a biofertilizer shall be assured to guarantee the success of the inoculation and to promote acceptance by the farmers. Normally, the term "quality" refers only to the density and viability of the available micro-organisms and their preservation. However, to assure a proper quality of the product for the final users, the legislation should introduce other parameters to be controlled at production level, which can later be reflected in the labeling requirements. The production standard and the label shall include the defini-tion of parameters such as the microbial density at the time of manufacture and at the expiry date, the expiry period, the permissible contamination, the pH, the moisture, the identification of the microbial strain, the specification of the kind of carrier utilized, and most important is efficiency of the micro-bial culture used.

10.2 MASS PRODUCTION OF BIOFERTILIZERS

10.2.1 BACTERIAL INOCULANTS

There are several steps involved in production of biofertilizers/microbial inoculants (Fig. 10.1). Product being live or latent microbes' utmost care is required in every step. There are three types of formulations: carrier/powder, liquid and granular available in market; among them, liquid is gaining more popularization due to several advantages. Although, India boasts to be the largest biofertilizers producer in the world, the creditability of biofertilizers is slow among farmers. Due to limitations of the carrier-based inoculums, formulation of liquid based biofertilizer was developed. This technology is an alternative solution to carrier-based biofertilizers, microbial pesticides, and bio-control agents. There are four basic concepts of formulations. They are as follows:

1. To stabilize the organisms during production, distribution, and storage.
2. To easily deliver in the field in the most appropriate manner.
3. To protect the microorganisms from harmful environmental factors at the target site (field), thereby increasing persistence.
4. To enhance activity of the organism at the target site by increasing its activity, reproduction, contact, and interaction with the target crops.

10.2.1.1   PREPARATION OF INOCULUM OR BROTH

- Prepare the respective broth media for *Rhizobium*, *Azospirillum*, *Azotobacter*, *Acetobacter*, phosphate solubilizing bacteria (PSB), potash mobilizing bacteria (KMB), in 250 mL, 3 L, and 5 L, flask and sterilize.
- Inoculate the media in 250 mL flask with efficient strains of the concerned organism under aseptic conditions.
- Incubate the flasks under shaking condition in gyratory shakers at room temperature for 2 days (Phosphobacteria), 3–5 days (*Rhizobium*), and 5–7 days (*Azotobacter, Azospirillum,* and *Acetobacter*).
- Check the flask for growth and determine the population load. This will serve as the starter culture.
- Using this starter culture (at log phase), inoculate the large flask (3 L and 5 L) containing media for the respective organisms.
- Incubate the large flasks at room temperature with occasional shaking for the growth period mentioned above.
- Prepare the concerned media in large quantities in fermenter, sterilize well, cool, and keep ready.
- Inoculate the media in fermenter with the log phase culture grown in large flasks (usually 1–2% inoculum is sufficient; however, inoculation may be done up to 5% depending on the growth and time requirement).
- Prior to inoculation, determine the population load of the organism in the broth used for inoculation.
- Grow the cells in fermenter by providing aeration, passing sterile air through compressor, and sterilizing agents like glass wool, cotton wool, acid, and others and steaming.
- Check the population of inoculated organism and contamination if any at growth period.
- Harvest the cell (spent broth) after incubation period and use for inoculant preparation.
- Determine the population load at harvest and check bacterial/fungal contamination in the spent broth.
- The population of the organisms should be more than 10^9 cells per ml of the fermented broth.
- There should not be any fungal or any other bacterial contamination at 10^{-5} dilution level.

- It is not advisable to store the broth after fermentation for periods longer than 24 h. Even at 4°C number of viable cells begins to decrease.

10.2.1.2 CARRIER OR POWDER FORMULATIONS

10.2.1.2.1 Preparation of Carrier Materials

- Grind the carrier material—peat soil or lignite to a fine powder so as to pass through 212 μm IS sieve.
- Neutralize the pH of the carrier material with the help of calcium carbonate powder as peat soil and lignite are acidic in nature.
- Sterilize the neutralized carrier material in an autoclave to eliminate the contaminants.
- Spread the neutralized sterilized carrier material in clean dry sterile metallic trays.
- Pour the appropriately grown bacterial (*Rhizobium*, *Azospirillum*, *Azotobacter*, *Acetobacter*, Phosphobacteria, KMB, etc.) cultures on the carrier and mix well manually by wearing sterile gloves or by mechanical mixer. The culture suspension is to be added to a level of 40% water holding capacity depending upon the population. The ratio of broth:carrier is to be maintained 2:3.
- Pack the inoculant in 200 g quantities in polythene bags. Seal with electric sealer and allow for curing for 2–3 days at room temperature. Curing can also be done by spreading the inoculants on a clean floor/ polythene sheet by keeping in open shallow tubs/trays with polythene covering for 2–3 days at room temperature before packet making.
- Check the initial population load of organism after curing in the inoculant packet and record.

10.2.1.2.2 Storage of Inoculant Packets

- Store the inoculant packets in a cool place away from heat or direct sunlight.
- Store packets at room temperature or in cold storage conditions in lots in plastic crates, boxes, or polythene/gunny bags.
- Determine the population of the organism in the inoculant packet at 15 days interval, which should not be less than 10^7 cells/g.

10.2.1.2.3   Adhesives

The use of a sticker to glue the inoculant onto the seed is a general recommendation for postproduction field applications. A number of stickers such as wall paper glue (5%), gum Arabic (4%), carboxy methyl cellulose (4%), honey (10%), powdered milk (10%), gur, or sugar (10%) are available and serve not only as glue for the bacteria to stick on to the seed but also serve as food for the bacteria until the infection process is complete in the case of rhizobia. The inclusion of adhesives within the prepared inoculant is desirable and practical but requires careful selection as some adhesives would merely set within the packet after impregnation. In India, generally gum Arabic or gum acacia powder is used as an adhesive. These gums are viscous exudates from Acacia (babul) trees. Acacia has number of species like *A. arabica*, *A. nelotica*, *A. Auriculiformis*, and others. Gum Arabic belongs to *A. arabica*.

10.2.1.2.4   Specifications for Raw Materials

- **Wood charcoal:** 50–200 mesh, good quality, pH 6.5–7.5, 120–140% water holding capacity, moisture percentage, 40% (w/w), and ash content 5–10%.
- **Raw lignite powder:** 150–200 mesh, 35% moisture percentage, finely powdered, duly packed in HDPE bag of 40–50 kg each.
- **Gum Arabic:** 150–200 mesh dry powder.
- **HDPE polythene bag:** made of high quality, 1st-grade milky white polythene tube having thickness of 300 ga and sealed from one end. Individual bag printed on both sides (Flexo printing) as per the design and size. Bag sizes range from 8×16 cm to 30×40 cm to contain 40 to 2880 g inoculant, respectively. In India, the standard packet size is 15×25 cm for 200 g biofertilizers.
- **Instafix gummed plastic tape:** 5 cm wide tape for fixing and sealing the corrugated cardboard boxes.
- **LDPE plastic strips:** For packing the cardboard boxes made of first grade LDPE, 5 cm wide and 200 m thick rolls.
- **Iron clips:** For packing plastic strip of 5 cm sizes.
- **Aluminum foil:** 0.03 mm thick.

These specifications can change with the development of new materials and changing needs of the production and marketing system.

10.2.1.3 LIQUID FORMULATIONS

Liquid biofertilizers are the microbial preparations containing specific beneficial microorganisms, which are capable of fixing or solubilizing or mobilizing plant nutrients by their biological activity. Liquid biofertilizers are special liquid formulations containing not only the desired microorganisms and their nutrients, but also special cell protectants or substances that encourage formation of resting spores and cyst for longer shelf life and tolerance to adverse condition.

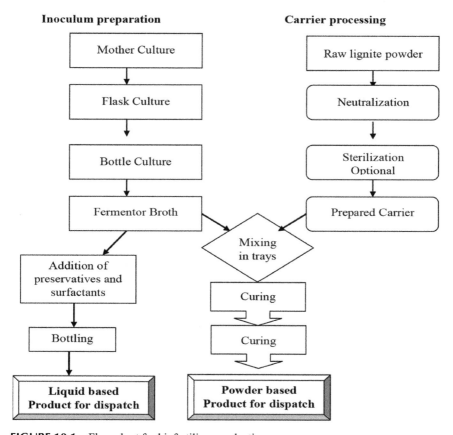

FIGURE 10.1 Flow chart for biofertilizer production.

10.2.1.4   LIMITATIONS OF CARRIER-BASED FORMULATIONS VS. ADVANTAGES OF LIQUID FORMULATIONS

Carrier formulations Limitations	Liquid formulations Advantages
• Low shelf life 1/2 year	• Longer shelf life, >1 year
• Automation difficult	• Easier to produce
• Temperature sensitive	• Temperature tolerant
• Low cell count	• High cell count
• More prone to contamination	• Contamination free
• Less effective	• More effective
	• Easy transport
	• Suitable for drip irrigation and protected cultivations

10.2.1.5 GRANULAR FORMULATIONS

Bead like formulations of biofertilizer which slowly release microorganism are known as granular formulations. For preparation of granular formulations, unconventional synthetic materials, viz. sodium alginate and clay are used. The main purpose of this formulation is to entrap the target organism for a longer period for better survival. They are applied as 5–10 beads/plant in the soil and give better growth as compared to carrier-based formulations.

10.3 MASS PRODUCTION OF MYCORRHIZA

The commercial utilization of mycorrhizal fungi has become difficult because of the obligate symbiotic nature and difficulty in culturing on laboratory media. Several researches in different parts of the world resulted in different methods of production of AM fungal inoculum using soil-based culture as well as carrier-based inoculum. Root organ culture and nutrient film technique provide scope for the production of soilless culture. As a carrier-based inoculum, pot culture is widely adopted method for production. The AM inoculum was prepared by using sterilized soil and wide array of host crops were used as host.

10.3.1 CULTIVATION TECHNIQUE

- Prepare a trench (1×1×0.3 m) lined with black polythene sheet to be used as a plant growth tub.
- Mix 50 kg of vermiculite and 5 kg of sterilized soil and pack in the trench up to a height of 20 cm. Spread 1 kg of AM inoculum (mother culture), 2–5 cm below the surface of vermiculite.
- Surface sterilizes maize seeds with 5% sodium hypochlorite for 2 min.
- Apply 2 g urea, 2 g super phosphate, and 1 g muriate of potash for each trench at the time of sowing maize seeds. Further, apply 10 g of urea twice on 30 and 45 days after sowing for each trench.
- Quality test on AM colonization in root samples to be carried out on 30th and 45th day.
- Stock plants are grown for 60 days (8 weeks). The inoculum is obtained by cutting all the roots of stock plants.
- The inoculum produced consists of a mixture of vermiculite, spores, pieces of hyphae, and infected root pieces.

- Thus, within 60 days, 55 kg of AM inoculum could be produced from 1 m^2 area. This inoculum will be sufficient to treat 550 m^2 nursery area having 11,000 seedlings/m^2.

10.4 MASS PRODUCTION OF BLUE GREEN ALGAE

The blue green algal inoculum may be produced by several methods, viz., in tubs, galvanized trays, small pits, and also in field conditions. However, the large-scale production is advisable under field condition which is easily adopted by farmers.

10.4.1 MULTIPLICATION IN TRAYS

- Big metallic trays (72×36×72 cm lbh) can be used for small-scale production.
- Take 10 kg of paddy field soil, dry powder well and spread.
- Fill water to a height of 3″.
- Add 250 g of dried algal flakes (soil based) as inoculum.
- Add 150 g of super phosphate and 30 g of lime and mix well with the soil.
- Sprinkle 25 g carbofuran to control the insects.
- Maintain water level in trays.
- After 10–15 days, the blooms of BGA (blue green algae) will start floating on the water sources.
- At this stage, stop watering and drain. Let the soil to dry completely.
- Collect the dry soil based inoculum as flakes.
- Store in a dry place. By this method, 5–7 kg of soil-based inoculum can be obtained.

10.4.2 MULTIPLICATION UNDER FIELD CONDITION

- Select an area of 40 m^2 (20×2 m) near a water source which is directly exposed to sunlight.
- Make a bund all around the plot to a height of 15 cm and give it a coating with mud or cow dung to prevent loss of water due to percolation.

- Plot is well prepared and leveled uniformly, and water is allowed to a depth of 5–7.5 cm and left to settle for 12 h.
- Apply 2 kg of super phosphate and 200 g lime to each plot uniformly over the area.
- The soil-based composite starter culture of BGA containing 8–10 species at 5 kg/plot is powdered well and broadcasted.
- Carbofuran at 200 g is also applied to control soil insects occurring in BGA.
- Water is let in at periodic intervals so that the height of water level is always maintained at 5 cm.
- After 15 days of inoculation, the plots are allowed to dry up in the sun, and the algal flakes are collected and stored.

10.5 MASS PRODUCTION OF *AZOLLA*

Azolla can be maintained in a nursery round the year, and from this, azolla can be broadcasted in rice fields. A simple *Azolla* nursery method for mass multiplication of *Azolla* has been evolved for easy adoption by the farmers.

- Select a wetland field and prepare thoroughly and level uniformly.
- Mark the field into plots of 20×2 m or any suitable size by providing bunds and irrigation channels.
- Maintain water level to a height of 5–10 cm.
- Mix 10 kg of cattle dung in 20 l of water and sprinkle in the field.
- Apply 100 g super phosphate as basal dose.
- Inoculate fresh *Azolla* biomass at 8 kg to each pot.
- Apply super phosphate at 100 g as top dressing fertilizer on 4th and 8th day after *Azolla* inoculation.
- Apply carbofuran (furadan) granules at 100 g/plot on 7th day after *Azolla* inoculation to control protozoa infestation.
- Maintain the water level at 5–10 cm height throughout the growth period of 2 or 3 weeks.

10.6 QUALITY CONTROL OF BIOFERTILIZERS

There are several specifications drawn by fertilizer control order (FCO) for quality control of various biofertilizers.

10.6.1 SPECIFICATIONS OF INDIVIDUAL MICROBIAL BIOFERTILIZER

S. no.	Rhizobium	Azotobacter	Azospirillum	Acetobacter	Phosphate solubilizing bacteria (PSB)	Potash mobilizing bacteria (KMB/KSB)	Zinc solubilizing bacteria
(i) Base	Carrier based* in form of moist/dry powder, granules, or liquid based						
(ii) Viable cell count	CFU minimum 5×10^7 cell/g of powder, granules or carrier material or 1×10^8 cell/ml of liquid						
(iii) Contamination level	No contamination at 10^5 dilution						
(iv) pH	6.5–7.5			5.5–6.0 for moist/ dry powder, granulated or carrier based and 3.5–6.0 for liquid		6.5–7.5	
(v) Particles size in case of carrier-based material	All material shall pass through 0.15–0.212 mm IS sieve						
(vi) Moisture percent by weight, maximum in case of carrier based	30–40%						
(vii) Efficiency character	Should show effective nodulation on all the species listed on the packet	The strain should be capable of fixing at least 10 mg of nitrogen per g of sucrose consumed	Formation of white pellicle in semisolid N-free bromothymol blue media	Formation of yellowish pellicle in semisolid medium N free medium	The strain should have phosphate solubilizing capacity in the range of minimum 30% when tested spectro-photometrically. Minimum 5 mm solubilization zone in prescribed media having at least 3 mm thickness		

*Types of carrier: The carrier material such as peat, lignite, peat soil, humus, wood, charcoal, or similar material favoring growth of the organism.

10.6.2 SPECIFICATION FOR MYCORRHIZAL BIOFERTILZERS

S. no.	Specification	Remarks
i	Form/Base	Fine powder/tablets/ granules/ root biomass mixed with growing substrate
ii	Particle size for carrier based	90% should pass through 250 micron IS sieve (60 BSS)
iii	Moisture content	8–12%
iv	pH	6–7.5
v	Total viable propagules/g of product	100/g of finished product
vi	Infectivity potential	80 infection points in test root/g of mycorrhizal inoculum used

10.6.3 SPECIFICATIONS FOR NPK CONSORTIA OF BIOFERTILIZERS

Department of Agriculture and Cooperation, Ministry of Agriculture, Government of India issued Gazette notification vide S.O.1181 (E) dated April 30, 2014 for the introduction of NPK consortia Biofertilizers in FCO, wherein a mixture of any two or more of the microorganisms, (a) *Rhizobium/Azospirillum/Azotobacter*, (b) PSB, and (c) KMB/KSB, may be used in the formulation of liquid and carrier-based Biofertilizer consortia. Accordingly, the name of the consortia shall be decided and printed on the product.

S. no.	Microbial strains for biofertilizer used	Consortia name
i	N, P, and K strains	NPK consortia or BIO-NPK consortia
ii	N and P strains	NP consortia or BIO-NP consortia
iii	N and K strains	NK consortia or BIO-NK consortia
iv	P and K strains	PK consortia or BIO-PK consortia

Consortia Specifications as per FCO

Specifications	Particulars	Carrier based consortia in the form of moist powder or granules	Liquid based consortia
(i) Individual viable cell count	CFU minimum in a mixture of any 2 or more of following microorganisms CFU minimum		
	Rhizobium, Azotobacter, or *Azospirillum*	1×10^7 per g	1×10^8 per ml
	PSB	1×10^7 per g	1×10^8 per ml
	KSB	1×10^7 per g	1×10^8 per ml
(ii) Particle size in case of carrier based moist powder	All materials that pass through	0.15–0.212 mm IS sieve	–
(iii) Moisture percent by weight		30–40%	–
(iv) Total Viable count of all the biofertilizer organisms in the product	CFU minimum	5×10^7 per g	5×10^8 per ml
(v) Contamination level		No contamination at 10^5 dilution	No contamination at any dilution
(vi) Efficiency character	*Azotobacter*	The strain should be capable of fixing at least 10 mg of nitrogen per g of C source	
	Azospirillum	The strain should be capable of fixing at least 10 mg of nitrogen per g of malate applied	
	Rhizobium	Nodulation test positive	
	PSB	Minimum 5 mm solubilization zone on prescribed PSB media having at least 3mm thickness	
	KMB	Minimum 5 mm solubilization zone on prescribed KMB media having at least 3 mm thickness	

10.6.4 PROCEDURE OF SAMPLING AND QUALITY ASSESSMENT

Biofertilizer sampling can be done at different stages of production/sell by drawing samples by appointed inspectors who may be graduated in agriculture/science with chemistry/microbiology. He/She may draw the samples as per the procedure laid down in Schedule II, prepare the sampling details in form J and handover one copy to the dealer/representative from whom the sample been drawn. The samples may be sent within 7 days and be analyzed within 30 days in accordance with the instructions contained in Schedule III in the NCOF/RCOF or any other laboratories notified by Central or State Government. The said report be dispatched within 15 days to the organization from the sample was drawn.

10.6.5 QUALITY CONTROL AGENCY OF BIOFERTILIZER IN INDIA

- NCOF, Sector-9, Hapur Road, Ghaziabad-201002 (U.P.) India.
- RCOF situated at Bangalore, Bhubaneswar Hisar, Imphal, Jabalpur, and Nagpur as a nodal agency for testing and maintaining quality of biofertilizer.
- BIS, ICAR Institutes & SAUs.

10.7 SCOPE OF BIOFETILIZER

- LBF being live microbial cultures, their promotion in organic cultivation have tremendous scope to reduce imports on fertilizers and their raw materials.
- Reduction in 25% NPK, cut down cultivation cost at great extent with improvement in farm, and rural economy with nurturing soil by plant probiotics.

10.8 MARKET OF BIOFERTILIZER

Most of the production of biofertilizers in the country is being done in the public sector by research institutions, universities, and National Centre for organic farming (Fig. 10.2). Few states and co-operative fertilizer units also have put a step forward into this field. The participation of private sector is limited in spite of it being a low investment and high-benefit technology. Almost all the production units in the country are manufacturing mainly the bacterial biofertilizers, and the present day annual production is estimated to be around 50,000 MT, which is lower than the demand. The chemical fertilizer (NPK) production and requirement, projected, estimates 27–31 million metric tons of India (Table 10.1).

TABLE 10.1   Fertilizer Requirement and Production in India Projections (Million Ton).

Year	2031	2051
Requirement	27.3	31.3
Production	20.9	23.9
Gap*	6.4	7.2

*There is a gap of approx 6–7 million ton, which can be bridged by alternative to chemical fertilizers like biofertilizers. Biofertilizer production and sale among farmers is promoted by Government of India. Recent data, published in Biofertilizer statistics 2013–2014, show that there is an increase in production of biofertilizers to the tune of 2 and 20 times in the year 2003–2004 and 2013–2014, respectively, as compared to 1993–1994 (Chanda et al., 2014). Thus, in last decades, biofertilizer production is advanced.

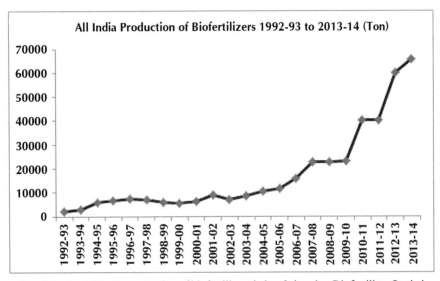

FIGURE 10.2   All India production of biofertilizers in last 2 decades, Biofertilizer Statistics, FAI, New Delhi (2013–2014).

On a conventional estimate, even a 10% saving through the use of biofertilizers is estimated to result in a huge savings on NPK fertilizers (Tables 10.2 and 10.3). The world-wide market for biofertilizers in relation to income was estimated to be worth about 5 billion USD in 2011, and according to its current market value and development scenario in the different continents, it is predicted to double by 2017 (http://www.marketsandmarkets.com/).

TABLE 10.2 Economics of Biofertilizer Use (Liquid).

Biofertilizer/Crop	Quantity required	Cost of application	Amount of nutrient mobilized
	l/ha	Rs/ha	kg/ha
Rhizobium in legumes	0.2–1.0 l	40–200	25–35 kg N
Azotobacter/Azospirillum in nonlegumes	0.5–2.0 l	80–400	20–25 kg N
Azoto + *Azosp* + PSB	0.5–2.0 l	80–400	20 kg N + 12 kg P
Mixed inoculants	0.5–2.0 l	80–400	25 kg N +15 kg P*
Mycorrhiza	2.0–5.0 kg	200–500	20–25 kg P + micronutrients + moisture

Source: Biofertilizer statistics 2013–2014, FAI, New Delhi, pp. 75.

*25 kg N 15 kg P (~1 Bag UREA + 1 Bag SSP).

TABLE 10.3 Economics of Biofertilizer Use (Carrier Based).

Quantity of biofertilizer	Equivalent quantity of fertilizer	Savings in nutrients
1 t *Rhizobium*	100–400 t (urea)	50–200 t N (minimum fixation of 50 kg ha[-1])
1 t *Azotobacter/Azospirillum*	50–100 t (urea)	20 t N (minimum fixation of 50 kg ha[-1])
1 t *Acetobacter*	80–120 t (urea)	40–60 t N (minimum fixation of 200–300 kg ha[-1])
1 t PSM	100 t (DAP)	40–50 t of P (minimum solubilization of 40 kg ha[-1] of P_2O_5)

Source: Biofertilizer statistics 2013–2014, FAI, New Delhi, pp. 75.

10.9 TECHNOLOGY

Unlike chemical fertilizers, all biofertilizers contain viable population of selected microbes, which are able to colonize the soil ecosystem. Their production on large scale needs favorable growth conditions and defined nutrient media. Bacteria and fungi being heterotrophs, need organic substrates, whereas algae are photolithotrophs and require inorganic nutrients and light.

10.10   SWOT ANALYSIS AND MANAGEMENT

There are many industries in India, which are producing biofertilizers. The major strength of using such microbial inoculants is reduction in the use of chemical fertilizers at least by 20–30%, and at the same time making the soil vital in an eco-friendly manner. Future approach in this area has to be carved out considering possible constraints, which may crop up in the large-scale application of this technology. The constraints can be:

(a) Nonavailability of region specific microbes backed by an authenticated germplasm collection center.
(b) Improper handling at production, distribution, and consumer levels.
(c) Absence of strict quality control legislation and casual concern about quality management at the production level.
(d) Neglect of an appropriate research and development support.

10.11   FACTORS WHICH LEAD TO LOW EFFECTIVE DEMAND

• Lack of awareness among the consumers and promoters about the application of the merits of technology: due to ready and quick response of chemical fertilizers, farmers are hesitant to use such microbial inoculants.
• Off-the-shelf invisibility of biofertilizers: knowledge of cropping pattern in an area or state at different time is required to make them timely available to the farmers which will also fetch good price as demand and supply will be met through.
• Lack of proper quality control measures: the innocent consumer being duped by unscrupulous manufactures interested in making a fast buck. Production of bioinoculants starting from their mother cultures to starter cultures and mass multiplication followed by downstream processing requires sterile environment and is to be operated upon by skilled personal and assured titer value before packaging. Good quality of carrier material is also required for longer shelf life.

Proper legislation with quality control and specifications laid down by Bureau of Indian Standards for certain products have to be strictly followed.

- Absence of federal support in the form of a dependable germplasm collection of region specific microbes of agricultural importance. So that they function both as depositories and service centers for selected microbes. Such facilities are absolutely necessary at state level where microbes are made available from time to time to the manufacturers at nominal price.
- Information gap on microbe–microbe interaction and fate of bioinoculants in soil.
- Varying response of inoculants at different locations due to agroecological factors and the quality of the inoculum. Application of consortium of microbes can combat such situations.
- Chemical fertilizers available at subsidized rates: In spite of heavy subsidy on chemical fertilizers, which has increased substantially during the last decade, the cost of fertilizer nitrogen is prohibitive and the gap between demand and supply is widening. By 2000 A.D., the total fertilizer consumption is targeted at 240 lakh tons, of which nitrogen alone would be 140 lakh tons. To bridge the gap between demand and supply of chemical nitrogen, cheaper sources of nitrogen for sustaining agricultural production are needed.

10.12 PROBLEMS IN BIOFERTILIZERS ADOPTION

- Unfamiliarity and inaccessibility of biofertilizers to the farmers.
- Poor quality biofertilizers.
- Expired date inoculant reach to farmers which are useless.
- Unsuitable transport and storage system may create high temperature, which destroy the microbial population in inoculants.

10.13 PROJECTIONS

A unit, capable of manufacturing about 1000 t (5 million packets) of bacterial biofertilizers, supported by the required research and development wing, would require an investment of about Rs. 10 million. Land requirement for bacterial biofertilizer unit will be around 3000–5000 m^2 with 10–12 skilled and unskilled workers. Algal biofertilizer unit will require another investment of about Rs. 1 million and additional 7000 m² land with 10–12 skilled and unskilled workers. One can begin with the low investment algal biofertilizer unit and then take up bacterial biofertilizers and possibly microbial pesticides.

Marketing can initially be entrusted to agencies like National Centre for organic farming, National Dairy Development Board, and State Department of Agriculture, who have the awareness as well as marketing channels developed for the agricultural inputs. This will appreciably cut down the initial investment and provide time to feel the market pulse. The selling price of a packet of biofertilizer sufficient to inoculate an acre is just double of the manufacturing cost; the breakeven point is 40% with internal rate of return being 18% and 5 year break even period.

At present, there are around 250 biofertilizer manufacturing units all over country run by State Agricultural Department, chemical fertilizer industries and agro-industries in the Govt. sector and in the private sector. To fully utilize the production potential of these units, intensive efforts are to be made and substantial demand from farmers is to be generated.

10.14 RECOMMENDATIONS

- Entrepreneur will come forward to establish biofertilizer industry, only if, there is demand with good market potential and quick returns. Demand has been envisaged, but technicalities have never been worked out properly. A large mass of consumers is aware of the biofertilizers and aggressive popularization will speed up the market diffusion rate. This will be further expedited if various government agencies, industrial units, and public and private R&D institutions join hands. Attentive attitude toward public/user awareness, quality control mechanisms for products, technical and fiscal support to small-scale businesses along with R&D support is required to develop entrepreneurship.
- There is a developing trend to shift from chemical-based agriculture to organic agriculture. Effective execution and acceptance of biofertilizers at large scale will depend on timely accessibility of high value products with sufficient shelf life, application at proper and precise time, use of specified equipments, with applicable education, and placement to trainers, distributors, and users.

10.15 CONCLUSION

- To obtain more effectiveness of biofertilizers, producers should make sure that farmers should get excellent product.
- Suitable efficient biofertilizer should be available to the farmers at appropriate time in all spell.

KEYWORDS

- **biofertilizers**
- **sustainable**
- **mass production**
- **FCO**

REFERENCES

Chanda, T. K.; Sati, K.; Robertson, C.; Arora, C. *Biofertilizer Statistics 2013–14*; The Fertilizer Association of India: New Delhi, 2014.

http://www.marketsandmarkets.com. Accessed 10.11.14

The Fertilizer (Control) Order, 1985 (As Amended up to June 2010); The Fertilizer Association of India: New Delhi, 1985.

Vora, M. S.; Shelat, H. N.; Vyas, R. V. *Handbook on Biofertilizers and Microbial Pesticides*; Satish Serial Publishing House, Delhi, 2008; pp 251.

Vyas, R. V.; Shelat, H. N.; Vora, M. S. Biofertilizers Techniques for Sustainable Production of Major Crops for Second Green Revolution in Gujarat—An Overview. *Greenfarming.* **2008,** *1,* 68–72.

CHAPTER 11

MICROALGAE: A PROMISING FEEDSTOCK AS SOURCE FOR THIRD-GENERATION RENEWABLE ENERGY

KAMLA MALIK*, RAMESH CHANDER ANAND, DEEPIKA KADIAN, and AMRITA NARULA

Department of Microbiology, Chaudhary Charan Singh Haryana Agricultural University, Hisar 125004, Haryana, India

Corresponding author. E-mail: kamlamalik@rediffmail.com; kamlamalik06@gmail.com

CONTENTS

ABSTRACT

Due to depleting conventional energy resources, environmental degradation, and hike prices of oil in worldwide, there is a dire need to look for sustainable, greener fuels which are economical as well as environment friendly. At present scenario, microalgae appear to be the only biosolution that can replace conventional fossil fuels completely. This chapter briefly overview the use of microalgae as future feedstock for biodiesel production and critically described their downstream processes. The challenges and future aspects for commercialization of biodiesel production from microalgae through genetic modification and metabolic engineering can also be used to enhance photosynthetic efficiency and higher microalgae oil yield.

11.1   INTRODUCTION

In recent years, the energy consumption is increasing drastically due to rapid depletion of fossil fuels throughout the world. The increasing costs of conventional energy sources and climate change due to this had diverted the interest of scientists toward cleaner production technology (Aransiola et al., 2014). Under these circumstances, a novel approach is to use renewable fuels like biodiesel as substitute of diesel. Biodiesel production from traditional feedstocks is based on expensive edible biomass as raw material (Goureia and Oliveira, 2009); however, the problem in storage poses threat to current biodiesel production. Therefore, there is a dire need to develop and explore of new feedstocks which are cost effective, nontoxic, eco-friendly, renewable, economically sustainable, and not directly linked to human food chain (Kumar et al., 2013). Biodiesel is a green fuel which does not contribute to the CO_2 saddle and produces drastically reduced engine emission and contributes 40–50% of the oxygen to the atmosphere (Um and Kim, 2009). It has been identified as one of the notable options for at least complementing conventional fuels. Biodiesel is safe, nontoxic, biodegradable, and renewable that contains no sulfur; and it is a better lubricant. Moreover, its use engenders numerous social benefits: rural revitalization, creation of new jobs, and reduced global warming (Lang et al., 2001; Bastianoni et al., 2008; Aransiola et al., 2012). The microalgae as renewable source for biodiesel is capable for fulfilling demand of transport fuels globally (Chisti, 2007). Microalgae are the largest primary producers of any aquatic ecosystem. It has been potentially used for biodiesel production due to a number of advantages including higher photosynthetic efficiency, higher biomass production, and higher growth rate besides short life span because of their simple structures compared to other energy crops. In addition, biodiesel from microalgae oil is similar in properties and is also more stable according to their flash point values to the standard biodiesel (Rajvanshi and Sharma, 2012).

In an Indian economy, the essentially diesel driven and the consumption of diesel fuel is four to five times more than motor-gasoline which is characteristically different from several developed economies. In this situation, there is an urgent need to identify and commercialize renewable alternative fuels for diesel substitution (Singla et al., 2010). The present chapter is focused on the overview on microalgae used as potential substitute of biodiesel production and its characteristics, cultivation, harvesting, and extraction methods.

11.2 GENERAL CHARACTERISTICS OF MICROALGAE

Microalgae are a large and microscopically (colonial or single-cell) diverse group of photosynthetic eukaryotic or prokaryotic microorganisms. They produce carbohydrates, proteins, and lipids as a result of photosynthesis. The majority of microalgae biomass is comprised of proteins, carbohydrates, and lipids. In general, algae biomass contains 20–30% carbohydrate, 10–20% lipid, and 40–60% protein (Singh et al., 2011). These microorganisms convert sunlight, water, and CO_2 to sugars, from which macromolecules (lipids and triacylglycerols) can be obtained. These triacylglycerols (TAGs) are used as a feedstock for biodiesel production (Khan et al., 2009). They can grow rapidly and have ability to live in harsh environments due to their cellular structure. They can be found anywhere, that is, in water, soils, ice, rivers, lakes, hot springs, and oceans. They have the ability to capture carbon dioxide and convert sunlight to chemical energy. These are primary producers that are found in marine, fresh water, desert sands, and hot spring (Guschina and Hardwood, 2006). More than 50,000 species are known, and of around 30,000 have been studied (Mata et al., 2010; Richmond, 2004). Some species of microalgae can be both phototrophic and/or heterotrophic depending on environmental conditions. Microalgae utilize sunlight more efficiently than terrestrial plants and also consume harmful pollutants, have required minimal resource and will not interfere with the products derived from crops such as production of food, fodder, etc. (Brennan and Owende, 2010). The microalgae are classified on the basis of cell wall constituents, kinds of pigments, and chemical nature of storage products (Alam et al., 2012). They are important producers of a large spectrum of pigments. Microalgae produce different kinds of hydrocarbons, lipids, and other complex oils that vary as per species levels. The microalgal strains, with high oil or lipid content, are summarized in Table 11.1. Commonly, the microalgae contain 20–50% (dry weight basis) of oil content, sometimes even up to 80% under certain circumstances. At present, all over the world, microalgae are cultivated commercially for human nutritional products in small- to medium-scale production systems and producing a few tons to a several hundreds of tons of biomass annually (Chisti, 2007).

TABLE 11.1   Oil Content of Various Microalgae Species.

Microalgae species	Oil content (% dry weight)	Reference
Amphora coffeaeformis	19.7	Renaud et al. (1991)
Anabaena cylindrical	4–7	Sydney et al. (2010)
Chlorella spp.	28–32	Chisti (2007)
Chlamydomonas reinhardtii	21	Um and Kim (2009)
Chlorella emersonii	25–63	Mata et al. (2009)
Chlorella protothecoides	14–57	Chisti (2007)
Chlorella vulgaris	5–58	Sialve et al. (2009)
Chaetoceros spp.	17	Renaud et al.(1991)
Botryococcus braunii	25–75	Chisti (2007)
Crypthecodinium cohnii	20	Chisti (2007)
Cryptomonas spp.	22	Renaud et al. (1991)
Dunaliella primolecta	23	Chisti (2007)
Dunaliella salina	6–25	Mata et al. (2009)
Dunaliella spp.	17–67	Mata et al. (2009)
Dunaliella tertiolecta	11	Um and Kim (2009)
Dunaliella bioculata	8	Chisti (2007)
Isochrysis spp.	25–33	Chisti (2007)
Isochrysis galbana	7–40	Mata et al. (2009)
Cylindrotheca spp.	16–37	Chisti (2007)
Monallanthussalina	20–22	Chisti (2007)
Monodus subterraneus	16.1	Rodolfi et al. (2009)
Euglena gracilis	14–20	Mata et al. (2009)
Haemato coccuspluvialis	25	Mata et al. (2009)
Hormidium spp.	38	Kumar et al. (2013)
Nitzschia spp.	45–47	Kumar et al. (2013)
Nannochloris spp.	31–68	Chisti (2007)

TABLE 11.1 *(Continued)*

Microalgae species	Oil content (% dry weight)	Reference
Nannochloropsis oculata	22–29	Mata et al. (2009)
Nannochloropsis spp.	12–53	Mata et al. (2009)
Tetraselmis sueica	15–23	Chisti (2007)
Tetraselmis spp.	12–14	Mata et al. (2009)
Schizochytrium spp.	50–77	Chisti (2007)
Phaeodactylum tricornutum	20–30	Chisti (2007)
Neochloris oleoabundans	35–54	Luisa (2011)
Ankistrodesmus spp.	24–31	Mata et al. (2009)
Pyrrosia laevis	69	Mata et al. (2009)
Pavlova salina	30.9	Mata et al. (2009)
Pavlova lutheri	35.5	Mata et al. (2009)
Prymnesium parvum	22–39	Um and Kim (2009)
Skeletonema costatum	13.5–51.3	Mata et al. (2009)
Scenedesmus obliquus	11–55	Mata et al. (2009)
Scenedesmus dimorphus	16–40	Kumar et al. (2013)
Thalassiosira pseudonana	20.6	Mata et al. (2009)
Tetraselmis suecia	15–23	Luisa (2011)
Rhodomonas spp.	18.7	Renaud et al. (1991)
Skeletonema spp.	13.3	Renaud et al. (1991)
Nephroselmis spp.	13.8	Sobczuk and Chisti (2010)
Gymnodium spp.	8–30	Mata et al. (2009)
Spirulina maxima	6–7	Sydney et al. (2010)
Spirulina platensis	4–16	Mata et al. (2009)
Oocystis pusilla	10.5	Mata et al. (2009)

11.3 ADVANTAGE OF MICROALGAE FOR BIODIESEL PRODUCTION

Nowadays, microalgae are being promoted as an ideal feedstock for third generation of biofuels. Microalgae could be a vital renewable fuel sources due to its fast growth rate, high oil content, higher biomass productivity, CO_2 fixation ability, easy to handle and grow in water, do not compete with food or feed crops, and can be produced on nonagricultural land. Considering all these advantages, the microalgae produced biodiesel 10–20 times higher/unit area than the biodiesel produced from oleaginous seed/vegetables oils (Chisti, 2007). The microalgae biomass is an alternative feedstock for biodiesel production (Chen et al., 2010). It has various advantages over conventional oil crops:

- Microalgae have higher photon conversion efficiency that represents higher biomass and fast growth rates with higher CO_2 sequestration capacity.
- These can be used to solve pollution problems and reduce the greenhouse emissions because microalgae use CO_2 to grow, which sourced out directly from high-emitting industries and may further reduce emissions by displacing existing fossil fuel sources.
- Microalgae are a very efficient biological system for harvesting solar energy for organic compounds production. It can effectively remove nutrients (nitrogen, phosphorus, and heavy metals) from wastewaters.
- Microalgae are easy to grow in a liquid medium, with better handling and can utilize salt, saline/brackish water/coastal seawater, and waste water streams. Production is not seasonal.
- Microalgae also produced valuable products such as carbohydrates, proteins, lipids, fertilizers, pigments, oil, starch, animal feed, etc.

Microalgae require specific methods for isolation, screening, harvesting, extraction, and conversion of biodiesel (Sheehan et al., 1998; Alabi et al., 2009).

11.4 METHODS FOR CULTIVATION/CULTURING OF MICROALGAE

Cultivation of microalgae is one of the most important steps for biodiesel production (Chen et al., 2010). The culture of microalgae is produced in different cultivation modes (photoautotrophic, heterotrophic, and mixotrophic). Generally, the biodiesel production from microalgae have been cultivated on photoautotrophic mode, and mixotrophic cultivation mode has rarely used (Chen et al., 2010). It can be cultivated in either open systems or closed systems, and sometimes both systems have a significant effect on the biodiesel production. For commercial scale, the microalgae culture system must have following characteristics: high productivity, high area productivity, cost effective, reliability, and easy to control the cultural conditions such as temperature, pH, turbulence, and oxygen (Kumar et al., 2013).The different methods for cultivation of microalgae used are as mentioned in Figure 11.1.

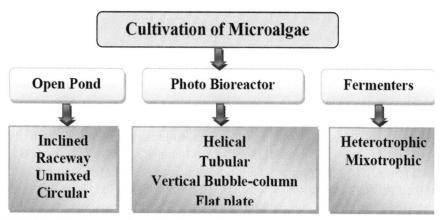

FIGURE 11.1   Methods for cultivation of microalgae for biodiesel production.

11.4.1   OPEN POND CULTURE METHOD

Microalgae cultivation in an open system is usually conducted in an open pond or open tank that is exposed to the environment. The open ponds culture method is simple, oldest, and commonly used for mass cultivation of microalgae. It can be categorized into natural waters (ponds, lakes, and lagoons) and artificial ponds, that is, containers. In open pond system, the shallow pond is about one-foot deep, and microalgae are cultured under natural environmental conditions. Control of temperature is very difficult and usually fluctuates in a diurnal cycle. Another significant issue with open ponds system is evaporative water loss due to atmosphere, CO_2 utilization is also much less efficient than in Photo bioreactors (PBRs). The biomass concentration is generally low (0.1–1.5 g/l) due to poor mixing (Chisti, 2007). In addition to the contamination by native species, and thus, are suitable for only a small number of microalgae species those can survive extreme environments such as high saline or alkaline conditions (Laws et al., 1988; Singh et al., 2011). Open pond culture methods are easy to operate, less expensive with a high production capacity, and can utilize sunlight. The nutrients can also be provided through runoff water from nearby land areas or by channeling water treatment plants/ water from sewage (Carlsson et al., 2007). The most commonly used open ponds methods are further categorized as inclined, raceway, unmixed, and circular ponds (Suali and Sarbatly, 2012).

11.4.1.1   INCLINED POND

The inclined pond method is open-air system and achieves high sustainable cell densities up to 10 g/l. It is consisting of slightly inclined shallow trays and turbulence is created by gravity. The thin layers microalgae culture suspension is prepared that flows from the top to the bottom of a sloping surface, thus achieved high turbulent flow rates which facilitates maximum cell concentrations with a higher surface-to-volume ratio. The productivity of inclined ponds is 31 g/m²/d (Doucha and Livansky, 2006). This method is used for only few microalgae (*Chlorella* and *Scenedesmus*) species because of their difficult operation system. The cultures are exposed to sunlight and operated in a continuous manner with CO_2. The nutrients are constantly fed to the pond with circulation of the remaining microalgae containing water at the other end. This system requires high energy input for mixing (Borowitzka, 2005).

11.4.1.2   RACEWAY POND

The raceway pond method is most commonly used for both the pilot scale as well as commercial scale for microalgae cultivation. It is the most popular cultivation system due to scalability and low cost of building. It is usually consisted of a closed loop with oval-shaped recirculation channels that is typically about 0.2–0.5 m deep. The proper mixing and circulation are required to stabilize microalgae growth and productivity. The microalgae culture is continuously fed in front of the paddlewheel and circulated through the loop to the harvest extraction point. The paddlewheel is in a continuous process that prevents sedimentation (Spolaore et al, 2010). The raceways are commonly made from concrete; sometimes, they are dug into the earth and lined with white plastic. In this system, cooling is done by evaporation. Evaporative water loss can be significant because of losses to atmosphere. The productivity is 14–50 g/m²/d. However, these systems have various drawbacks, including rapid process than closed systems and biological contamination by fungi, amoeba, algae grazers, and other algae. These methods are low-cost with low biomass productivity because of poorly mixing and cannot sustain an optically dark zone. The major advantage of this method is that it is cheap to provide water and nutrients using paddle wheels on regular frequency for better circulation of microalgae. It is commonly used for the large scale cultivation of *Spirulina* (*Arthrospira platensis*), *Dunaliella salina*, *Haematococcus pluvialis*, and *Chlorella vulgaris* (Putt et al., 2011).

11.4.1.3 UNMIXED POND

This method is commonly used for the cultivation of *D. salina*. Unmixed pond method has low productivities (<1 $g/m^2/d$) (Brennan and Owende, 2010). These ponds are not suitable for commercial scale cultivation of microalgae species (Doucha and Livansky, 2006).

11.4.1.4 CIRCULAR POND

The circular ponds are used for high productivity (21 $g/m^2/d$) of microalgae (Benemann and Oswald, 1993). The addition of organic carbon supports the respiration process in dark as well as supports microalgae growth at the bottom of the pond where sunlight exposure is less (Doucha and Livansky, 2006). It is mostly used for cultivation of *Chlorella* spp. The open ponds culture methods are less expensive, susceptible to contaminates and fast-growing heterotrophic organisms. Some drawbacks of methods are inefficient stirring mechanisms, evaporation, environment regulation, high land requirements, low biomass productivity, and also uncontrollable conditions for a large-scale cultivation system such as light intensity, duration of sunlight, and temperature (Sanchez et al., 2011).

11.4.2 CLOSED POND CULTURE METHOD

To overcome the major drawbacks with open-pond culture methods, the closed ponds are most popular method for culturing of microalgae. Cultivation of microalgae in a closed pond methods can be operated in photobioreactor (Carvalho et al., 2006; Suali and Sarbatly, 2012).

11.4.2.1 PHOTO BIOREACTORS (PBR)

PBR methods are commonly used for high productivity of microalgae (Tredici et al., 1999; Molina et al., 2003; Pulz, 2001; Carvalho et al., 2006). PBR is a translucent closed container that make use of light source and can be located both indoor provided with artificial light and outdoor with natural light. PBRs are characterized by new advance technology with controlled cultivation conditions as well as less contamination risk and no CO_2 losses (Pulz, 2001). It can be operated in batch culture with a continuous stream of sterilized water that contains air, nutrients, and CO_2. It is suitable for single culture of microalgae species for longer durations. These methods have been successfully used for large quantities of microalgae biomass production (Rajvanshi and Sharma, 2012). These are mainly of four types: (1) helical PBRs, (2) tubular PBRs, (3) vertical bubble column or air lifter PBRs, and (4) flat plate PBRs.

11.4.2.1.1   Helical PBRs

These PBRs consists of parallel flexible transparent tubes which coiled around a cylinder. The helical shape is effective in increasing the surface area of sunlight that can increase productivity (113.7 $g/m^2/d$) of biomass of microalgae (Carvalho et al., 2006).

11.4.2.1.2   Tubular PBRs

Tubular PBRs can be horizontal, vertical, inclined, helical, α, and conical in shape. It consists of straight transparent tubes which are made of either glass or plastic. The microalgal culture is circulated through the tubes and exposed to light for photosynthesis and returns to a reservoir. The tubes are usually 10 cm in diameter that allows for sunlight penetration. The biomass is prevented from settling by maintaining high turbulent flow in the reactor with either a mechanical pump or an airlift pump (Chisti, 2007). This system requires a gas exchange chamber to reduce the elevated dissolved O_2 levels in the liquid. These PBRs are cost effective, high biomass productivities, and have a large illumination surface. It is mostly used for cultivation of *Spirulina* species and *Chlorella pyrenoidosa* with maximum productivity of 30 $g/m^2/d$ (Chiu et al., 2008).

11.4.2.1.3   Vertical Bubble Columns or Air Lifter PBRs

This reactor is one of the most common types of PBR that has high surface-to-volume ratio, low cost, easy to operate, low shear forces, high efficiency of CO_2 use, absence of wall growth, and the ability to use sunlight. The columns are simple, aerated from the bottom, placed vertically, and illuminated through transparent walls or internally. It offers the proper mixing, high rate of volumetric gas transfer, and optimized growth conditions. On the bottom of the tube is an air inlet which bubbles air through the column, which provides mixing and gas exchange.

11.4.2.1.4 Flat Plate PBRs

Flat plate PBRs are made of rectangular thin boxes which are composed of translucent glass or plastic, that are open at one end, and ribs are running vertically from bottom to top. Air is bubbled from the bottom that provides sufficient mixing and gas transfer. These reactors have baffles running horizontally inside the reactor to aid mixing and gas exchange efficiencies. It has increased surface area for light to reach microalgae cells that can have significantly enhanced productivity up to 1.09 $g/m^2/d$ with *Spirulina platensis* (Tredici and Materassi, 1992; Carvalho et al., 2006). Due to the low accumulation of dissolved oxygen and the high photosynthetic efficiency, these PBRs are suitable for mass cultures of microalgae.

The flat plate PBR is most popular method for microalgae cultivation due to low cost, low shear forces, easier to control the process conditions, high surface–to-volume ratio, high biomass productivity, high efficiency of CO_2 conversion, better CO_2 consumptions ability, low operating cost, and minimum risk of contamination as compared to other systems.

11.4.3 HYBRID PBRS

In hybrid PBRs method, combination of different growth stages of two types of PBRs (open and closed system) exists. It is a well-designed bioreactor and conditions are optimized with minimum contamination of the other microorganisms and to promote continuous cell division. In α-shaped hybrid reactor, the culture is lifted 5 m by air to a receiver tank and culture flows down (Brennan and Owende, 2010).

11.4.4 FERMENTERS

Microalgae are cultivated by using two types of fermenters, they are heterotrophic and mixotrophic.

11.4.4.1   HETEROTROPHIC

Mostly, microalgae grow phototrophically, some microalgae species (heterotrophic) are able to utilize organic substrates as sole sources of carbon and energy. The heterotrophic culture of microalgae is cultivated in fermenters where a high degree of culture manipulation can be performed. This method is well established. Fermentation technologies, high degree of process control, high production, the independence from weather and climate conditions, elimination of the light requirement; oxygen is required for the catabolism of the organic substrates and lower harvesting costs (Barclay et al., 1994; Abayomi, 2009).

11.4.4.2   MIXOTROPHIC

This method is mainly dependent upon the nutritional requirement, photo assimilation CO_2, and simultaneously the oxidative catabolism of organic carbon sources that increases the productivity. Maximum productivity of 127 $g/m^2/d$ (day time) and 79 $g/m^2/d$ (night time) has been reported in marine microalgae *Phaeodactylum tricornutum* (Garcia et al., 2000). The mixotrophic cultures contribute too much higher cell densities than phototrophic systems (Wang et al., 2014).

The prerequisition of cultivation process of microalgae depends on selection of microalgal strain, production facility, and product of interest. The basic principle in all of the photobioreactor designs is to reduce the light path and to provide the light to each micro algal cell. The photobioreactor designs are more complicated as compared to an open pond culture system. At present, there is no information available on the designs, operations, yields, and other various aspects of commercial cultivation of microalgae production.

The overall advantages and disadvantages of various microalgae culture systems are mentioned in Table 11.2 (Dragone et al., 2010; Rajvanshi and Sharma, 2012).

11.5   HARVESTING OF MICROALGAE

After cultivation, microalgae should be recovered or separation of biomass from the culture medium should constitute about 20–30% of the total biomass production cost and drying. The desired product can be obtained;

TABLE 11.2 Advantage and Disadvantages of Microalgae Culture Systems.

Microalgae culture system	Advantages	Disadvantages
Open pond culture	Inexpensive, simple, easy to clean up, easy maintenance, low energy inputs, use of nonagricultural land	Poor light utilization, large evaporative losses, poor stability, ponds are susceptive to weather conditions, difficult to grow algal cultures for long periods, diffusion of CO_2 to the atmosphere, requires large areas, low productivity due to lack of proper stirring, limited to only few algal strains, and cultures are easily contaminated by predators and other fast growing heterotrophs and extremophilic spp.
Closed pond culture Photo bioreactors (PBRs) Tubular PBRs	Cost effective, increase of illumination surface area, high biomass productivities, suitable for outdoor cultures	Requires controlled conditions (temperature, pH, dissolved oxygen, and CO_2), fouling, requires large land space, photo inhibition, costly
Column photobioreactors	Low energy inputs, readily tempered, high mass transfer, proper mixing, good exposure to light–dark cycles, low shear stress, easy to sterilize, less photo inhibition, high photosynthetic efficiency	Small illumination surface area, sophisticated construction materials, low illumination surface area, expensive, modest scalability
Flat photobioreactors	Easy to operate, cheap, high illumination surface area, low energy consumption, high biomass productivities, low oxygen build-up, suitable for outdoor cultures	Problem in scale-up which requires many compartments and support materials, difficult temperature control, hydrodynamic stress to some algal strains, low photosynthetic efficiency, some degree of wall growth, scale-up

Sources: Dragone et al. (2010); Rajvanshi and Sharma (2012).

this process is known as harvesting (Mata et al., 2010). Microalgae cells are typically small in size (2–200 μm), and their separation from culture media is more challenging. So, the harvesting is difficult and expensive process due to the high water content (99.98%), posing a huge operation cost for dewatering. So the harvesting of microalgal cultures has been considered as a major step toward the commercial scale processing of microalgae for biodiesel production (Chen et al., 2010). Microalgal biomass harvesting can be done by physical, chemical, or biological methods such as centrifugation, filtration, flocculation, ultrasound, positively charged surface, gravity sedimentation, electrophoresis, dissolved air flotation, etc. Most commonly, harvesting consists of two processes: bulk and thickening. In bulk harvesting process, for large-scale production, the biomass is separated from bulk culture through flocculation and flotation or gravity sedimentation (increasing cell concentration to 1–5% w/v, dry weight basis). The thickening process is mainly done by centrifugation, filtration, and ultrasonic aggregation (final cell concentration to 10–25% w/v, dry weight basis). Hence, it is a more energy requiring step (Li et al., 2008; Rawat et al., 2013). Therefore, it is necessary to select most reliable harvesting method depending up on the characteristic of microalgae, that is, cell size, density, and also the desired specification of the final product. Once the algae are harvested, then it is dried (Rajvanshi and Sharma, 2012). After harvesting, the microalgal cultures can be concentrated by both chemical as well as physical flocculation and oil is extracted. Some other processes are proposed like cell breakage, solvent extraction, and three-phase centrifugation.

11.6 EXTRACTION OF OIL FROM MICROALGAE BIOMASS

Before oil extraction, microalgae must first need to be dried. Sun drying is one of the cheapest methods for microalgal biomass drying. However, this method takes more time, requires large surface, and less risk of losses of bioreactive products. Nowadays, advance drying technologies have been developed for drying microalgae biomass such as drum drying, fluidized bed drying, spray drying, and freeze drying. These methods are more efficient and expensive. After that, microalgae cell is disrupted by using high-pressure homogenizers, autoclaving, and addition of sodium hydroxide, hydrochloric acid, and alkaline lysis.

11.6.1 MECHANICAL METHOD

11.6.1.1 MECHANICAL PRESS METHOD

In this method, the microalgae biomass is subjected to high pressure that result in rupturing the cell wall and release the oil. This method is easy to use and solvent is not required. This method can extract oil to the tune of 70–75% from microalgal biomass and special skill is not required (Suali and Sarbatly, 2012).

11.6.1.2 ULTRASONIC ASSISTED EXTRACTION METHOD

It is a most promising method that is to be used for extraction of oil from microalgae. In this method, microalgae are exposed to a high intensity ultrasonic wave and create small cavitation bubbles around the cells. Collapsing of bubbles emits shockwaves that shattered the cell wall, and desired compounds are released into solution (Harun et al., 2010). This method is commonly used at laboratory scale, requires less solvent, less time, higher penetration of solvent into cellular materials, and release of cell contents into bulk medium. It can extract 76–77% yield of oils from microalgae (Suali and Sarbatly, 2012).

11.6.2 CHEMICAL METHOD

11.6.2.1 SOLVENT EXTRACTION

In this method, microalgal oil can be extracted by using different organic solvents such as hexane, chloroform, acetone, benzene, cyclo-hexane, etc. These organic solvents are added to microalgae paste and damage cell wall, and oil is extracted from aqueous medium due to their higher solubility in organic solvents as compared to water. These solvents are low cost, recycled, high extraction capability in less time, and the yields obtained are about 60–70% (Galloway et al., 2004; Nautiyal et al., 2014).

11.6.2.2 SUPERCRITICAL FLUID EXTRACTION (SFE)

Supercritical fluid extraction method has been widely used for microalgae oil extraction and purification. This method is more efficient that can replace organic solvent extraction method. In SFE, high pressure and temperature are used to rupture the microalgae cells and produce highly purified extracts which are free from potentially harmful solvent residues and safe for thermally sensitive products. In this method, quick extraction and separation is done. It can extract 100% yield of oils from microalgae. Use of CO_2 as supercritical fluid which is liquefied under pressure and heat is to attain the properties of both liquid as well as gases that act as solvent for the extraction of oil (Halim et al., 2012). This method requires high pressure equipment. However, it is expensive and energy intensive. The advantage of this method is nontoxic, easy to operate, and nonflammable (Xiaodan et al., 2012).

From the above methods, the solvent extraction and supercritical fluid extraction methods are the most popular methods for oil extraction as compared to mechanical methods.

11.7 CONVERSION OF MICROALGAL OIL TO BIODIESEL

After the oil extraction processes, the microalgal oil can be converted into biodiesel. Biodiesel is one of the most common biofuel which is produced from microalgae. Microalgal oil contained high carbon and hydrogen contents with low oxygen content as compared to plant-based oils. These characteristics make microalgae as an attractive biodiesel feedstock because they lead to a fuel with high energy content, low density, and low viscosity (Gong and Jiang, 2010). There are different methodologies that can be utilized to process microalgae into biodiesel. Biodiesel in the fatty acid methyl esters forms are prepared from microalgae biomass by different methods: (1) oil extraction from microalgal biomass followed by transesterification and (2) direct transesterification of microalgae biomass. In this case, both dry as well as wet biomasses are used as feed stocks for biodiesel production. The biodiesel production mainly consists of four primary ways, that is, direct use and blending, microemulsions, thermal cracking (pyrolysis), and transesterification. The most popular used method is transesterification of oils. Transesterification of alcohols and lipids are the chemical reaction required to produce biodiesel, with glycerol being produced as a by-product. Transesterification is a multiple-step process which occurs in a reactor where it is blended with alcohol (methanol) and NaOH used as catalyst that react with the triglyceride present in the microalgae oil. Then upstream product is pumped into a separator tank. When a base or acid used as a catalyst, the upper layer is contains methyl ester, catalyst, and excess alcohol, and the lower level is glycerol (Guan et al., 2004; Kouzu et al., 2008). Biodiesel conversion from microalgae consists of five major steps as depicted in Figure 11.2.

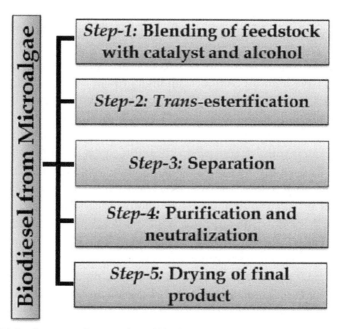

FIGURE 11.2   Step toward conversion of biodiesel from microalgae.

Nevertheless, most technologies for the production of biodiesel using low-quality feedstock are still in their infancy, and research efforts are needed in the future. But the development of these technologies in the future will lead to benefits for the biodiesel industry and also to less environmental impact, safer processes, high glycerol purity, reduction in land use, and noninterference in the food chain (Luque and Melero, 2012).

11.8   BIOTECHNOLOGICAL AND GENETIC ENGINEERING IMPLICATIONS FOR MICROALGAL BIODIESEL PRODUCTION

A microalga has been recognized as promising biomass feedstocks for biodiesel production (Chen et al., 2010). Microalgae are natural resources which synthesize high lipid content that further increase either TAGs production or enhance biomass and biodiesel production which has possible through heterotrophic cultivation, genetic engineering, and metabolic engineering

(Guan Hua et al., 2010). Today, cultivation, harvesting, and extraction of microalgae biomass are major problems, so new techniques and strategies should be devised to lower the cost and improve the oil extraction efficiency. Genetic engineering is one of most potential technique to improve the overall microalgal biomass yield and lipid content. Discovery of new strains and genetically modified strains capable of micro algal biodiesel production is the need of the hour. Selection of superior microalgae strains that should have high oil content with high growth rate and wide environmental tolerances for biodiesel production has mainly focused on species selection and cultivation techniques. There is one possibility to increase microalgal oil yield by the development of microalgae molecular biology, genetic, or metabolic engineering. Therefore, genetic engineering techniques can be used for high oil content, increase photosynthetic efficiency, and higher yield of microalgae biomass. The photosynthesis is the first stage of biodiesel production, if increase in photosynthetic efficiency that will be benefitted for downstream process of biodiesel production (Schenk et al., 2008). The successfully down regulation of light harvesting complexes (LHCs) in *Chlamydomonas reinhardtii* for improved photosynthetic efficiency and light penetration in liquid culture was reported by Mussgnug et al. (2007). The LHC mutant was more efficient conversion of solar energy to biomass. Uses of genetic engineering have increased microalgae biosynthesis of oil by the activity of the enzyme which catalyzes the rate-limiting step in lipid production. Till date, in microalgae, the expression of genes is involved in fatty acid synthesis, and TAG biosynthesis is not well understood. Based on biotechnological processes, that is, transgenic microalgae are still in their infancy, the complete genome sequences from the red alga *Cyanidios chyzonmerolae* have been studied by Nozaki et al. (2010), the diatoms *Thalassiosira pseudonana* (Armbrust et al., 2004), *P. tricornutum* (Bowler et al., 2008), and the unicellular green alga *Ostreoco ccustauri* (Derelle et al., 2006). Various genetic transformation systems have been developed in green algae such as *Chlamydomonas reinhardtti* and *Volvox-carteri* (Nguyen et al., 2011). Characterization of nine genes (LHCBM 1–9) encode for the major light system of green alga *C. reinhardtti* have been reported by Natali and Croce (2015). Transcriptomic and proteomic analyses have shown that those genes are all expressed in different amounts and some of them only in certain conditions.

Till date, there is no success story with respect to over production of lipid by microalgae using the genetic engineering techniques, only understanding the TAG biosynthesis path-way, which is to be identical in all species except the differences in the location of reactions and the structure of some key enzymes, has been studied. However, researches have also been carried out at different species for the enhancement of lipid production using the genetic engineering techniques. Similarly, lipid production was enhanced by using biochemical, genetic, and transcription factor engineering approaches (Courchesne et al., 2009; Ramasamy et al., 2011). Extensive research should be done before we utilize recent techniques like genomics, proteomics, lipidomics, and metabolomics that increases microalgal oil production. At present, only a few algae species have been successfully genetically manipulated and by using of induced random mutagenesis for identification of new genes which play important roles in processes of interest. Some mutagenic strategies have been developed including isolation of tagged mutant genes and insertional mutagenesis of the nuclear genome. These strategies have strong preference for specific genomic regions. Modifications in biological function of algae have been transformed by heterologous gene expression and nuclear genome. A variety of reporter genes and drug-resistance genes have been identified. The analysis of transgene expression has only been performed in *Chlamydomonas spp.* But problems with transgene silencing in *Chlamydomonas* have encountered as well as in identification of strains which have mutated in the silencing pathways. Recently, the nuclear genome and the plastid genome of *Chlamydomonas* have been successfully transformed by using heterologous protein expression method. There is a dire need to isolate, screen, select superior microalgae strains by genetic engineering for higher oil content and overall productivity (e.g., high photosynthetic efficiency in mass culture), resistance to contaminants (predators, grazers, heterotrophs, and extremophilic spp.), temperature and other environmental factors, etc. (Nguyen et al., 2011; Yu et al., 2011; Misra et al., 2012).

11.9   MICROALGAE BIOMASS DOWNSTREAM PROCESSING

In the last few years, research and development concerning microalgae paid much attention on strain development and bioreactor design, but their biomass downstream processing into biodiesel has been neglected. When aiming to produce biodiesel on a commercial/ industrial scale, a very large volume of both biomass and culture medium should be processed. The overall biomass downstream processing sequence is represented in Figure 11.3, showing the main steps such as cultivation, separation of the cells from the liquid growth medium, and oil extraction by transesterification process for biodiesel production.

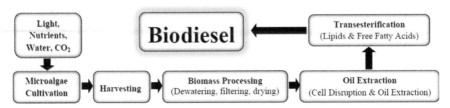

FIGURE 11.3   The overall integrated process of biodiesel from microalgae.

Microalgae for biodiesel production have some proposed advantages as compared to conventional crops. In reality, biodiesel production from microalgae has number of challenges to overcome in cultivation, productivity, extraction, and conversion of oil to biodiesel. However, each potential advantage is counterbalanced by a potential limiting downside to microalgae for biodiesel production as mentioned in Table 11.3.

TABLE 11.3   Advantages and Downsides to Microalgae for Biodiesel Production.

Proposed advantages	Cautionary advice
1. Can be grown on marginal lands, nonarable lands, urban areas, industrial parks, and no competition with food products	Cultivation much more technologically difficult as compared to traditional crops terrestrial plants
2. Rapid growth under optimal conditions	Optimal conditions like pH, temperature, CO_2, light intensity difficult to maintain
3. High lipid contents	Only few spp. under optimized conditions
4. Not seasonal, it means production throughout the year	Low productivity during winter months, heating is necessary

11.10   CHALLENGES AND FUTURE PERSPECTIVE FOR COMMERCIALIZATION OF BIODIESEL FROM MICROALGAE

The fast growth rate, high light intensity, and high oil content have been considered as reasons to more emphasize on microalgae for biodiesel production. However, there are a number of challenging barriers to overcome that range from how and where to grow these microalgae, isolation, screening, sources of nutrient and utilization, production management, harvesting, oil extraction, refining, residual biomass utilization, and other bioactive compounds. There are a number of issues regarding the challenges for engineers to either design PBRs that requires low cost for large-scale production or for engineers and biologists to combine work to develop species that will be grown efficiently in open systems. The current technologies such as genetic engineering, proteomics, lipidomics, and metabolomics can enhance growth rate, cultivation, harvesting, and oil extraction from improved strains of microalgae.

The main issue that must be faced in the future prospective that is how we can reach at commercial-scale production systems with low-cost, higher productivities, and maximum oil content is currently achieved by using modern techniques. Bioprospecting is of great importance to identify microalgal species that have high lipid content, growth rate, and growth densities, whereas growing on inexpensive media with potential strategy on present scenario is like that genetic engineering and breeding will be required for these strains to be economic viable. The potential use for genetic engineering for microalgae is just beginning which has to be realized for improving lipid biogenesis. Modern technologies such as genetic engineering, lipidomics, metabolomics, genomics, proteomics, and mutagenesis offer immense opportunities to improve microalgal oil production in future.

11.11 CONCLUSIONS

Microalgae have been considered as a potential feedstock for biodiesel because of their high productivities as compared to other conventional energy crops. A number of constrains related to cultivation, harvesting, drying, and extraction of oils have hindrance in the large-scale production of biodiesel from microalgae. However, several important scientific and technical hurdles remain to be overcome before the commercial-scale production of microalgae-derived biodiesel. Producing cost-effective microalgal biodiesel requires mainly improvements to microalgae characteristics through genetic and metabolic engineering and further lowering the cost of production by modification in photobioreactor design, microalgal biomass harvesting, extraction, drying, and processing are the important areas that may lead to enhanced cost-effectiveness. Therefore, there is a need to have better technique to separate and concentrate microalgal cell biomass from liquid medium rapidly with low energy consumption. Production of low-cost microalgae biodiesel through genetic, metabolic engineering, genetic modification, nutrient management, metabolic pathways, and cultivation conditions has been optimized by researchers. The main challenges are to identify high oil-content producing microalgae strains that can produce maximum biodiesel at commercialization scale and reduce the consumption of fossil fuel. One way to make feasibility for biodiesel production from microalgae is by increasing the lipid production to recover the overall production cost. The scientific efforts increase lipid production by genetic manipulation of the culture system in terms of the engineering and design low cost photobioreactor and also modification in the phycological metabolism of the microalgal cell. These technologies can lead to increase in the biomass as well as lipid production and microalgae production at industrial scale. Microalgae can be used as feedstock for third-generation energy source.

KEYWORDS

- **biodiesel**
- **microalgae**
- **biomass**
- **oil content**
- **renewable energy**

REFERENCES

Abayomi, O. A. Microalgae Technology and Processes for Bio Fuel Bio Energy Production in British Colombia. Seed Science Ltd: Blackjack Drive, Lanlzville, Brilish Columbia, 2009.

Alabi, A. O.; Tampier, M.; Bibeau, E. Microalgae Technologies and Processes for Biofuels/ Bioenergy Production in Columbia: Current Technology, Suitability and Barrier to Implementation. Seed Science Ltd. Report. The British Columbia Innovation Council: British Columbia, 2009.

Alam, F.; Datea, A.; Rasjidin, R. Biofuel from Algae—Is it a Viable Alternative. *Procedia Eng.* **2012**, *49*, 221–227.

Aransiola, E. F.; Betiku, E.; Ikhuomoregbe, D.; Ojumu, T. V. Production of Biodiesel from Crude Neem Oil Feedstock and Its Emissions from Internal Combustion Engines. *Afr. J. Biotechnol.* **2012**, *11*, 6178–6186.

Aransiola, E. F.; Ojumu, T. V.; Oyekola, O. O.; Madzimbamuto, T. F.; Omoregbe, D. I. O. A Review of Current Technology for Biodiesel Production: State of the Art. *Biomass Bioenergy.* **2014**, *61*, 276–297.

Armbrust, E. V.; Berges, J. A.; Bowler, C. The Genome of the Diatom *Thalassiosirapseudonana*: Ecology, Evolution and Metabolism. *Science.* **2004**, *306*, 79–86.

Barclay, W. R.; Meager, K. M.; Abril, J. R. Heterotrophic Production of Long Chainomega-3 Fatty Acids Utilizing Algae and Algae-Like Microorganisms. *J. Appl. Phycol.* **1994**, *6*(2), 123–129.

Bastianoni, S.; Coppola, F. T.; Colacevicin, A.; Borghini, F.; Focardi, S. Biofuel Potential Production from the Orbetello Lagoon Macroalgae: A Comparison with Sunflower Feedstock. *Biomass Bioenergy.* **2008,** *32,* 619–628.

Benemann, J. R.; Oswald, W. J. Systems and Economic Analysis of Microalgae Ponds for Conversion of Carbon Dioxide to Biomass. Final Report: Grant No.DE-FG22-93PC93204. Pittsburgh Energy Technology Center, Pittsburgh, PA, US Department of Energy, 1993.

Borowitzka, M. A. Culturing Microalgae in Outdoor Ponds. In *Algal Culturing Techniques;* Andersen, R. A., Eds.; Elsevier Academic Press: Burlington, MA, 2005; pp 205–218.

Bowler, C.; Allen, A. E.; Badger, J. H. The Phaeodactylum Genome Reveals the Evolutionary History of Diatom Genomes. *Nature.* 2008, *465,* 239–244.

Brennan, L.; Owende, P. Biofuels from Microalgae-A Review of Technologies for Production, Processing and Extraction of Bio Fuels and Co-Product. *Renewable Sustainable Energy Rev.* **2010,** *14,* 557–577.

Carlsson, A. S.; van Beilen, J. B.; Moller, R.; Clayton, D. *Micro- and Macro-Algae: Utility for Industrial Applications*; 1st ed. CPL Press: Newbury, 2007.

Carvalho, A. P.; Meireles, L. A.; Malcata, F. X. Microalgal Reactors: A Review of Enclosed System Designs and Performances. *Biotechnol. Prog.* **2006,** *22,* 1490–1506.

Chen, F.; Dong, W.; Zhang, X.; Chen, G. Biodiesel Production by Microalgal Biotechnology. *Appl. Energy.* **2010,** *87,* 38–46.

Chisti, Y. Biodiesel from Microalgae. *Biotechnol. Adv.* **2007,** *25,* 294–306.

Chiu, S. Y.; Kao, C. Y.; Chen, C. H.; Kuan, T. C.; Ong, S. C.; Lin, C. S. Reduction of CO_2 by a High-Density Culture of *Chlorella* spp. in a Semicontinuous Photo Bioreactor. *Bioresour. Technol.* **2008,** *99,* 3389–3396.

Courchesne, N. M.; Parisien, A.; Wang B.; Lan, C. Q. Enhancement of Lipid Production using Biochemical, Genetic and Transcription Factor Engineering Approaches. *J. Biotechnol.* 2009, *141,* 31–41.

Derelle, E.; Ferraz, C.; Rombaut, S. Genome Analysis of the Smallest Free-Living Eukaryote *Ostreococcustauri* Unveils Many Unique Features. *Proc. Natl. Acad. Sci. U.S.A.* 2006, *103,* 11647–11652.

Doucha, J.; Livansky, K. Productivity CO_2/O_2 Exchange and Hydraulics in Outdoor Open High Density Microalgal (*Chlorella* sp.) Photobioreactors Operated in a Middle and Southern European Climate. *Appl. Phycol.* 2006, *18,* 811–826.

Dragone, G.; Fernandes, B.; Vicente, A. A.; Teixeria, A. A. Third Generation Biofuels from Microalgae. In *Current Research, Technology, and Education Topics in Applied Microbiology and Microbial Biotechnology*; Mendez-Vilas, A. Eds.; Formatex: Spain, 2010; pp 1355–1366.

Galloway, A.; Koester, K. J.; Paasch, B. J. Effect of Sample Size on Solvent Extraction for Detecting Continuity in Polymer Blends. *Polymer.* **2004,** *45,* 423–428.

Garcia, M. C.; Sevilla, J. M.; Fernandez, F. G. Mixotrophic Growth of *Phaeodactylumtricornutum* on Glycerol Growth Rate And Fatty Acid Profile. *J. Appl. Phycol.* **2000,** *12,* 239–248.

Gong, Y.; Jiang, M. Biodiesel Production with Microalgae as Feedstock: From Strains to Biodiesel. *Biotechnol. Lett.* **2010,** *33,* 1269–1284.

Goureia, L.; Oliveira, A. C. Microalgae as a Raw Material for Biofuels Production. *J. Ind. Microbial. Biotechnol.* **2009,** *36,* 269–274.

Guan Hua, H.; Feng, C.; Dong, W.; Xue, Zhang, W.; Gu, C. Biodiesel Production by Micro Algal Biotechnology. *Appl. Energy.* **2010,** *8,* 8738–8746.

Guan, Y.; Deng, M.; Yu, X.; Zhang, W. Two-Stage Photo-Biological Production of Hydrogen by Marine Green Alga *Platymonassubcordiformis. Biochem. Eng. J.* **2004,** *19,* 69–73.

Guschina, I. A.; Harwood, J. L. Lipid and Lipid Metabolism in Eukaroyotic Alga. *Progress Lipid Res.* **2006,** *45,* 160–186.

Halim, R.; Danquah, M. K.; Webley, P. A. Extraction of Oil From Microalgae for Biodiesel Production: A Review. *Biotechnol. Adv.* **2012,** *30,* 709–732.

Harun, R.; Singh, M.; Forde, G. M.; Danquah, M. K. Bioprocess Engineering of Microalgae to Produce a Variety of Consumer Products. *Renewable Sustainable Energy Rev.* **2010,** *14,* 1037–1047.

Khan, S. A.; Rashmi, Hussain, M. Z.; Prasad, S.; Banerjee, U. C. Prospects of Biodiesel Production From Microalgae in India. *Renewable Sustainable Energy Rev.* **2009,** *13,* 2361–2372.

Kouzu, M.; Kasuno, T.; Tajika, M.; Yamanaka, S.; Hidaka, J. Active Phase of Calcium Oxide used as Solid Base Catalyst for Transesterification of Soybean Oil with Refluxing Methanol. *Appl. Catal. A: Gen.* **2008,** *334*(1–2), 357–365.

Kumar, M.; Sharma, M. P.; Dwivedi, G. Algae Oil as Future Energy Source in Indian Perspective. *Int. J. Renewable Energy Res.* **2013,** *3,* 913–921.

Lang, X.; Dalai, A. K.; Bakshi, N. N. Preparation and Characterization of Biodiesels from Various bio Oils. *Bioresour. Technol.* **2001,** *80,* 53–62.

Laws, E. A.; Taguchi, S.; Hirata, J. Optimization of Micro Algal Production in a Shallow Outdoor Flume. *Biotechnol. Bioeng.* **1988,** *32*(2), 140–147.

Li, W.; Zhang; W. W.; Yang, M. M.; Chen, Y. L. Cloning of the Thermostable Cellulase Gene from Newly Isolated *Bacillus subtilis* and Its Expression in *Escherichia coli. Mol. Biotechnol.* **2008,** *40,* 195–201.

Luisa, G. *Microalgae as a Feedstock for Biofuels*; Springer Heidelberg Dordrecht: London, New York, 2011.

Luque, R.; Melero, J. A. *Advances in Biodiesel Production: Processes and Technologies*; Published by Woodhead Publishing Limited: Cambridge, UK, 2012; pp 1–8.

Mata, T. M.; Martins, A. A.; Caetano, N. S. Microalgae for Biodiesel Production and Other Applications: A Review. *Renewable Sustainable Energy Rev.* **2009,** *7,* 17–22.

Mata, T. M.; Martins, A. A.; Caetano, N. S. Microalgae for Biodiesel Production and Other Applications: A Review. *Renewable Sustainable Energy Rev.* **2010,** *14,* 217–232.

Misra, N.; Panda, P. K.; Parida, B. K.; Mishra, B. K. Phylogenomic Study of Lipid Genes Involved in Microalgal Biofuel Production—Candidate Gene Mining and Metabolic Pathway Analyses. *Evol. Bioinform.* 2012, *8,* 545–564.

Molina Grima, E.; Belarbi, E. H.; Acien Fernandez, F.; Robles Medina, A.; Chisti, Y. Recovery of Micro Algal Biomass and Metabolites: Process Options and Economics. *Biotechnol. Adv.* **2003,** *20,* 491–515.

Mussgnug, J. H.; Thomas-Hall, S.; Rupprecht, J.; Foo, A.; Klassen, V.; McDowall, A.; Schenk, P. M.; Kruse, O.; Hankamer, B. Engineering Photosynthetic Light Capture: Impacts on Improved Solar Energy to Biomass Conversion. *Plant Biotechnol. J.* 2007, *5,* 802–814.

Natali, A.; Croce, R. Characterization of the Major Light Harvesting Complexes (LHCBM) of the Green Alga *Chlamydomonas Reinhardtti.* PLoS. **2015,** *10*(2), e0119211, doi: 10.1371/journal.pone.0119211.

Nautiyal, P.; Subramanian, K. A.; Dastidar, M. G. Production and Characterization of Biodiesel from Algae. *Fuel Process Technol.* **2014,** *120,* 79–88.

Nguyen, H. M.; Baudet, M.; Cuine, S. Proteomic Profiling of Oil Bodies Isolated from The Unicellular Green Microalga *Chlamydomonasreinhardtii:* With Focus on Proteins Involved in Lipid Metabolism. *Proteomics.* 2011, *11,* 4266–4273.

Nozaki, H.; Nakada, T.; Watanabe, S. Evolutionary Origin of *Gloeomonas* (Volvocales, Chlorophyceae), based on Ultrastructure of Chloro-Plasts and Molecular Phylogeny. *J. Phycol.* **2010**, *46*, 195–200.

Pulz, O. Photobioreactors: Production Systems for Phototrophic Microorganisms. *Appl. Microbiol. Biotechnol.* **2001**, *57*, 287–293.

Putt, R.; Singh, M.; Chinnasamy, S. An Efficient System for Carbonation of High-Rate Algae Pond Water to Enhance CO_2 Mass Transfer. *Bioresour. Technolnol.* **2011**, *102*, 3240–3245.

Rajvanshi, S.; Sharma, M. P. Microalgae: A Potential Source of Biodiesel. *J. Sustainable Bioengergy Syst.* **2012**, *2*, 49–59.

Ramasamy, S.; Sanniyasi, E.; Mohommad, M. Microalgae Lipid Research, Past, Present: A Critical Review for Biodiesel Production, in the Future. *J. Exp. Sci.* **2011**, *2*(10), 29–49.

Rawat, I.; Kumar, R. K.; Mutanda, T. Biodiesel from Microalgae: A Critical Evaluation from Laboratory to Large Scale Production. *Appl. Eng.* **2013**, *103*, 444–467.

Renaud, S. M.; Parry, D. I.; Thinh, L. V.; Kuo, C.; Padovan, A.; Sammy, N. Effect of Light Intensity on the Proximate Biochemical and Fatty Acid Composition of *Isochrysis* sp And Nannochloropsis Oculata for use in Tropical Aquaculture. *J. Appl. Phycol.* **1991**, *3*, 43–53.

Richmond, A. *Handbook of Microalgal Culture: Biotechnology and Applied Phycology*; Blackwell Science Ltd: Oxford, 2004.

Rodolfi, L.; Zittelli, G. C.; Bassi, N.; Padovani, G.; Biondi, N.; Bonini, G. Microalgae for Oil: Strain Selection, Induction of Lipid Synthesis and Outdoor Mass Cultivation in a Low-Cost Photobioreactor. *Biotechnol. Bioeng.* **2009**, *102*(1), 100–112.

Sanchez, A.; Gonzalez, A.; Maceiras, R.; Cancela, A.; Urrejola, S. Raceway Pond Design for Microalgae Culture for Biodiesel. *Chem. Eng. Trans.* **2011**, *25*, 845–850.

Schenk, P.M.; Thomas-Hall, S.R.; Stephens, E.; Markx, U. C.; Mussgnug, J. H.; Posten, C.; Kruse, O.; Hankamer, B. Second Generation Biofuels: High-Efficiency Microalgae for Biodiesel Production. *Bioenergy Res.* 2008, *1*, 20–43.

Sheehan, J.; Dunahay, T.; Benemann, J.; Roessler, P. A Look Back at the U.S. Department of Energy's Aquatic Species Program-Biodiesel from Algae. Golden (CO): National Renewable Energy Laboratory; 1998 July. Report No.: NREL/TP-580-24190. Contract. No: DE-AC36-83CH10093. Sponsored by the U.S. Department of Energy, 1998.

Sialve, B.; Bernet, N.; Bernard, O. Anaerobic Digestion of Microalgae as a Necessary Step to Make Micro Algal Biodiesel Sustainable. *Biotechnol. Adv.* 2009, *27*, 409–416.

Singh A.; Nigam, P. S.; Murphy, J. D. Mechanism and Challenges in Commercialization of Algal Biofuels. *Bioresour. Technol.* 2011, *102*, 26–34.

Singla, A.; Kumar, N.; Sharma, P. B. Biodiesel Production from Microalgae—A Sustainable Fuel. *J. Biofuels.* 2010, *1*, 37–45.

Sobczuk, T. M.; Chisti, Y. Potential Fuel Oils from the Microalga *Choricistis Minor*. *J. Chem. Technol. Biotechnol.* 2010, *85*, 100–108.

Spolaore, P.; Joannis-Cassan, C.; Duran, E.; Isambert, A. Commercial Applications of Microalgae. *J. Biosci. Bioeng.* 2010, *101*, 87–96.

Suali, E.; Sarbatly, R. Conversion of Microalgae to Bio Fuels. *Renewable Sustainable Energy Rev.* 2012, *16*, 4316–4342.

Sydney, E. B.; Sturm, W.; de Carvalho, J. C.; Thomaz-Soccol, V.; Larroche, C.; Pandey, A.; Soccol, C. R. Potential Carbon Dioxide Fixation by Industrially Important Microalgae. *Bioresour. Technol.* 2010, *101*, 5892–5896.

Tredici, M. R.; Materassi, R. From Open Ponds to Vertical Alveolar Panels: The Italian Experience in the Development of Reactors for the Mass Cultivation of Phototrophic Microorganisms. *J. Appl. Phycol.* **1992**, *4*, 221–231.

Tredici, M. R.; Carlozzi, P.; Zittelli, G. C. A Vertical Alveolar Panel (VAP) for Outdoor Mass Cultivation of Microalgae and Cyanobacteria. *Bioresour. Technol.* 1999, *38*, 153–159.

Um, B. H.; Kim, Y. S. Review: A Chance for Korea to Advance Algal-Biodiesel Technology. *J. Ind. Eng. Chem.* 2009, *15*, 1–7.

Wang, J.; Yang, H.; Wang, F. Mixotrophic Cultivation of Microalgae for Biodiesel Production: Status and Prospects. *Appl. Biochem. Biotechnol.* 2014, *172*(7), 3307–3329.

Xiaodan, W.; Rongsheng, R.; Zhenyi, D.; Yuhuan, L. Current Status and Prospects of Biodiesel Production from Microalgae. *Energies.* 2012, *5*, 2667–2682.

Yu, W. L.; Ansari, W.; Schoepp, N. G.; Hannon, M. J.; Mayfield, S. P.; Burkart, M. D. Modifications of the Metabolic Pathways of Lipid and Triacylglycerol Production in Microalgae. *Microb. Cell Fact.* **2011**, *10*, 91–102.

INDEX

Printed and bound by CPI Group (UK) Ltd, Croydon, CR0 4YY

23/10/2024

01777701-0015